BUS

HANDBOOK OF PLASTIC FOAMS

HANDBOOK OF
PLASTIC FOAMS

Types, Properties,
Manufacture and Applications

Edited by

Arthur H. Landrock (ret.)

Plastics Technical Evaluation Center (PLASTEC)
Picatinny Arsenal
Dover, New Jersey

 NOYES PUBLICATIONS
Park Ridge, New Jersey, U.S.A.

Library of Congress Catalog Card Number: 94-15236
ISBN: 0-8155-1357-7
Printed in the United States

Published in the United States of America by
Noyes Publications
Mill Road, Park Ridge, New Jersey 07656

10 9 8 7 6 5 4 3 2 1

Library of Congress Cataloging-in-Publication Data

Handbook of plastic foams : types, properties, manufacture, and
    applications / [edited by] Arthur H. Landrock.
            p.   cm.
        Includes bibliographical references and index.
        ISBN 0-8155-1357-7
        1. Plastic foams. I. Landrock, Arthur H.
    TP1183.F6H35 1995
    668.4'93--dc20   94-15236
                                                            CIP

Dedication

To my wife, Rose–Marie,
for her unfailing support and understanding

PREFACE

This book is intended to be useful to anyone working with plastic foams (cellular plastics), and to a lesser extent, elastomeric foams. The emphasis is on practical, rather than theoretical aspects. The books should prove helpful to materials engineers, chemists, chemical engineers, sales personnel. It may also find use as a textbook or reference source in materials engineering courses. The book is a comprehensive technical treatment of plastic foams and covers information not available in any other single source.

A brief description of the books contents may be helpful. Chapter 1 by M. Okoroafor and K.C. Frisch is an introduction and also covers the subject of foam formation. The chapter includes a discussion of the Montreal Protocol mandating the development of foams with substantially reduced CFC content by 1995. Chapter 2 is a comprehensive discussion of thermosetting foams of all types, with the emphasis on urethane and phenolic foams. The authors, K. Ashida and K. Iwasaki, are recognized authorities in their fields of specialization. This chapter presents extensive discussions of such fields as composites and syntactic foams. There is some overlap between this chapter and topics covered in later chapters, but the treatments are different.

Chapter 3 and all subsequent chapters were prepared by the editor, A.H. Landrock. Chapter 3 covers all types of thermoplastic foams, including rigid, semi-rigid, and structural foams. Chapter 4 briefly discusses elastomeric foams. Chapter 5 discusses a number of miscellaneous and specialty foams, many of which were also covered in Chapter 2.

Chapter 6 covers solvent cementing of thermoplastic foams and adhesive bonding of all foam types, including thermosets. Bonding of plastic foams to non-plastic substrates is also covered. Additives, fillers and reinforcements are covered in Chapter 7, a chapter of 37 pages. This chapter covers antistatic agents, slowing agents, catalysts, fire retardants, mold-release agents, nucleating agents, reinforcements, and stabilizers.

The problem of ozone depletion considered in the Montreal Protocol is discussed in detail. Chapter 8 considers methods of manufacture, including molding methods, spraying, frothing, laminating, structural foam molding, syntactic foam preparations, and foam–in–place techniques. Chapter 9 on sources of information covers journals, manufacturers' bulletins, technical conferences and their published proceeding, seminars and workshops, standardization activities, trade associations, consultants, and information centers, and books.

Chapter 10 on test methods is an expansion of the listings of standard test methods presented in Chapter 11. The first section of this chapter lists, in alphabetic order, 130 properties of cellular plastics and elastomers and tabulates the standard test methods used for each. Only number designations are given. The second section is a somewhat detailed discussion, also in alphabetical order, of 22 foam properties tested by standard test methods. The third section is a brief invited presentation of several non–standardized test methods currently in use. Chapter 11 on standardization documents lists published specifications, test methods and other related standards used in the U.S., in addition to British standards and ISO International Standards. A total of 361 standards are covered. A glossary of 221 terms is included.

I wish to express my appreciation to Messrs. William C. Tanner and Harry S. Katz for their helpful suggestions in the planning of this book. Mr. George Narita, Vice President and Executive Editor of Noyes Publications, has been most helpful in the development of the book. Lastly, I must express appreciation to my wife, Rose–Marie, for her patience and understanding during the many months spent on the book.

Sparta, New Jersey   Arthur H. Landrock
November 1994

CONTRIBUTORS

Kaneyoshi Ashida
Polymer Institute
University of Detroit Mercy
Detroit, Michigan

Kurt C. Frisch
Polymer Technologies, Inc.
University of Detroit Mercy
Detroit, Michigan

Kadzuo Iwasaki
Iwasaki Technical Consulting
  Lab., Ltd.
Ohta City
Gumma–Prefecture, Japan

Arthur H. Landrock (ret.)
PLASTEC
Picatinny Arsenal
Dover, New Jersey

Michael O. Okoroafor
Technical Center
PPG Industries, Inc.
Monroeville, Pennsylvania

Notice

To the best of our knowledge the information in this publication is accurate; however, the Publisher does not assume any responsibility or liability for the accuracy or completeness of, or consequences arising from, such information. This book is intended for informational purposes only. Mention of trade names or commercial products does not constitute endorsement or recommendation for use by the Publisher. Final determination of the suitability of any information or product for use contemplated by any user, and the manner of that use, is the sole responsibility of the user. We recommend that anyone intending to rely on any recommendation of materials or procedures mentioned in this publication should satisfy himself as to such suitability, and that he can meet all applicable safety and health standards.

CONTENTS

xx   Contents

1

INTRODUCTION TO FOAMS AND FOAM FORMATION

Michael O. Okoroafor and Kurt C. Frisch

INTRODUCTION

Cellular plastics or plastic foams, also referred to as expanded or sponge plastics, generally consist of a minimum of two phases, a solid–polymer matrix and a gaseous phase derived from a blowing agent. The solid–polymer phase may be either inorganic, organic or organometallic. There may be more than one solid phase present, which can be composed of polymer alloys or polymer blends based on two or more polymers, or which can be in the form of interpenetrating polymer networks (IPNs) which consist of at least two crosslinked polymer networks, or a pseudo– or semi–IPN formed from a combination of at least one or more linear polymers with crosslinked polymers not linked by means of covalent bonds.

Other solid phases may be present in the foam in the form of fillers, either fibrous or other–shaped fillers which may be of inorganic origin, e.g. glass, ceramic or metallic, or they may be polymeric in nature. Foams may be flexible or rigid, depending upon whether their glass–transition temperatures are below or above room temperature, which, in turn, depends upon their chemical composition, degree of crystallanity, and degree of crosslinking. Intermediate between flexible and rigid foams are semi–rigid or semi–flexible foams. The cell geometry, i.e. open vs. closed cell, size and shape, greatly affect the foam properties.   Thus, closed–cell foams are most suitable for thermal insulation, while open–cell foams are best for acoustical insulation.

Plastic foams can be produced in a great variety of densities, ranging from about 0.1 lb/ft^3 (1.6 kg/m^3) to over 60 lb/ft^3 (960 kg/m^3) (1). Since the mechanical–strength properties are generally proportional to the foam densities, the applications of these foams usually determine which range of foam densities should be produced. Thus, for rigid foam, load-bearing applications require high densities and (or) fiber–reinforced foams, while low densities are usually used for thermal insulation.

The production of polymeric–foam materials can be carried out by either mechanical, chemical, or physical means. Some of the most commonly used methods are the following (2):

1. Thermal decomposition of chemical blowing agents gen-erating either nitrogen or carbon dioxide, or both, by application of heat, or as the result of the exothermic heat of reaction during polymerization.

2. Mechanical whipping of gases (frothing) into a polymer system (melt, solution or suspension) which hardens, either by catalytic action or heat, or both, thus entrapping the gas bubbles in the polymer matrix.

3. Volatilization of low–boiling liquids such as fluoro-carbons or methylene chloride within the polymer mass as the result of the exothermic heat of reaction, or by application of heat.

4. Volatilization of gases produced by the exothermic heat of reaction during polymerization such as occurs in the reaction of isocyanate with water to form carbon dioxide.

5. Expansion of dissolved gas in a polymer mass on reduction of pressure in the system.

6. Incorporation of hollow microspheres into a polymer mass. The microspheres may consist of either hollow glass or hollow plastic beads.

7. Expansion of gas–filled beads by application of heat or expansion of these beads in a polymer mass by the heat of reaction, e.g. expansion of polystyrene beads in a polyurethane or epoxy resin system.

The production of foams can take place by many different techniques. These may include (3):

1. Continuous slab–stock production by pouring or impingement, using multi–component foam machines.
2. Compression molding of foams.
3. Reaction–injection molding (RIM), usually by impingement.
4. Foaming–in–place by pouring from a dual– or multi–component head.
5. Spraying of foams.
6. Extrusion of foams using expandable beads or pellets.
7. Injection molding of expandable beads or pellets.
8. Rotational casting of foams.
9. Frothing of foams, either by introduction of air or of a low–boiling volatile solvent (e.g. dichlorodifluoromethane, F–12).
10. Lamination of foams (foam–board production).
11. Production of foam composites.
12. Precipitation foam processes where a polymer phase is formed by polymerization or precipitation from a liquid which is later allowed to escape.

It should be recognized that almost every thermoplastic and thermoset resin may be produced today in cellular form by means of the mechanisms and processes cited above. The physical properties of the foams reflect in many ways those of the neat polymers, taking into account the effects of density and cell geometry.

There are numerous books, chapters in books, and reviews published on foams, covering a wide spectrum of cellular plastics. Some of these are listed in references 1–15.

In addition, two journals (in English) deal exclusively with plastic foams. These are the "Journal of Cellular Plastics" (Technomic Publishing Co.) and "Cellular Polymers" (RAPRA Technology Ltd.). A valuable source of information for foamed plastics has been the annual proceedings of various technical organizations such as the Society of the Plastics Industry (SPI); the German FSK and others.

CFC EFFECTS AND ALTERNATIVES

A search for alternate blowing agents for urethane foams became necessary in 1987 following the Montreal Protocol, which mandated the development of foams with substantially reduced CFC content by 1995.

CFC's or chlorofluorocarbons are chemicals that cause ozone depletion in the stratosphere as well as the "Greenhouse Effect". They have been typically employed as blowing agent in foams. Since the initial proclamation, the mandate has been revised several times to accelerate the CFC phaseout schedule, with the latest revision resulting from the Copenhagen agreement in November 1992 where 87 nations resolved to move up total CFC phaseout by four years in January 1996. The recent Copenhagen revision induced major CFC manufacturers to accelerate their phaseout time table. DuPont announced recently that it plans to stop CFC production by 1994, almost 2 years ahead of plan.

Some countries have independently banned CFC use. For example, Sweden banned the use of CFC's in 1991, followed by Switzerland in 1992. In Europe *Rigid Foam* manufacturers are using hydrocarbon blowing agents, such as cyclopentane as an alternative to CFC.

The U.S. Environmental Protection Agency issued a final rule banning the use of CFC's in flexible plastics and packaging foams, among other uses, after February 15, 1993. Exceptions are CFC–11 and CFC–13 which can be used, temporarily, in mold release agents and the production of plastic and elastomeric materials. However, in 1994, no CFC's will be allowed in flexible foams in the U.S., and a tax will be levied on other CFC uses. Total CFC phaseout is mandated in the U.S. for 1995.

Users of CFC's in foam applications are, for the time being, able to employ alternative blowing agents (ABA's) available to them. CFC–11, the workhorse of the foam industry can now be replaced with a hydrochlorofluorocarbon, namely HCFC–141b. Although HCFC–141b and other HCFC's are not considered drop–ins for CFC–11, the use of foam additives, such as surfactant and softening agents, has made it possible to achieve comparable insulation value in rigid foams blown with HCFC's.

Total U.S. HCFC use is to be phased out by 2005; however current trends indicate HCFC's may be dropped in some industries as early as 1997. Even though eventual phaseout of all HCFC substitutes is expected to start by the year 2003, because of its ozone depleting potential, foam manufacturers, especially, rigid foam blowers, are committed to it in the short term.

For flexible foams in which CFC's have typically been employed as auxiliary blowing agents, entirely water–blown foams can be achieved with the performance additives.

FUNDAMENTALS OF FOAM FORMATION

The preparation of a polymeric foam involves first the formation of gas bubbles in a liquid system, followed by the growth and stabilization of these bubbles as the viscosity of the liquid polymer increases, resulting ultimately in the solidification of the cellular resin matrix.

Foams may be prepared by either one of two fundamental methods. In one method, a gas such as air or nitrogen is dispersed in a continuous liquid phase (e.g. an aqueous latex) to yield a colloidal system with the gas as the dispersed phase. In the second method, the gas is generated within the liquid phase and appears as separate bubbles dispersed in the liquid phase. The gas can be the result of a specific gas-generating reaction such as the formation of carbon dioxide when isocyanate reacts with water in the formation of water–blown flexible or rigid urethane foams. Gas can also be generated by volatilization of a low–boiling solvent (e.g. trichlorofluoromethane, F–11, or methylene chloride) in the dispersed phase when an exothermic reaction takes places. (e.g. the formation of F–11 or methylene chloride–blown foams).

Another technique to generate a gas in the liquid phase is the thermal decomposition of chemical blowing agents which generate either nitrogen or carbon dioxide, or both.

Saunders and Hansen (3) have treated in detail the colloidal aspect of foam formation utilizing blowing agents. The formation of internally blown foams takes place in several stages. In the first stage the blowing agent generates a gas in solution in a liquid phase until the gas reaches a saturation limit in solution, and becomes supersaturated. The gas finally comes out of solution in the form of bubbles. The formation of bubbles represents a nucleation process since a new phase is formed. The presence of a second phase which may consist of a finely divided solid, e.g. silica, or some finely dispersed silicone oils, or even an irregular solid surface such as an agitator or wall of a vessel, may act as a nucleating agent.

The factors affecting the stability and growth of bubbles in aqueous foams have been reviewed in depth by deVries (3). In order to disperse a given volume of gas in a unit volume of liquid, one must increase the free energy of the system by an amount of energy ΔF as follows;

$$\Delta F = \gamma A$$

where γ is the surface tension and A is the total interfacial area. When the surface tension of the liquid is lowered, either by heat or by the addition of a surfactant, the free–energy increase associated with the dispersion of the gas will be reduced and will aid in the development of fine cells which corresponds to a large value of A.

According to classical theory, the gas pressure in a spherical bubble is larger than the pressure in the surrounding liquid by a difference Δp, as shown in the following equation:

$$\Delta p = \frac{2\gamma}{R}$$

where R is the radius of the bubble. Hence, the gas pressure in a small bubble is greater than that in a large bubble.

In the case of two bubbles of radii R_1 and R_2, the difference in pressure Δp^2, is given by the equation:

$$\Delta p^2 = 2\gamma \left(\frac{1}{R_2} - \frac{1}{R_2} \right)$$

Therefore, in a liquid system, a diffusion of gas takes place from the small bubbles into the large bubbles, resulting in the disappearance of the small bubbles, while the large bubbles grow in size with time. It is also apparent that low values of γ, e.g. by addition of a surface–tension depressant such as a silicone surfactant, reduce the pressure differences between bubbles of different sizes and hence lead to better bubble stability and small average cell size.

In the formation of polymeric foams, a number of the relationships described below are applicable, at least to some extent, when the polymer phase is still a liquid. In order to form a stable foam, there must be at least two components, one which is preferentially absorbed at the surface. The Gibbs theorem teaches that the surface tension is dependent upon the type and amount of absorbed solute, as follows:

$$d\gamma = \Sigma \Gamma d\mu$$

where Γ is the surface excess of a component with a chemical potential μ. This relationship explains the resistance to an increase in the surface area or a thinning of the cell membrane. Due to the fact that membranes tend to rupture more easily the thinner they are, this resistance to thinning helps to stabilize the cell.

When a membrane expands and the concentration of a surfactant at the interface decreases, there exist two mechanisms to restore the surfactant surface concentration. The first mechanism, termed the "Marangoni effect" (16), refers to the fact that the surface flow can drag with it some of the underlying layers, i.e. the surface layer can flow from areas of low surface tension, thus restoring the film thickness. It is also a source of film elasticity or resilience.

In the second mechanism, the "Gibbs effect," the surface deficiency is replenished by diffusion from the interior and the surface tension is lowered to obtain a desirable level. For the best stabilization of a foam, an optimum concentration of surfactant as well as an optimum rate of diffusion is desirable (3).

Another factor which affects the bubble stability is temperature, since an increase in temperature reduces both surface tension and viscosity, which results in thinning of the cell membrane and may promote cell rupture.

Still another factor in cell stability is the drainage of the liquid in the bubble walls which is due to gravity and capillary action. This drainage from both capillary action and gravity can be retarded by an increase in viscosity, especially at the film surface. This is particularly important in primarily thermoset systems which involve simultaneous polymerization and foaming of the liquid components. A balance between the viscosity and gas evolution must be provided in order to obtain not only a stable foam, but also one with the highest foam volume possible. It is obvious that if the viscosity increases too rapidly (as the result of too fast a polymerization) the gas evolution will eventually cease before reaching its desired foam volume, especially for the production of low–density foams. On the other hand, if the viscosity is too low, when most of the foam evolution occurs, foam stabilization may be very difficult and may result in foam collapse (3).

The proper balance between viscosity and gas evolution can be controlled by a number of factors such as a suitable type and concentration of catalyst and surfactant, the presence of a nucleating agent (not always necessary) (17,18) and control of reaction temperature (or exotherm). Additional factors that must be considered are the use of a suitable chemical blowing agent, which is especially important for the production of thermoplastic foams, and the formation of oligomers (prepolymers) which exhibit higher viscosities than monomers in the preparation of thermoset foams (e.g. polyurethane foams).

The "electrical double layer" effect, i.e. the orientation of electrical charge on each film surface due to the use of ionic emulsifiers, is generally more important in aqueous foams than in organic polymeric foams. The stability effect arises from the repulsion of the electrical charges as the two surfaces approach each other, thus limiting the thinning of the film (cell walls) (3).

The morphology of cellular polymers has been studied in great detail by numerous investigators, in particular by Hilyard (5), Gent and Thomas (19), Harding (5), Meinecke and Clark (20), and others.

The markets for plastic foams have been growing worldwide with North America, the E.E.C., and Japan as the leading producers and consumers of foams. However, the Comecon countries, Latin America, especially Brazil, Argentina, and Mexico, and Asian countries, (other than Japan) such as Taiwan, South Korea and India, are rapidly developing foam markets and production facilities. Many developing countries in all continents are using foams at an ever-increasing rate by starting foam production employing either imported or locally produced raw materials, with major efforts being expended in utilizing certain domestic plant or forest products, especially for foam composites.

The major industries which utilize flexible or semi-flexible foams are:

> Furniture
> Transportation
> Comfort cushioning
> Carpet underlay
> Packaging
> Textiles
> Toys and novelties
> Gasketing
> Sporting goods
> Shock (vibration) and sound attenuation
> Shoes

Rigid foam markets include the following industries:

> Thermal insulation
> Building and construction
> Appliances
> Tanks/pipes
> Transportation
> Packaging

Furniture
Flotation
Moldings (decorative)
Business–machine housings
Food–and–drink containers
Sporting goods
Sound insulation

Surveys of foam markets are frequently prepared by raw–material suppliers as well as various marketing–research organizations. A very useful publication is the U.S. Foamed Plastics Markets & Directory, published annually by Technomic Publishing Co., Lancaster, PA 17604.

REFERENCES

1.   *Plastics Engineering Handbook*, Society of the Plastics Industry, Inc., 5th Edition, ed. M.L. Berins, Van Nostrand Reinhold, N.Y., 1991.

2.   Frisch, K.C., in *Plastic Foams*. Vol. 1, eds. by Frisch, K.C. and Saunders, J.H., Marcel Dekker, N.Y., (1976).

3.   Saunders, J.H., and R.H. Hansen in *Plastic Foams*, Vol. 1, eds. Frisch, K.C. and Saunders, J.H., Marcel Dekker, N.Y., (1972) Chapter 2.

4.   Benning, C.J., *Plastic Foams*, Vols. 1 and 2, Wiley–Interscience, New York (1969).

5.   *Mechanics of Cellular Plastics*, ed. by Hilyard, N.C., Macmillan, New York, (1982).

6.   Shutov, F.A., *Integral/Structural Polymer Foams*, eds. Henrici–Oliv, G., and Oliv, S., Springer, Berlin, (1985).

7.   Berlin, A.A., Shuto, F.A., and Zhitinkina, A.K.. *Foams Based on Reactive Oligomers*, Technomic Publishing Co., Lancaster, PA (1982).

8.   *Polyurethane Handbook*, ed. by Oertel, G., Hanser, and distrib.

by Macmillan Co., Munich, N.Y., (1985).

9.   Ferrigno, T.H., *Rigid Plastic Foams*, 2nd edition, Reinhold, New York, (1967).

10.   Buist, J.M.,   and Gudgeon, H., *Advances in Polyurethane Technology*, Maclaren & Sons, London, (1968).

11.   *Developments in Polyurethanes*, ed. by Buist, J.M., Elsevier, London and New York, (1978).

12.   *Handbook of Foamed Plastics*, ed. Bender, R.J., Lake Publishing, Libertyville, Illinois, (1965).

13.   Saunders, J.H., and Frisch, K.C., *Polyurethanes*, Wiley–Interscience, Part I, 1962, Part II, 1964, reprinted.   Krieger, Malabar, Florida (1983).

14.   Woods, G., *Flexible Polyurethane Foams*, Applied Science, London & Englewood, New Jersey, 1982.

15.   Landrock, A.H., *Handbook of Plastic Foams*, PLASTEC Report, R52, PLASTEC, Picatinny Arsenal, Dover, New Jersey, (1985).

16.   deVries, A.J., *Rubber Chem. & Technol.*, *31*, 325 (1965).

17.   Marangoni, C., *Nuovo Cimento*, 2, 5 (1871)

18.   Hansen, R.H., and Martin, W.M., *Ind. Eng. Chem. Prod. Res. Dev. 3*, 137 (1964).

19.   Hansen, R.H., and Martin, W.M., *J. Polym. Sci., 3B*, 325 (1965).

20.   Gent, A.N., and Thomas, A.G., *Rubber Chem. & Technol., 36*, 597 (1963).

21.   R.H. Harding, *J. Cell Plastics, 1*, 385 (1965).

22.   Meinecke, E.A., and Clark, R.C., *Mechanical Properties of Polymeric Foams*, Technomic Publishing Co., Lancaster, Pennsylvania (1973).

2

THERMOSETTING FOAMS

Kaneyoshi Ashida and Kadzuo Iwasaki

INTRODUCTION *(by Kaneyoshi Ashida)*

A new era of plastics began with the appearance of plastic foams. This period might be called the "Plastic Foam Age."

Plastic foams can be called expanded plastics, cellular plastics or foamed plastics, and include both thermoplastic and thermosetting plastics.

Thermosetting foams can be defined as foams having no thermo-plastic properties. Accordingly, thermosetting foams include not only cross-linked polymer foams, but also some linear polymeric foams having no thermoplastic properties, e.g., carbodiimide foams and polyimide foams. These foams do not melt and turn to char by heating.

Most thermosetting foams are prepared by the simultaneous occurrence of polymer formation and gas generation. This is the principle of preparation of thermosetting plastic foams, as shown in Figure 1.

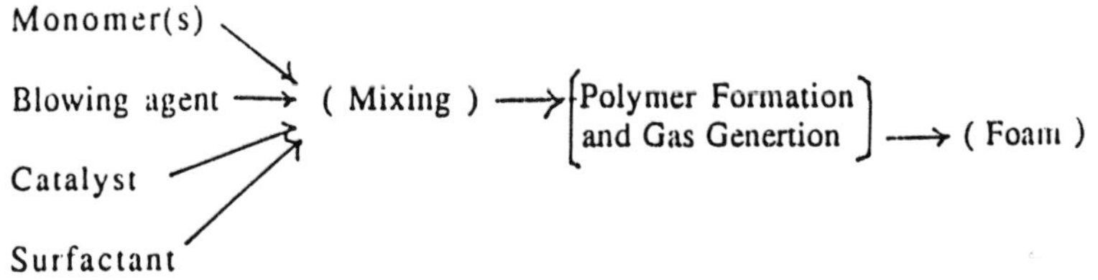

```
Monomer(s)
Blowing agent ──→  ( Mixing ) ──→ ⎰Polymer Formation⎱
                                   ⎱and Gas Genertion⎰ ──→ ( Foam )
Catalyst
Surfactant
```

Figure 1. Mechanism of thermosetting foam preparation

In principle any kind of polymer–forming reactions can be employed for foam preparation. Accordingly, all kinds of thermosetting polymers can theoretically lead to foamed materials.

Table 1 shows a classification of thermosetting foams. Among the foams listed in this table, polyurethane foams have the largest market share in the thermosetting–plastic–foam market.

Table 1: Classification of Thermosetting Foams

Foam	Reaction	Property
Polyurethane	Polyaddition	Flexible and Rigid
Polyisocyanurate	Cyclotrimerization	Rigid
Polyamide	Polycondensation	Flexible and Rigid
Polyimide	Polycondensation	Semi–Rigid
Pyranyl	Radical polymerization	Rigid
Polyurea	Polyaddition	Flexible and Rigid
Epoxy	Ring–opening polymerization	Rigid
Phenolic	Polycondensation	Rigid
Urea–formaldehyde	Polycondensation	Rigid
Polycarbodiimide	Polycondensation	Rigid
Polyoxazolidone	Ring–opening polyaddition	Semi–rigid
Unsaturated polyester (or vinyl ester)	Radical polymerization	Rigid
Rubber (natural & synthetic)	Vulcanization	Flexible
Viscose	Regeneration of cellulose	Flexible
Polyvinyl alcohol	Formal formation	Flexible

Polyisocyanurate foams, polyurea foams and phenolic foams are growing rapidly in recent years. Urea–formaldehyde foams disappeared recently from the U.S. market. Rubber foams and pyranyl foams are no longer available in the worldwide market.

Other foams listed in Table 1 are still of interest to polymer chemists, although their usage is still small.

Many professional books on isocyanate–based plastic foams are available (1–24, 36, 115, 116, 227, 228, 229).

However, only few primers on polyurethane and other thermosetting foams are available to students, beginners and sales and marketing people. This chapter, therefore, is intended as an introduction to thermosetting plastic foams for materials engineers, design engineers, fabricators, chemists, chemical engineers and students.

ISOCYANATE–BASED FOAMS *(by Kaneyoshi Ashida)*

Introduction

Polyisocyanates are very reactive compounds and produce various polymers such as fibers, resins, elastomers, foams, coatings and adhesives by the reaction of polyaddition, polycondensation or stepwise polymerization.

Nylon 66 was invented by W. Carothers of E.I. du Pont de Nemours & Co. in the mid–1930's. The invention became a stimulus to chemists to make synthetic polymers such as fibers and elastomers. Since then, many exploratory studies in polymer syntheses have been carried out worldwide.

For example, polyurethanes were investigated independently by O. Bayer and his collaborators of I.G. Farbenindustrie A.G. in Germany (24, 206), by T. Hoshino and Y. Iwakura of Tokyo Institute of Technology in Japan (200), and by a research group at E.I. du Pont de Nemours & Co. in the United States.

Similarly, polyureas were synthesized by the reaction of diiso–cyanate with aliphatic diamines carried out by three research groups, I.G. Farbenindustrie A.G., E.I. du Pont de Nemours & Co. and Tokyo Institute of Technology, mentioned above.

The first method of making isocyanate–based foams was based on the reaction of a carboxyl–terminated polyester with an organic diisocyanate, e.g., toluene diisocyanate. The simultaneous reactions resulting in carbon dioxide generation and polyamide formation produced cellular plastics.

The foams obtained were polyamide foams and not polyurethane foams, as shown by the model reaction A. This method was invented by Hoechtlen and Dorste in 1941 (121).

The first patent involving polyurethane foams was assigned to Zaunbrecher and Barth in 1942 (22). The method involved simultaneous and competitive reactions comprising polyurethane formation (reaction of an organic diisocyanate with a hydroxyl–terminated polyester), and gas generation (reaction of the diisocyanate with water to form carbon dioxide and polyurea). Equations B and C show the model reactions.

(A) $R—NCO + R'COOH = R—NH—CO—R' + CO_2$

(B) $2 R—NCO + H_2O = R—NH—CO—NH—R + CO_2$

(C)
$$R—NCO + R'OH = R—NH—\overset{\overset{\displaystyle O}{\|}}{C}—R'$$

Both methods could be recognized as a new combination of a known gas–generation and a known polymer–formation reaction. Since then, a variety of isocyanate–based foams were developed as described below. New combinations of new gas–generation reactions and a known polymer formation lie outside of the prior art. This new area was studied by Aahida and his collaborators, and the work was reviewed (33).

Isocyanate–based foams include polyurethane, polyisocyanurate, polyurea, polycarbodiimide, polyamide, polyimide, and polyoxazolidone foams.

In addition to these unmodified foams, many modified or hybrid foams have appeared in the literature, e.g., urethane–modified isocyanurate foams, isocyanurate–modified urethane foams, urea–modified isocyanurate foams, unsaturated polyester–polyurethane hybrid foams, etc.

A number of isocyanate reactions for making isocyanate–based polymers are listed in Table 2 which is expressed by model reactions.

The polyisocyanate–based polymer–forming reactions can be classified into three types of reactions: addition reactions, condensation reactions, and cyclotrimerization reactions. Among the isocyanate reactions shown in Table 2, the addition reaction is the major isocyanate reaction in polyurethane foam preparation. A model addition reaction is shown below:

$$A—N{=}C{=}O + B—H \qquad A—\underset{\underset{\displaystyle H}{|}}{N}—\overset{\overset{\displaystyle O}{\|}}{C}—B$$

where A is residual radical of isocyanate; H is active hydrogen; and B is residual radical of active hydrogen compound.

Table 2: Model Isocyanate Reactions for Foams

Reaction	Material	Product	Linkage
Addition	A-NCO + B-OH	$\overset{O}{\overset{\|}{A-NH-C-O-B}}$	Urethane
"	2 A-NCO + H_2O	$\overset{O}{\overset{\|}{A-NH-C-NH-A}} + CO_2$	Urea
"	A-NCO + B-NH$_2$	$\overset{O}{\overset{\|}{A-NH-C-NH-B}}$	Urea
"	A-NCO + B-COOH	$\overset{O}{\overset{\|}{A-NH-C-O-B}} + CO_2$	Amide
"	A-NCO + $\overset{O}{\overset{\|}{A-NH-C-NH-B}}$	$\overset{O}{\overset{\|}{A-N-C-NH-B}}$ $\underset{O=C-NH-A}{}$	Biurette
"	A-NCO + $\overset{O}{\overset{\|}{A-NH-C-O-B}}$	$\overset{O}{\overset{\|}{A-N-C-O-B}}$ $O=C-NH-A$	Allophanate
"	A-NCO + CH$_2$—CH-B (epoxide, O bridge)	A-N——CH$_2$ $O=C$ CH-B (O bridge)	2-Oxazolidone
Cyclotrimerization			Isocyanurate
"	3 A-NCO	(isocyanurate ring structure) A-N, N-A, O=C, C=O, N-A	
Polycondensation	2 A-NCO	A-N=C=N-A + CO$_2$	Carbodiimide
"	A-NCO + (phthalic anhydride structure)	A-N (imide ring with benzene) + CO$_2$	Imide

Raw Materials for Isocyanate–Based Foams

The major raw materials for making isocyanate–based foams include the following compounds: polyisocyanates, polyols, catalysts, blowing agents, surfactants, epoxides, and flame retardants.

Polyisocyanates.   The polyisocyanates which can be used for preparing isocyanate–based foams are mainly aromatic compounds and some aliphatic or aralkyl polyisocyanates. Major polyisocyanates in the market are listed in Table 3. TDI is widely used for flexible foams. Pure MDI is used for elastomers and coatings. Modified TDI and modified MDI are used for high–resilience flexible foams. Polymeric isocyanates (polymeric MDI or oligomeric MDI) are mostly used for preparing rigid urethane and isocyanurate foams, and in part, for preparing flexible and semi–flexible foams.

Table 3: Important Polyisocyanates

TDI : Toluene diisocyanate (Tolylene diisocyanate)

2,4-TDI   2,6-TDI

MDI : Diphenylmethane diisocyanate

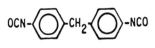

4,4'-MDI   2,4'-MDI

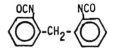

2,2'-MDI

(continued)

Table 3: (continued)

Polymeri MDI (Oligomeric MDI)

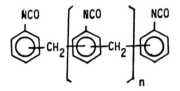

Tables 4 and 5 (48) show typical properties of both TDI and MDI.

Table 4: Typical Physical Properties of TDI

	TDI Isomer Ratio (2,4–/2,6–)		
	100	**80:20**	**65:35**
Physical state at normal temp.	Liquid/ Solid	Liquid	Liquid
Viscosity (mPa·s at 25°C)	(3–6)	3–6	(3–6)
Color		*	*
Odor		**	**
Specific Gravity (g/ml) (at 25°C)	1.21	1.21	1.21
Boiling temp. (°C)	251	251	251
Flash temp. (°C)	(135)	135	(135)
Fire temp. (°C)	(142)	142	(142)
Freezing temp. (°C)	22	15	8
Vapor density (Air=1)	6.0	6.0	6.0
Vapor pressure (mbar at 25°C)	0.03	0.03	0.03
Molecular Weight	174.2	174.2	174.2

  * Colorless to pale yellow
 ** Characteristic pungent

NOTE: ( ) : expected value from result on 80:20 material

Table 5: Typical Physical Properties of MDI

	Typical Physical Properties	
	MDI (monomeric)	**MDI (polymeric)**
Physical State at normal temperatures	solid	liquid (oily)
Viscosity (mPa·s at 25 °C)	–	100–800
Color	white to light yellow	dark brown (opaque)
Odor	–	earthy, musty (characteristic)
Specific Gravity (@ 25 °C)	1.22 (43 °C)	1.23
Boiling Point (°C)	171 at 1,33 mbar 200 at 6,66 mbar 230 Decomposition	polymerizes about 260 °C with evolution of carbon dioxide
Flash Point (°C)	199	over 200
Fire Point (°C)	232	over 200
Freezing Point (°C)	38	below 10
Vapor Density (air = 1)	8.5	8.5
Vapor Pressure mbar at 25 °C	$<10^{-5}$	$<10^{-5}$

The conventional method of producing organic isocyanates is based on phosgenation of aromatic or aliphatic amines, as shown by the following model reaction.

$$R\text{–}NH_2 + COCl_2 \rightarrow R\text{–}NCO + 2\,HCl$$

Very recently, phosgene–free methods for producing organic isocyanates have been developed. One method involves reductive carbonylation of a nitro compound in the presence of a monoalcohol to produce a urethane compound, followed by thermal dissociation of the resulting urethane compound, as shown below:

$$R\text{–}NO_2 \xrightarrow{\quad 3CO + R\text{—}OH \quad} R\text{—}NH\text{—}CO\text{—}O\text{—}R' + 2\,CO_2$$

$$R\text{—}NH\text{—}CO\text{—}O\text{—}R' \xrightarrow{\quad \Delta \quad} R\text{—}NCO + R'\text{—}OH$$

This method was developed for producing TDI by ARCO Chemical

(145), Mitsubishi Chemical Corp. (30,148), Mitsui Toatsu Chemicals, Inc. (146), and Bayer AG (147), respectively.

Oxidative carbonylation was developed by Asahi Chemical Industry Co., Ltd. for producing MDI (31). The process consists of steps (1) oxidative carbonylation, (2) condensation, and (3) decomposition of the condensation reaction product, as shown below:

1. Carbonylation

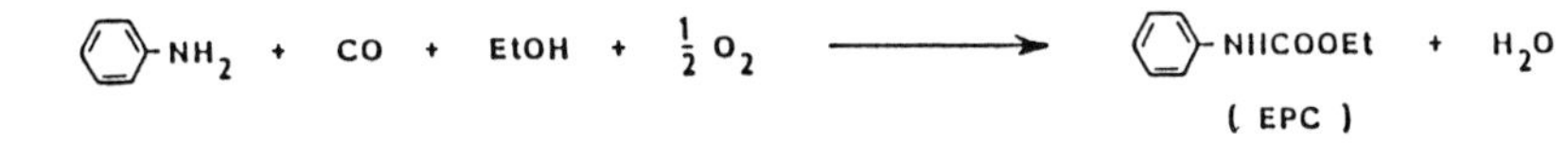

2. Condensation

1st Step: Condensation

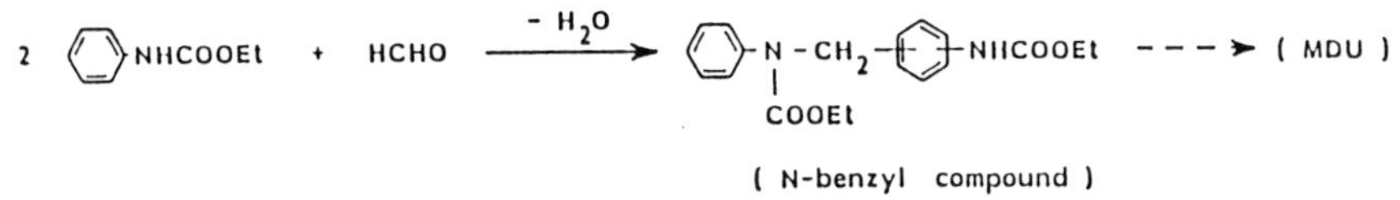

(N-benzyl compound)

2nd Step: Intermolecular Transfer Reaction

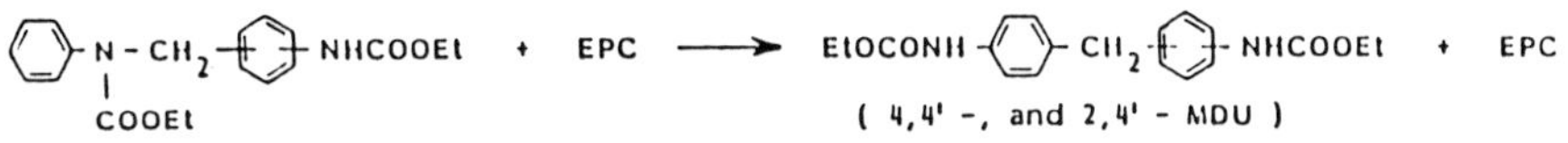

3. Decomposition

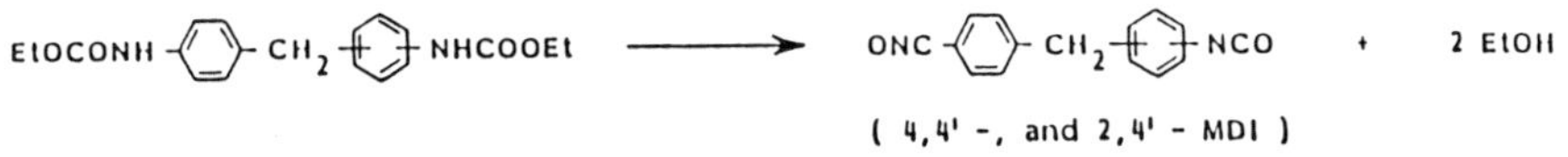

Another phosgene–free method was developed by Akzo Co. to product p–phenylene diisocyanate (32).

$$(1) \quad R{-}COOR \xrightarrow{NH_3} R{-}CO{-}NH_2 \xrightarrow{NaOCl} R{-}CO{-}NH{-}Cl$$
$$[I]$$

$$(2) \quad [I] \xrightarrow[-\ NaCl]{NaOH\ /\ HN(C_2H_5)_2} R{-}NH{-}CO{-}N{-}(C_2H_5)_2$$
$$[II]$$

$$
(3) \qquad [II] \xrightarrow[- (C_2H_5)_2 \text{ NH HCl}]{HCl} R\text{—}NCO
$$

$$
[III]
$$

In addition, American Cyanamid Co. has commercialized tetra-methyl xylene diisocyanate (TMXDI) in meta and para forms (117). The synthetic method is shown by the model reactions (1) to (3) as shown below. The NCO–groups are produced by the thermal dissociation of urethane groups. Recently, another similar method was disclosed by the same company (149). The NCO groups are also produced by the thermal dissociation of urethane groups.

$$
(1) \qquad H_2N.CO.NH_2 + CH_3OH \rightarrow H_2N.CO.O.CH_3 + NH_3
$$

$$
(2) \quad H_2N.CO.O.CH_3 + R\text{—}\underset{\underset{CH_3}{|}}{\overset{\overset{CH_3}{|}}{C}}{=}CH_2 \rightarrow R\text{—}\underset{\underset{CH_3}{|}}{\overset{\overset{CH_3}{|}}{C}}\text{—}NH.CO.O.CH_3
$$

$$
[I]
$$

$$
(3) \qquad [I] \rightarrow R\text{—}\underset{\underset{CH_3}{|}}{\overset{\overset{CH_3}{|}}{C}}\text{—}NCO + CH_3OH
$$

Modified polyisocyanates are prepared by incorporating at least one linkage into monomeric polyisocyanates. Such linkages include urethane, carbodiimide, allophanate, biuret, amide, imide, isocyanurate, and oxazolidone. These modifications provide some advantages, e.g., lower vapor pressure, increased viscosity, and controlled reactivity.

Some examples of modified polyisocyanates are isocyanate–terminated quasi–prepolymers (semi–prepolymers), urethane–modified MDI, carbodiimide–modified MDI, isocyanurate–modified TDI, and isocyanurate–modified isophorone diisocyanate.

Blocked polyisocyanates, which are addition compounds of labile hydrogen–containing compounds with polyisocyanates, are shown below:

$$\underset{\underset{H}{|}}{A-N-\overset{\overset{O}{\|}}{C}-B} \quad \rightleftharpoons \quad A-NCO \; + \; B-H$$

B–H in the above equation is a "blocking" agent, that is, a labile active hydrogen-containing compound, which includes phenol, nitrophenol, and epsilon-caprolactam.

Blocked polyisocyanates are inert compounds at ambient temperature, and they generate free polyisocyanates at elevated temperatures by thermal dissociation (208). Blocked polyisocyanate technology is used in one-component urethane coatings.

Polyols. The polyols for urethane foams are oligomers or polymeric compounds having at least two hydroxyl groups. Such polyols include polyether polyols, polyester polyols, hydroxyl-terminated polyolefins and hydroxyl-containing vegetable oils.

Polyether Polyols. The major polyols for preparing various urethane foams are polyether polyols. Polyester polyols are used only in specific applications. The advantages of polyether polyols are: choice of functionality and equivalent weight; the viscosities are lower than those of conventional polyesters; production costs are cheaper than for aliphatic polyesters; and resulting foams are hydrolysis-resistant.

Polyether polyols are prepared by the anionic polymerization of alkylene oxides, such as propylene oxide and/or ethylene oxide, in the presence of an initiator and a catalyst, as shown in the following equation:

$$R-(OH)_f \; + \; (f \; x \; n) \; \underset{\underset{O}{\diagdown\diagup}}{CH_2-\overset{\overset{R'}{|}}{CH}} \; \rightarrow \; R[-OCH_2-CH-)_n-OH]_f$$

where $R-(OH)_f$ represents initiators which are low-molecular-weight polyols having 2 to 8 functionality, shown as follows (functionality is in parenthesis): ethylene glycol (2), glycerol (3), trimethylolpropane (3), 1,2,6-hexanetriol (3), triethanolamine (3), pentaerythrytol (4), aniline (2), toluenediamine (4), alpha-methyl glucoside (4), sorbitol (6), sucrose (8). The initiators can be other types of active hydrogen-containing compounds, such as aliphatic or aromatic amines.

The functionality and equivalent weight of polyether polyols can be widely varied. This is a big advantage of polyether polyols over polyester polyols, and, for this reason, polyether polyols are used for producing various polyurethanes, e.g., rigid, flexible and semi–flexible foams, elastomers, coatings, adhesives, and resins.

The most widely used catalyst for the stepwise ring–opening polymerization of alkylene oxides is potassium hydroxide. This reaction (KOH catalyst), however, is accompanied by side reactions, e.g., the formation of allyl alcohol brought about by the isomerization of propylene oxide.

The allyl alcohol then yields vinyl–terminated polyether monols and the presence of monols results in many problems. Hence, the maximum molecular weight available as commercial products is limited to less than 5,000.

Very recently, a novel method for producing high–molecular–weight polyols without the formation of the monols has been disclosed by ARCO Chemical Co. (118, 150), and Asahi Glass Co. (223). The catalysts employed for this procedure were double metal cyanide salts, e.g., zinc hexacyanocobaltate complex, e.g., $Zn_3[Co(CN)_6]_2 \cdot xZnCl_2 \cdot y$ Glyme$\cdot zH_2O$. This catalyst was discovered in the 1960's by Herold and his co–workers at the General Tire and Rubber Co. (now GenCorp.) (205).

The catalyst makes it possible to product outstandingly high–equivalent–weight polyether polyols, e.g., about 10,000. In other words, polyether diols of 20,000 molecular weight and polyether triols of 30,000 molecular weight can be produced.

Another method of producing polyether polyols is the ring–opening polymerization of cyclic ethers, such as tetrahydrofuran, to produce polytetramethylene ether glycols or poly(oxytetramethylene) glycols, (PTMEG), as shown below.

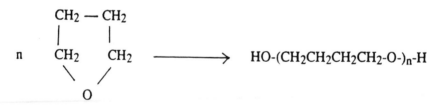

Modified polyether polyols have appeared in recent years, i.e., graft polyether polyols (polymer polyols, copolymer polyols) which were first developed by Union Carbide Corp. in the mid–1960's.

Polyurea dispersion polyols (PHD polyols, Polyharnstoff Dispersion polyols) were developed by Mobay Corp. (151). PHD polyols are usually produced by adding TDI into hydrazine–containing polyether polyols under vigorous stirring. These polyols are preferably used for producing molded flexible foams and high–resilience foams having high load-bearing properties.

Graft polyols include acrylonitrile–grafted, as well as acrylonitrile- and styrene–grafted polyether polyols. The percent of grafting was about 20 to 21% when these materials were first introduced commercially. Recently, however, polyether polyols having higher percentages of grafting, e.g., about 40 to 50%, have become available as commercial products (126, 193).

Polyester Polyols.   Polyester polyols for urethane and related polymer foams include: (a) aliphatic polyesters prepared by the reaction of dibasic acids, such as adipic acid, phthalic acid, and sebacic acid, with glycols such as ethylene glycol, propylene glycol, diethylene glycol, 1,4–butanediol and 1,6–hexanediol; (b) aliphatic polyesters prepared by the ring–opening polymerization of lactones, e.g., epsilon–caprolactone; and (c) aromatic polyesters prepared by the transesterification of reclaimed polyethylene terephthalate or distillation residues of dimethylterephthalate. The polyesters (a) and (b) are used for making flexible foams, elastomers, coatings and adhesives, and (c) is used for producing rigid urethane foams and urethane–modified isocyanurate foams.

Other Polyols.   Hydroxyl–containing vegetable oils such as castor oil were used for producing semi–flexible foams in the initial stage of the urethane foam industry, but they have not been used much in recent years.

New polyols, such as polycarbonate polyols (Duracarb, PPG Ind. Inc.), hydantoin–containing polyols (Dantocol DHE, Lonza Inc.), polyolefinic polyols (Poly bd, Atochem Co.) and its hydrogenated polyols, i.e., Polytail (Mitsubishi Chemical Corp.) are now available as commercial products. An application of polyolefinic polyols for foams has recently been reported (119). The chemical structures of the above polyols are shown below:

Polycarbonate Polyols (Duracarb):

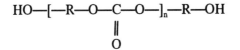

Dimethylhydantoin Polyols (Dantocol DHE):

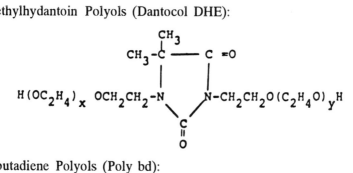

Polybutadiene Polyols (Poly bd):

$$HO \left[\left(CH_2-CH=CH-CH_2 \right)_{0.2} \left(CH_2-\underset{\underset{CH=CH_2}{|}}{CH} \right)_{0.2} \left(CH_2-CH=CH-CH_2 \right)_{0.6} \right]_N OH$$

Blowing Agents. Gas generation is an essential part of plastic–foam formation. In preparing thermosetting foams two kinds of blowing agents are used, i.e., chemical and physical blowing agents, and combinations of both.

Major blowing agents appearing in the literature are listed in Table 6, and Tables 7A through 7D.

Chemical Blowing Agents. The conventional gas–generation reaction for flexible urethane foams is the water–isocyanate reaction which was first described in a German patent (122). Its chemical reaction is shown as follows:

$$2 \text{ R—NCO} + \text{H}_2\text{O} \rightarrow \text{CO}_2 + \text{R—NH—CO—NH—R}$$

This reaction is carried out in two stages, i.e., carbon dioxide gas–generation, with the simultaneous formation of the substituted urea linkage.

Unconventional gas–generation reactions have been reported by Ashida (33). The blowing agents used include the following compounds (a) enolizable compounds such as nitroalkanes (nitroethane, nitropropane) aldoximes (acetaldoximes), nitrourea, acid amides (formamide, acetamide), active methylene–containing compounds, (acetylactone, ethyl acetoacetate), and (b) boric acid. The mechanisms of their gas–generation reactions are also discussed (33).

Recently, Speranza disclosed new blowing agents—carboxyl–

terminated polyether oligomers. The carboxyl groups in these agents react with isocyanate groups to generate carbon dioxide (234).

Gills has found a new gas–generation reaction, i.e., an aldehyde and a ketone having an isocyanate–reactive center within 6 carbon atoms of the carbonyl group reacting with isocyanate groups to generate carbon dioxide gas. Examples include hydroxyl acetone and o–hydroxybenxaldehyde (235). Formic acid has also been reported to be used as a chemical blowing agent (25); one mole of formic acid reacts with two isocyanate groups to produce one mole of CO gas and one mole of CO_2.

Physical Blowing Agents. Different from chemical blowing agents, e.g., water, physical blowing agents have the following advantages: (a) the reaction exotherm is removed in part by the evaporation of the physical blowing agents, and the resultant foams have reduced discoloration, scorching and fire risk; (b) the system viscosity is lowered, and pour–in–place foaming is facilitated; (c) some physical blowing agents (e.g., CFC–11) provide higher thermal insulation properties in foam formation than those of water–blown foams.

Physical blowing agents may be classified as CFCs (chlorofluorocarbons), HCFCs (hydrochlorofluorocarbons), HFCs (hydrofluorocarbon ethers) and non–fluorine–containing organic liquids. These fluorinated blowing agents can also be used in foaming polyisocyanurate foams, polyoxazolidone foams, and polyurea foams.

CFCs. Table 6 shows major CFCs. CFC–11 has been the representative blowing agent for both flexible and rigid urethane foams (155). The use of CFCs brought significant advantages to both flexible and rigid polyurethane foams. However, ozone–depletion potential (ODP) in the stratosphere have led to the worldwide ban of the production and use of CFCs by 1995.

HCFCs. Alternative blowing agents for CFCs, for the time being, are HCFCs (hydrochlorofluorocarbons) as shown in Table 7A, in which HCFC–141b is the most feasible alternative blowing agent for CFC–11. However, HCFCs also have problems, i.e., higher costs, and a small ozone–depletion potential, and therefore, HCFCs are also expected to be phased out by 2020.

HFCs. Hydrofluorocarbons (HFCs) as blowing agents are shown in Table 7B. Each of these agents have no ODP. Rosbotham, et al. used HFC–134a as an alternative blowing agent (230). HFC–356 has almost the same boiling point and molecular weight as CFC–11, and therefore it is an attractive alternative for CFC–11 (231), but its cost reduction for commercial applications is not clear.

Table 6: Major CFCs*

Designation	Formula	Mol. Wt.	B.P., °C	Vapor at 60°C W/mK	Approx. at Life Years	O.D.P.**
CFC-11	$CCl_3 F$	138.	23.8	0.0093	60	1.0
CFC-12	$CCl_2 F_2$	121.	-29.8	0.0114	100	0.9
CFC-113	$CCl_2 F CCl F_2$	188.	47.6	0.0044	90	0.8
CFC-114	$CCl F_2 CCl F_2$	171.	3.6	0.0060	200	0.6
CFC-115	$CCl F_2 CF_3$	155.	-38.7	0.00724	400	0.3

* Chlorofluorocarbons
** Ozone depletion potential

Table 7A: Alternative Blowing Agents (HCFCs)*

Designation	Formula	Mol. Wt.	B.P., °C	Vapor at 60°C W/mK	Approx. at Life Years	O.D.P.
HCFC-21	$CH\ Cl_2\ F$	103	8.9	-	-	<0.05
HCFC-22	$CH\ Cl\ F_2$	86.5	-40.8	0.0130	-	0.05
HCFC-31	$CH_2\ Cl\ F$	68.5	-9.1	-	-	<0.05
HCFC-123	$CH\ Cl_2\ CF_3$	153	27.9	0.0138	1.4	0.02
HCFC-124	$CH\ Cl\ F\ CF_3$	137	-11.0	0.0159	-	0.02
HCFC-132b	$CH_2\ Cl\ C\ Cl\ F_2$	99.5	46.8	-	-	<0.05
HCFC-133a	$CH_2\ Cl\ CF_3$	116	6.1	-	-	<0.05
HCFC-141b	$CCl_2\ F\ CH_3$	117	32.1	0.0138	<1	0.15
HCFC-142b	$CClF_2CH_3$	101	-9.8	0.0159	-	0.06

* Hydrochlorofluorocarbons

Table 7B: Alternative Blowing Agents (HFCs)*

Designation	Formula	Mol. Wt.	B.P., °C	Vapor at 60°C W/mK	Approx. at Life Years	O.D.P.
HFC-134a	$CF_3\ CH_2\ F$	102	-26.2	0.01717	6	0
HFC-152a	$CHF_2\ CH_3$	66.0	-24.7	0.0182	-	0
HFC-356	$C_4\ H_4\ F_6$	166	24.6	0.0095	0.4	0

* Hydrofluorocarbons

Table 7C: Alternative Blowing Agents (PFCs)*

Designation	Formula	Mol. Wt.	B.P., °C	O.D.P.
PF-5040	C_4F_{10}	238	-2	0
PF-5050	C_5F_{12}	288	30	0
PF-5060	C_6F_{14}	338	56	0
PF-5070	C_7F_{16}	388	80	0

*Perfluorinated hydrocarbons.

Table 7D: Alternative Blowing Agents (HFEs)*

Designation	Formula	Mol. Wt.	B.P., °C	O.D.P.
E-245	$CF_3CH_2OCF_2H$	134	26	0
E-356	CF_3CHOCH_3	150	27	0

*Hydrofluorocarbon ethers

PFC's. Table 7C lists perfluorocarbons (PFC's). None of these agents have an ODP. Their boiling points lie in the appropriate range. A disadvantage is in their relatively high costs and incompatibility with urethane foam ingredients, e.g., polyisocyanates and polyols. Therefore, emulsion-type systems have to be developed (232). Other problems include their long atmospheric life, e.g., 100-10,000 years, which may affect the global warming potential (GWP).

HFEs. Hydrofluorocarbon ethers (HFEs) are shown in Table 7D. They have no ODP and suitable boiling points, nearly equal to CFC-11. Fishback and Reichel reported that the tested HFEs function as near drop-in replacements for CFCs and HCFCs for rigid PUR and PIR foams (233).

Non-fluorinated blowing agents. n-Pentane has been used in European countries, e.g., Germany, as a blowing agent for rigid urethane foams. According to Heiling and co-workers' test results, it has been concluded that there were no indications of higher risks in the case of a real fire. Specifically, the fear concerning explosive-gas mixtures of pentane and air was not confirmed (236). Explosion-proof dispensing machines have been developed by some companies. Cyclopentane can also be used as a physical blowing agent (241).

Recently n-pentane-based blowing agents of a blend type have been patented (196). This patent claims the use of a blend of liquid hydrocarbon and chlorinated hydrocarbon, e.g., a blend of n-pentane and methylene chloride. This method is a convenient way to produce various rigid foams, e.g., polyurethane foams, polyisocyanurate foams, and polyoxazolidone foams. Methylene chloride and pentane have nearly equal boiling points and their blends act like a single solvent. The use of methylene chloride alone results in foam collapse, but a blend of the two solvents does not result in such collapse. A blend of 80/20-90/10 wt % of methylene chloride/pentane is substantially non-combustible, and can be used as the blowing agent for polyisocyanurate-based foams. For rigid polyurethane foams, a blend of about 50/50 wt % is suitable. These blends could solve the disadvantages of 100% water-blown rigid foams mentioned above.

What will be the next generation of blowing agents? Decaire et al. (237) list the requirements for alternative blowing agents as follows: zero ozone depletion potential (ODP), non-flammable or moderately flammable, 50°C boiling point upper limit, and molecular weight below 180. In addition, the cost ($/mole) of a blowing agent is another important industrial factor.

The use of some azeotropic mixtures as blowing agents for rigid urethane foams have been proposed by Doerge. These blowing agents include CFC–11/methyl formate (238), and HCFC–141b/2–methyl butane (239). Ashida et al disclosed halogen–free azeotropes (228).

2–Chloropropane as blowing agent for rigid urethane foams has been developed by Recticel (240).

Mixed gas/liquid blowing agents for rigid urethane foams have also been proposed (245). The patent claims the use of hydrocarbons having boiling points (a) less than 10°C or (b) 20–30°C, or (c) an inert organic liquid having a boiling point of 35–125°C. Another mixed blowing agent for rigid urethane foams was proposed by a patent which claims the use of a mixture of cycloalkanes, e.g., cyclopentane and cyclohexane, and, if necessary, water (246). The non–fluorinated blowing agents described above can also be applied to polyisocyanurate foams, polyoxazolidone foams, polyurea foams, etc.

Methylene chloride has been used as an auxiliary blowing agent for a long period of time. In some countries, however, due to possible occupational and environmental problems, increased restrictions have been placed on the use of methylene chloride. Therefore, other types of auxiliary blowing agents have been proposed.

Liquefied carbon dioxide is proposed as an auxiliary blowing agent for water–blown flexible urethane foams (242). Hydrocarbons having a boiling point of 38–100°C are proposed for use in self–skin foam production (243). Blends of hydrocarbons having a boiling point above – 50°C and below 100°C have been proposed as auxiliary blowing agents for water–blown flexible foams (244).

Catalysts. Many kinds of catalysts for various isocyanate reactions have appeared in the literature. Some have relatively high selectivity for specified reactions and others have relatively low selectivity and act as catalysts for various isocyanate reactions.

In the case of urethane–foam formation, tin catalysts (Table 8) mainly promote the reaction between isocyanate and hydroxyl groups, i.e., the formation of the urethane linkage.

Table 8: Major Tin Catalysts

Stannous 2–ethylhexanoate	$[CH_3(CH_2)_3CH(C_2H_5)CO_2]_2$ Sn
Dibutyltin dilaurate	$[CH_3(CH_2)_{10}CO_2]_2$ Sn $[(CH_2)_3CH_3]_2$

In contrast, tertiary amine catalysts (Table 9) mainly accelerate the water–isocyanate reaction, that is, the gas–generation reaction, but are also catalysts for the hydroxyl–isocyanate reaction.

Table 9: Major Tertiary Amine Catalysts

Aliphatic Tertiary Amines

Bis-2-dimethylaminoethyl ether

$$(CH_3)_2NC_2H_4OC_2H_4N(CH_3)_2$$

N,N,N',N'-Tetramethylpropylene diamine

$$(CH_3)_2NC_3H_6N(CH_3)_2$$

N,N,N',N' -Tetramethylhexamethylene diamine`

$$(CH_3)_2NC_6H_{12}N(CH_3)_2$$

N,N,N'N' N"- pentamethyldiethylene triamine

$$(CH_3)_2N(C_2H_4)-N-(C_2H_4)N(CH_3)_2$$
$$|$$
$$CH_3$$

Aromatic & Alicyclic Tertiary Amines

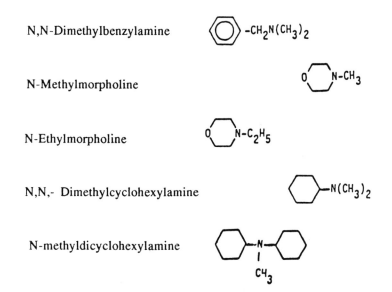

N,N-Dimethylbenzylamine

N-Methylmorpholine

N-Ethylmorpholine

N,N,- Dimethylcyclohexylamine

N-methyldicyclohexylamine

(continued)

Table 9: (continued)

Heterocyclic Amines

Triethylenediamine
(Dabco,) (1,4-diazabicyclo-2,2,2,octane)
[109]

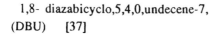

1,8- diazabicyclo,5,4,0,undecene-7,
(DBU)   [37]

2-Methyl-2 azanorbornane [125]

Recently, delayed–action catalysts have been used to adjust reactivity profiles for rigid and flexible molded foams (38). Some of these catalysts are shown in Table 10.

Table 10: Delayed–Action Catalysts

Polycat SA-1 [38]

Polycat SA·-102 [38]

Polycat SA-610/50   [38]

A comprehensive review of trimerization catalysts was prepared by Zhitankina et al (116). Isocyanate trimerization catalysts are shown in Table 11. 2–Oxazolidone catalysts and carbodiimide catalysts are shown in Tables 12 and 13 respectively.

The synthesis and properties of 2–oxazolidones and 2–oxazolidone-containing polymers have been reviewed by Pankroatov et al (115). It is interesting to note that some isocyanate cyclotrimerization catalysts also act as catalysts for urethane, oxazolidone and/or carbodiimide linkages.

Table 11: Cyclotrimerization Catalysts

Potassium 2-ethylhexanoate [39]	$CH_3(CH_2)_3CH(C_2H_5)COOK$
Potassium acetate [167]	CH_3COOK
Alkali metal salt of aminocarboxylic acid [189]	$(CH_3)_2\ N\ CH_2COOK$
Metal alkoxide [58,167, 169]	$NaOCH_3,\ KOC_4H_9$
Tertiary amine [172] [173]	$N(C_2H_5)_3\ ,\ C_6H_5CH_2N(CH_3)_2$ $[\ (CH_3)_2\ N\ C_2H_4]_2\ O$
Alkali hydrosulfide [173]	$NaSH$
Alkali cyanide [174]	$NaCN$
Alkali polysulfide [175]	Na_2S_2
Titanium (IV) n-butoxide [176]	$Ti[O(CH_2)_3CH_3]_4$
Sodium salicylate [177]	$2\text{-}(HO)C_6H_4COONa$

(continued)

Table 11: (continued)

Tributylphosphine [178   ] $[CH_3(CH_2)_3]_3$ P

Group Va Organometallics [169,170,171] Bu_2SbO, $(isoBu)_3AsO$

Choline [179] $HOCH_2CH_2N(CH_3)_3{}^+$ OH^-

Quaternary ammonium carboxylate

   (a) (Polycat SA-102) [37]

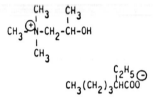

   (b) (Dabco TMR)   [187,188]

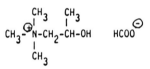

   (c) (Dabco TMR-2) [187, 188]

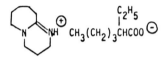

DBU salt of phenol. [37]
(Polycat SA-1)

Co-catalyst Combinations with Dabco

   Butanedione [180] $CH_3COCOCH_3$

(continued)

Table 11: (continued)

Alkylene Oxide [181]

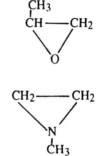

N-substituted
alkyleneimine [181]

Aldehyde [181]   C₄H₉ CHO

2,4,6-tris(dimethylaminomethyl)phenol
  [58]

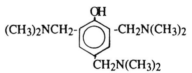

N,N',N"-Tris(dimethylaminopropyl)
sym-hexahydrotriazine [58]

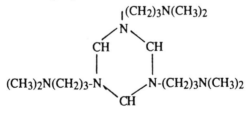

Titanates   [40]

                                    Ti [OC₂H₄N(CH₃)₂]₄

Silicates [40]   Si[OC₂H₄N(CH₃)₂]₄

Quaternary ammonium salt of tetrahedral boron-oxygen   complex.
[40]   ⊖   ⊕
                        (OC₄H₉)₄ B   (CH₃)₄ N

                                                  (continued)

Table 11: (continued)

Chelate Compounds

      Salicylaldehyde potassium [47]

      Quinizarin potassium [47]

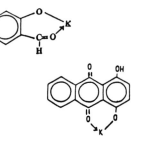

      Potassium acetylacetonate [47]

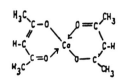

      Cobalt acetylacetonate [47]

Combination Catalyst

      Tertiary Amine/Alkylene Carbonate   [182]

      Tertiary Amine / Alcohol   [183,184]

Aminimides [185]

(continued)

Table 11: (continued)

Pyrazine Compounds   [186]

Table 12: Oxazolidone Catalysts

Tetraethylammonium bromide   [41]

$$(C_2H_5)_4N \cdot Br$$

Alkoxides   of II-a and III-a   metals   [42]

$$Al(OC_2H_5)_3$$

Complex of   a   magnesium halide and a phosphine oxide   [43]

$$MgCl_2 \ [(C_6H_5)_3 \ P(O)]_n$$

Alkali metal chelates   [44]

Complexes of Lewis acid   and   Lewis base.   [45]

$$AlCl_3 \cdot [(CH_3)_2N]_3 \ P(O)$$

Organoantimony iodide [225]

$$(C_6H_5)_3 \ Sb^{++} \ (I)^- \ (I_3)^-$$

Table 13: Carbodiimide Catalysts

Phospholenes [82]

Phospholanes [82]

Phospholene Oxides [82]

Pholpholane Oxides [82]

2,4,6-tris(diethanolamino)-s-triazine   [120]

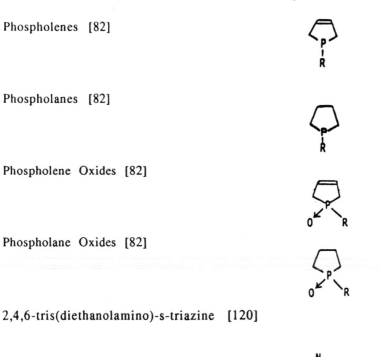

Surfactants. A surfactant is a major raw material for polyurethane foams. Surfactants play an important role in obtaining required cell structures, e.g., fine cells, coarse cells, closed cells, and open cells, and these cell structures then influence foam properties.

Surfactants for urethane and related polymer foams are usually silicone–surfactants. These surfactants generally are copolymers of poly(dimethylsiloxane) $[-Si(CH_3)_2-O-]_n$, oxyalkylene chains, e.g., polyethylene oxide chain $(EO)_n$, and polypropyene oxide chains, $(PO)_n$. The copolymers can be linear, branched or pendant types. The surfactants have different functions, i.e., emulsifying, foam stabilizing, and cell–size control.

The contents of EO, PO, and Si significantly affect the function of surfactants (226). For example, the relationship between molar ratios of

(EO + PO)/Si (or content of oxyalkylene chains) and molar percents of EO/(EO + PO) can influence surfactant properties such as meeting the needs of various applications.

Many choices of surfactants are commercially available for use in various foams such as flexible foams, HR foams, rigid foams, etc.

Epoxides. Epoxides are used as raw materials to make poly–2–oxazolidone foams. In foam preparation fast reaction is essential, and therefore aromatic epoxides are favored. Aromatic polyepoxides include bisphenol–A–based epoxides and novolac–based epoxides, as shown below.

Bisphenol–A–based epoxides:

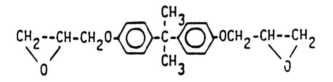

Novolac–based epoxides:

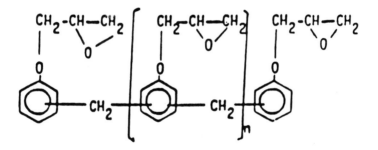

Flame Retardants. Among the isocyanate–based foams, polyurethane foams, both flexible and rigid, are flammable. Due to serious fire hazards of polyurethane foams, strict fire regulations have come out on the use of foams in the areas of furniture and public transportation. In addition, the use of rigid urethane foams in building insulation have resulted in stricter fire regulations.

For these reasons, studies to find effective flame retardants have been carried out by many companies over a long period, but these studies have not yet been fully successful. The problems encountered include ease of ignition, surface flame spread, heat release, smoke evolution, fire endurance and toxicity of fire gases.

Other types of isocyanate–based polymer foams, such as polyiso-cyanurate foams modified by oxazolidone, carbodiimide or imide linkages, have outstanding properties in flame retardance and fire endurance without the addition of any flame retardants

However, flame–retardant urethane foams must have flame retardants. For this reason many flame retardants have been developed for polyurethane foams and are readily available in the market (154).

Combustion is a radical reaction, and halogens generated from flame retardants act as radical scavengers for the radical reaction.

In contrast, phosphorus compounds act as char–forming agents which result in reduced generation of flammable gases. Both reaction mechanisms in flame retardance are different, and the combined use of the two is recommended.

Flame retardants for polyurethane foams can be classified into two categories, i.e., reactive and non–reactive. Reactive flame retardants include brominated polyethers, halogen– and/or phosphorus–containing polyethers, dibromopropanol, and dibromoneopentyl glycol. The effectiveness of halogenated flame retardants is in the following increasing order: fluorine, chlorine, bromine, and iodine.

Non–reactive flame retardants include both organic and inorganic compounds. Organic compounds include halogen–containing phosphates, e.g., tris(chloroethyl)phosphate, halogenated aromatic powders, e.g., chlorendic anhydride. Inorganic compounds include antimony trioxide, sodium borate, and aluminum hydroxide (alumina trihydrate). Antimony trioxide exhibits a synergistic effect when combined with halogenated compounds. Alumina trihydrate shows flame retardance by means of endothermic splitting of water, dilution of cracked gases formed by water vapor and protective–layer formation consisting of alumina.

Polyurethane Foams

Preparation. Polyurethane foams (often referred to as urethane foams) are prepared by the reaction of a polyisocyanate with a polyol in the presence of a blowing agent, a surfactant, and a catalyst without external heating of the foaming system. The principle of preparation of urethane foams is based on the simultaneous occurrence of two reactions, i.e., polyurethane formation and gas generation in the presence of catalyst and surfactant, as shown below:

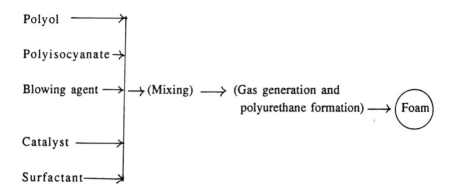

In flexible urethane foams, the major blowing agent is water and, at the same time, auxiliary blowing agents, i.e., low–boiling–point inert solvents such as CFC–11, methylene chloride, can be used. But in the case of rigid urethane foams, the major blowing agent has been CFC–11, which vaporizes due to the exothermic reaction of polyurethane formation. Model equations of polyurethane foam formation are shown in Figure 2.

a) Polyurethane Formation

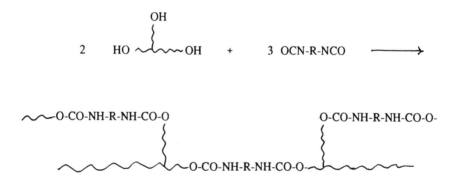

b) CO₂ Gas Generation

$$n \ OCN-R-NCO + n \ H_2O \longrightarrow (-R-NH-CO-NH-)_n + n \ CO_2$$

Figure 2. Schematic diagram of polyurethane foam formation.

Side reactions, e.g., formation of allophanate, biuret, isocyanurate, or carbodiimide linkages, may be formed, depending upon the reaction conditions.

In recent years, the ban on the use of CFCs resulted in major changes in foam formulations. A number of studies were carried out on the use of 100% water–blown foams for both rigid and flexible foams. These studies required modifications or improvements in raw materials, e.g., polyisocyanates, polyols, catalysts and surfactants.

Urethane foams can be classified into two principle types, i.e., flexible and rigid foams. In some cases, flexible foams can be further subdivided into flexible and semi–flexible (or semi–rigid) foams. The differences in physical properties of the two foams are mainly due to the differences in molecular weight per cross–link, the equivalent weight and functionality of the polyols, and the type and functionality of the isocyanate.

These foams can be prepared by the proper choice of equivalent weight and functionality of the polyols employed. Polyisocyanates can be considered as the joining agents of the polyols. A rough classification of the three kinds of foams based on the type of polyols used is shown in Table 14.

Table 14: Classification of Urethane Foams

Foam	Rigid Foam	Semi–Rigid Foam	Flexible Foam
Polyol:			
OH No.*	350–560	100–200	5.6–70
OH Equivalent**	160–100	560–280	10,000–800
Functionality	3.0–8.0	3.0–3.5	2.0–3.0
Elastic Modulus at 23°C			
MPa	>700	700–70	<70
lb/in^2	>100,000	100,000–10,000	<10,000

 * OH Number: mgKOH/g
** OH Equivalent: 56,110/OH Number

Processes of Urethane Foam Preparation. Urethane foams can be prepared by the one–shot process, semi–prepolymer (quasi–prepolymer) process or prepolymer process. The one–shot process is most commonly used. The semi–prepolymer process is sometimes preferred because of the advantages of easy processing, stabilized–foam rise and lower exotherm, but the prepolymer process is used only for limited purposes. Schematic diagrams of the three foaming processes are shown in Figure 3.

The one–shot process is used for flexible and rigid foams. In the case of slabstock foams, the ingredients are separately supplied to the mixing head. In order to adjust viscosity and mixing accuracy, some of the ingredients, such as polyol and tin catalyst, water and amine catalyst, are pre–mixed.

Two–component premix systems are widely used for different foam systems because these systems can be supplied in two kinds of drums. In addition, the machines for foam preparation can be of simple structure, and their handling and maintenance are easy.

The semi–prepolymer systems have the same advantages as the one–shot, two–component premix systems.

The above processes are conducted by mixing the components at ambient temperature for both flexible and rigid urethane foams. The above three foaming processes are classified as non–frothing, and the breakdown of the foaming phenomena is shown in Figure 4. The technical terms employed in the figure are explained in the following pages.

In contrast to the non–frothing process, the frothing process is conducted under pressurized mixing. The frothing process is further classified into the following three processes: (a) R–12 (CFC–12 process (25); (b) Thermal Froth process (157); and (c) Chemical Froth process (157).

These frothing processes are distinguished from conventional methods by the difference in foaming profile, i.e., the frothing processes have no cream time, and they begin to expand immediately from the mixing head like shaving cream. The difference between the conventional and frothing processes is also shown in Figure 5.

The technical terms employed in the figures are as follows: *Cream Time*—the time between the start of the mixing and the point at which the clear mixture turns creamy or cloudy and starts to expand; *Gel Time*—the time interval between the start of mixing and the start of gelation; *Rise Time*—the time interval between the start of mixing and the completion of expansion of the foaming mass; and *Tack–Free Time*—the time interval between the start of mixing and the time to reach a non–sticky state.

In the case of conventional urethane–foam processing, cream times are adjusted in the range of about 10 to 30 seconds, and rise times are adjusted in the range of about 60 to 120 seconds.

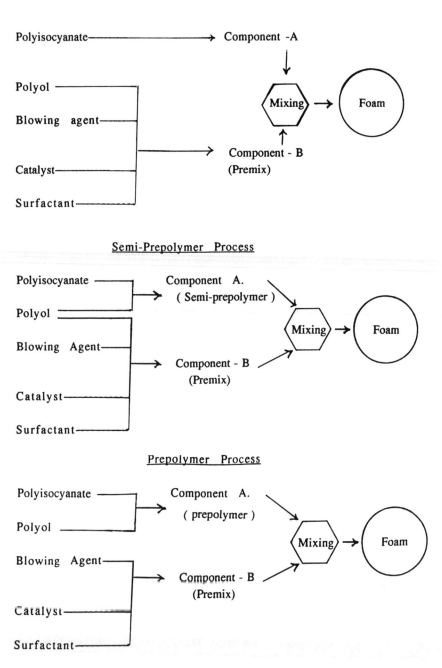

Figure 3. Schematic diagram of foaming processes.

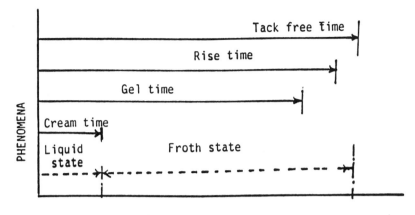

Figure 4. Breakdown of foaming phenomena.

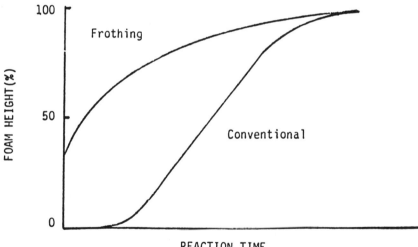

Figure 5. Comparison of rise curves.

In the case of flexible foams, gelling usually occurs before rise and tack–free times come after rise is completed. In contrast, tack–free times of rigid urethane foams sometimes are needed before rise is completed, although they are usually reached after rise.

The frothing process was developed by the Du Pont Co. (25). The process has the following advantages: isotropic physical properties and lower foaming pressure. The method is preferably used for large–panel production at in–plant production or pour–in–place foaming in field applications, e.g., building panels, chemical tanks, etc.

The CFC–12 frothing process is characterized by the co–use of CFC–12 (boiling point 29.8°C) with CFC–11, and occurs at ambient temperature.

The thermal frothing process is carried out by elevating the temperature of the foaming systems to a temperature above the boiling point of CFC–11.

The chemical frothing process is carried out by using an accelerated gas–generation reaction by means of a special chemical reaction at ambient temperatures. An example is the addition of a small amount of methanol in the foaming system (157).

Flexible Urethane Foams

Introduction

Classification. Flexible urethane foams have the largest market of all polyurethane products. The production properties and applications of various flexible urethane foams are described in the following sections. Flexible urethane foams are defined as open–cell urethane foams having the property of complete recovery immediately after compression. They can be classified into two kinds, i.e., polyether foams and polyester foams. Polyether foams are further classified as follows: conventional flexible foams, high–resilience flexible foams (HR foams), cold–molded foams, super–soft foams, and viscoelastic foams.

Semi–flexible foams (or semi–rigid foams) are sometimes classified as subdivisions of flexible foam because the foams have higher load–bearing properties and good compression recovery. Microcellular flexible foams and integral–skin flexible foams may also be classified in this category. In some classifications, however, microcellular foams are classified as elastomers.

Different foams can be prepared by the proper choice of polyols, which will be described later. Polyisocyanates are used as joining agents for the polyols, and therefore, polyols are considered to be the major components important in determining the physical properties of the resulting polymers.

Flexible urethane foams include slabstock foam, molded foam, and pour–in–place foam. In some cases, the latter two foams can be called flexible RIM foams (RIM is an abbreviation for reaction injection molding).

Hand–Mixing Process. In the research and development of urethane foams, the hand–mixing technique (or bench–mixing technique) is widely used in the research and development of various foams, not only for flexible foams, but also for rigid foams, rather than the use of miniature foaming machines. Well–trained technicians can prepare excellent foams with good reproducibility by means of the hand–mixing process.

Materials and Equipment

System Containers: paper cups (wax–coated) or polyethylene beakers: ca. 0.5 to 1.0 liter.

Foaming Containers: wooden box (about 15 x 15 x 15 cm or 30 x 30 x 30 cm, lined with kraft paper or polyethylene film), paper boxes or polyethylene buckets which are available as drum liners in about 2– to 5–liter capacity. The proper size of the containers is chosen based on the amount of foaming system to be used and on the estimated foam density. In bench foaming, about 1–liter (1–quart or 2–pint) containers can be used for screening experiments. Larger–volume containers are preferred for formulation studies, because such containers provide more uniform specimens for testing.

Tools for Hand Mix–Foaming: balance (electronic balance or triple–beam balance, capacity is 1 to 2 kg and sensitivity is 0.1 g); syringes for catalysts and surfactants (capacity is 5.0 ml, 1.0 ml and 0.1 ml); stirrer—spatula or glass rod; high–speed electric mixer; disposable wood sticks (ice–cream stick); and stop watch.

Foaming Procedures. The steps in the foaming procedure are as follows:

(a)   Adjust the temperature of raw materials to 20° to 25°C.

(b)   Preparation of Component A: Weigh polyisocyanate in a paper cup (the isocyanate component is normally called Component A).

(c)   Preparation of Component B: Weigh all other ingredients except polyisocyanate, e.g., polyols, catalysts blowing agents, surfactants, etc. in another paper cup and stir gently with a glass rod. Small amounts of ingredients, e.g., catalysts, surfactants, etc. are conveniently added by using syringes. Good accuracy can be obtained by choosing the proper size of the syringes. The volume of these ingredients can be calculated based on their density. The use of syringes is very simple, easy, and introduces no risk of over charge.

Trace amounts of catalysts, e.g., tin catalysts, are conveniently added by means of a pipette. The weight of one drop of catalyst can be determined in advance. Small amounts of catalyst can also be added by using an analytical balance. However, the use of such a balance is not recommended because of disadvantages, including time–consuming weighing procedures and risks of over charge.

Another important procedure is the method of addition of fluorocarbon blowing agent, e.g., CFC–11 (boiling point is 24°C)), into the foaming ingredients. Due to its low boiling point, it evaporates during mixing, and, therefore, the actual content of CFC –11 dissolved in the premixed system is not equal to the amount added. The resulting error sometimes exceeds 10%. For this reason it is necessary to repeat the addition and mixing procedures to obtain the exact calculated amount of CFC–11. After complete mixing CFC–11 does not evaporate in a short period of time, e.g., 1–2 minutes. The premix obtained is called Component B.

(d) *Foam Preparation:* The stirring of Component B in a paper cup is carried out for about 5 seconds. Immediately after the stirring, the total amount of Component A, already weighed in another paper cup, is poured immediately into Component B (the remaining amount of Component A sticking to the paper cup is determined in advance). Vigorous stirring is applied for about 5 to 20 seconds (the time should be shorter than the cream time), and then the mixture is immediately poured into a foaming container. In some cases, the mixed system is left and allowed to rise in the same container. The cream time, gel time, rise time, and tack–free time are measured and recorded. Modified methods also can be used, depending upon the sizes of the foam sample prepared, laboratory facilities, etc.

(e) *Post Cure:* After the completion of foam rise, the foam should be placed in an oven at about 60° to 80°C for several hours, preferably overnight, for the completion of the isocyanate reactions. Alternatively, room–temperature cure can be carried out by keeping samples for several days before testing or use.

Foam Properties and Testing Methods. In general, most physical properties of foams, both flexible and rigid, are proportional to the foam density. Therefore, at all times, the first physical property to be determined is foam density. Foam density can be calculated from the following equation:

$$\text{Density} = \frac{\text{Weight}}{\text{Volume}}$$

The unit of density is kg/m^3 (SI unit) and lb/ft^3 (pcf) (US unit). The conversion is as follows: $1.000 \ lb/ft^3 = 16.018 \ kg/m^3$.

Representative test methods of flexible foam properties, as defined by the ASTM D–3574 are: density, IFD (indentation force deflection), CFD (compression force deflection), sag factor, compression set, tensile and tear strengths, elongation, resilience, dry–heat aging and steam autoclave aging.

Flame retardance is another important property and is defined by different test methods. Some of the small–scale methods include horizontal flame spread (FMVSS 302, ASTM D–1692) vertical flammability (ASTM D–3014, so–called Butler Chimney Test), limiting oxygen index (ASTM D–2863), and smoke density (ASTM D–2840).

Very recently, stricter regulations in flame–retardant seatings for furniture have been issued. Representative testing methods are California Technical Bulletins 117 and 133, Boston Fire Department Chair Test, BS 5852: Part I and Part 2.

One of the most important properties of flexible urethane foams is cushioning. Figure 6 shows a comparison of hysteresis curves among polyester–, polyether– and HR–foams, respectively.

Sag factor (or comfort factor or support factor) is the ratio of the 65% compression force divided by the 25% compression force, and is used for evaluating cushioning ability. The sag factors of three kinds of flexible foams are shown below:

Foam	Sag Factor
Flexible polyester foam	1.2–1.7
Flexible polyether foam	1.4–1.8
HR foam	2.4–3.0

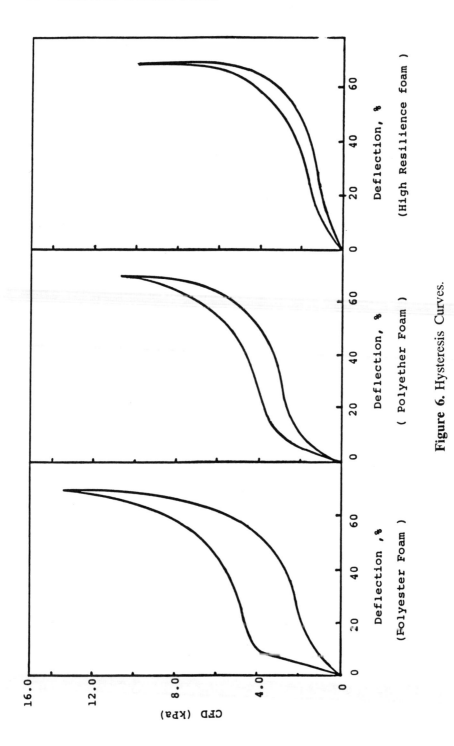

Figure 6. Hysteresis Curves.

Other important properties of flexible urethane foams include indentation force deflection (IFD) or compression force deflection (CFD), compression set, and humid aging.

Table 15 shows the unit conversion factors for five properties.

Table 15: Unit Conversion Factors — US to SI Unit

	US Unit	x	Multiplier	=	SI Unit
Density	lb/ft^3		16.018		kg/m^3
Stress or Pressure	lb/in^2		0.006895		MPa
Notched Izod Impact	ft–lb/in		53.38		J/m
Tensile Impact	ft–lb/in^2		2.103		KJ/m^2
Tear Strength	lb/in		0.175		KN/m
Thermal Conductivity	Btu–in/ft^2/°F		0.1445		W/(m.K)

Applications of Flexible Urethane Foams. Cushioning materials are the major application of flexible urethane foams. In 1980, the worldwide consumption of urethane foams was as follows (48): furniture and mattress applications, 37% and automotive applications, 18%. In addition to these applications, a wide variety of additional applications have been reported, including transportation, textiles, packaging, appliances, household materials, medical supplies, sound absorbents, sporting goods, cosmetics, agricultural applications such as artificial soil, and toys.

Slabstock Foams

The continuous loaf of foam made by the continuous pouring of liquid foaming components on a moving conveyor is called a slabstock foam, and a cut–off segment of the slabstock is called block foam or bun foam. The horizontal–conveyer process has been used widely since the beginning of the urethane foam industry. Recently, however, vertical production processes have been developed.

The cross section of slabstock foams produced by the horizontal conveyer process is not exactly rectangular, and has a slightly rounded top surface, which results in lowering the yield of fabricated foams. Accordingly, some improved processes for producing foams having rectangular cross sections have been utilized in the foam industry.

Figure 7 shows a scheme of a machine for continuous production of slabstock foams. Figure 8 shows a scheme of a machine for continuous production of laminates.

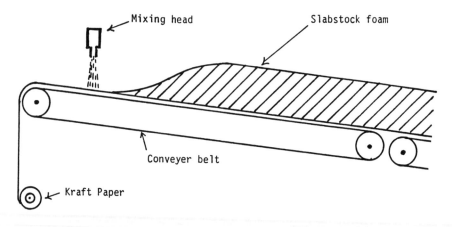

Figure 7. Continuous production of slabstock foams.

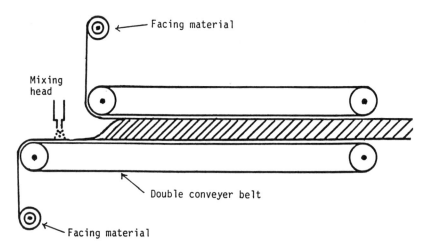

Figure 8. Continuous production of laminates.

Rectangular slabstock foams are more cost effective than conventional slabstock foams. The foams are produced by a horizontal process (36) and by a vertical process (123).

In the case of machine foaming, a one-shot process is widely used. All ingredients are pumped into the mixing head through separate streams. However, from the standpoint of viscosity, mixing ratio and ease of processing, some ingredients are pre-mixed, for example, water and amine catalyst or tin catalyst and a part of the polyol.

The foam densities of slabstock foams can be varied in the range of 10 to 30 kg/m³ (0.63 to 1.9 pcf), based on the market requirements. Higher-density foams also can be produced by the slabstock technique, if required.

A variety of formulations and their modifications are used for producing slabstock urethane foams. The formulation variations are based on available raw materials, foam properties required, product costs and production processes. Examples of formulations for producing free-rise slabstock foam are shown in Tables 16 and 17. A comprehensive list of slabstock formulations has been prepared by Iwata and his co-workers (191). The list is shown in Table 18.

Table 16: Formulation GM-21
(Slabstock foam with no flame retardant) (26)

	Weight Percent
Polyol — Glycerol-based poly(oxypropylene) poly(oxyethylene) polyol, 3500 mol.wt.	68.1
Isocyanate – TDI (80/20)	23.5
Blowing Agent – CO_2 produced by isocyanate reaction with water	1.7
Auxiliary blowing agent – Fluorocarbon	5.5
Catalyst – Organotin salt	0.2
Catalyst – Tertiary amines	0.3
Surfactant – Silicone copolymer	0.7

Physical Properties	US Unit	SI Unit
Density, lb/cu.ft. (kg/m³)	1.80	29
ILD		
25% R, lb/50 in² (N/323 cm²)	27.	120
65% R, lb/50 in² (N/323 cm²)	53	236
Sag factor	1.96	1.96
Tensile strength, lb/in² (N/m)	12.0	83.
Elongation %	200.	200.
Tear strength, lb/in (N/m)	1.6	280
Compression set, %	7.5	7.5
Rebound, %	50.	50.

Table 17: Formulation GM–23
(Slabstock foam with flame retardant) (26)

	Weight Percent
Polyol – Glycerol-based poly(oxypropylene) poly(oxyethylene)polyol, 3500 mol.wt.	62.9
Isocyanate – TDI (80/20)	23.5
Blowing agent – CO_2 produced by isocyanate reaction with water	1.7
Auxiliary blowing agent – Fluorocarbon	6.0
Catalyst – Organotin salt	0.2
Catalyst – Tertiary amines	0.3
Surfactant – Silicone copolymer	0.7
Flame retardant – Chlorinated phosphonate ester	4.7

Physical Properties	US Unit	SI Unit
Density, lb/ft^2 (kg/m^3)	1.75	28
ILD		
25% R, $lb/50$ in^2 ($N/323$ cm^3)	24	107
65% R, $lb/50$ in^2 ($N/323$ cm^3)	49	218
Sag factor	2.04	2.04
Tensile strength, lb/in^2 (kPa)	12	83
Elongation, %	200	200
Tear strength, lb/in (N/m)	1.6	280.
Compression set, %	7.5	7.5
Rebound, %	53	53

Table 18: Formulations and Properties of Various Flexible Foams (191)

Foams	Conventional	Super Soft	High Load-bearing	High Resilient
Formulation (phr)				
Polyether Triol	Conventioanl (OHV : 56) 100 phr	EO-capped (OHV:50, 75% EO) (70 phr) Conventioanl (30 phr)	AN-grafted (OHV: 31) (50 phr) Conventional (50 phr)	ST-grafteed (OHV:27) (50 phr) Conventional (50 phr)
TDI (80/20)	51.	41.	37.	37.
Water	3.9	3.5	2.8	2.7
Tertiary amine	0.1	0.2	0.25	0.18
Silicone surfactant	1.0	1.8	1.0	1.5
Stannous octoate	0.25	0.08	0.20	0.10
CFC-11	0	12.0	0	0
Properties				
Density kg / m^3 *	25.	20.	30.	35.
25% Hardness, kgf *	12.5	1.5	20.5	9.0
Elongation, % **	205.	380.	165.	160.
Ball rebound,% *	42.	40.	38.	60.

* JIS-K-6401,
** JIS-K-6301.(JIS stands for Japanese Industrial Standards.)

Molded Flexible Urethane Foams

Molded flexible urethane foams have been used for producing intricate foam products, such as automotive seats and furniture cushions. Molded foams are composed of high–density foam skin and low–density foam core. An example of density distribution of a 10–cm thick molded mattress foam is shown in Figure 9.

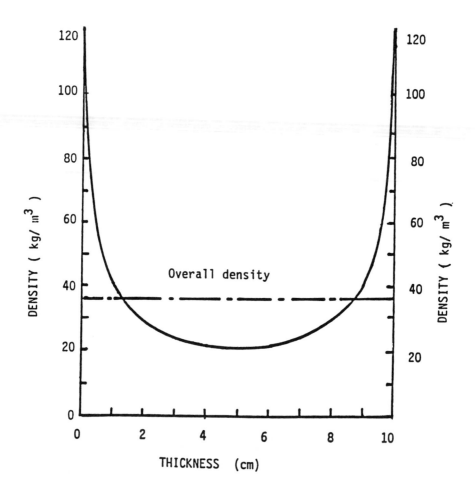

Figure 9. Density profile of molded flexible foam.

The CFD (compression force deflection) as well as IFD (indentation force deflection) curves of these foams are relatively linear in comparison with slabstock foams, as shown in Figures 10 and 11.

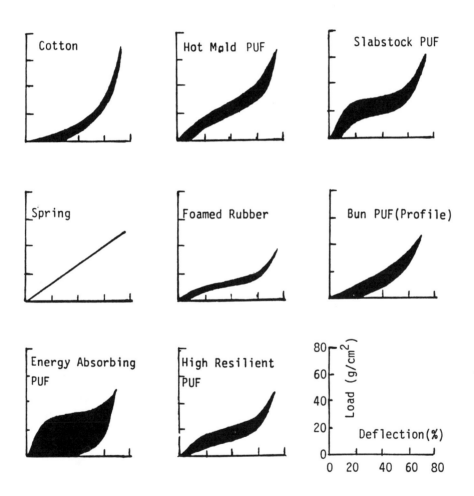

Figure 10. Hysteresis curves

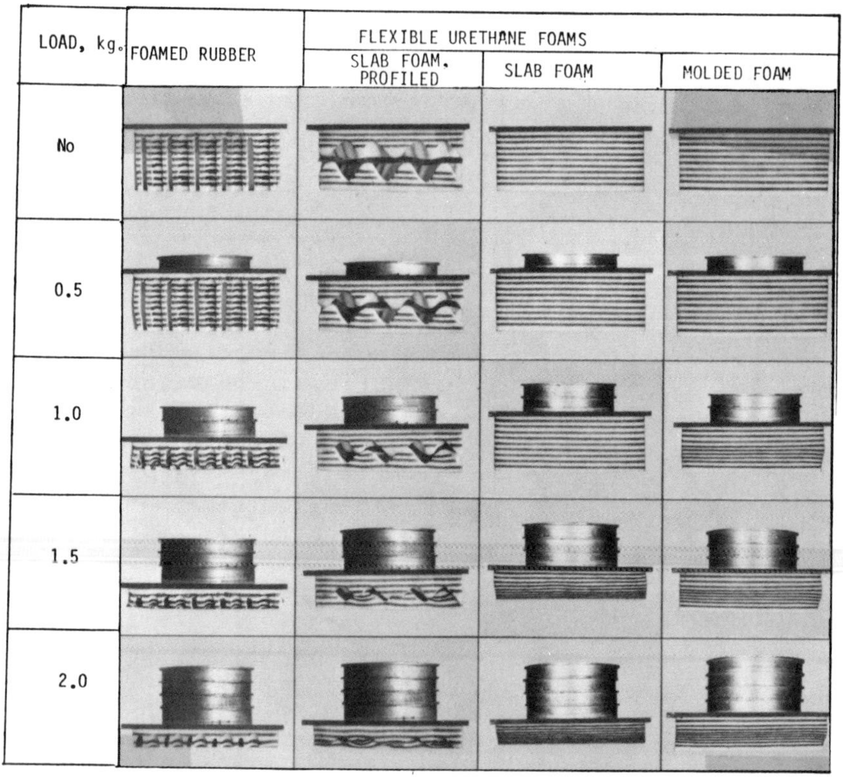

LOAD, kg.	FOAMED RUBBER	FLEXIBLE URETHANE FOAMS		
		SLAB FOAM, PROFILED	SLAB FOAM	MOLDED FOAM
No				
0.5				
1.0				
1.5				
2.0				

Figure 11. Compressive deformation of flexible foams.

In the molding process, mixed ingredients are poured through a mixing head into a pre-heated mold made of aluminum, steel, or epoxy resin. The mixed ingredients expand in the mold and the resulting foams are kept at the required curing conditions, followed by demolding.

The molded foam process can be classified into two types—the hot-molded foam process and the cold-molded foam process. The two processes are classified by the molding temperature and oven temperature.

HR foam (high-resilient foam) is sometimes classified as a cold-molded foam because it can be molded at low temperatures. However, HR foams have slightly different formulations than standard cold-molded foams. There are many kinds of molded-foam formulations in the literature. The following formulations are only a few examples.

Hot-Molded Foam and Cold-Molded Foam. Hot-molded foams are produced by using either conventional polyether polyols (3,000 MW,

3–functional propylene oxide adducts) or primary OH–terminated 3–functional polyether polyols with TDI, but not with MDI. For these reasons hot–molded foams can be considered as essentially all TDI–based foam.

Cold–molded foams, however, are prepared by using formulations based on blends of MDI and TDI, or 100% MDI, because the co–use of MDI with TDI makes it possible to use low oven and mold temperatures due to the faster reaction of MDI. The MDI can be either pure MDI, modified liquid MDI, or polymeric MDI. Some confusion may arise between cold–molded foam and high–resilience (HR) foam.

In recent years, HR foams have been prepared by using almost the same molding conditions as cold–molded foams, i.e., low mold temperature using the MDI–TDI blend formulation. However, in the development stage of HR foam, only TDI–based formulations were used, which required hot–mold conditions. For these reasons HR foams are not exactly equal to cold–cure foams. However, in recent years, HR foams have been produced by blends of MDI and TDI. The resulting foam systems can be molded at the same molding conditions as those for cold–molded foam systems.

In comparison with cold–molded foams, hot–molded foams have disadvantages of higher mold temperatures and longer mold–retention times, but have the following advantages: higher ratio of load–bearing/density, lower cost, and easier molding of complicated products. An example of a formulation of hot–molded foams is shown in Table 19.

Table 19: Formulation for Hot Molded Foam (4 inch thick)

Formulation:		Weight Percent
3000 MW Triol (Sec.OH)		100
Silicone surfactant		2.0
Dibutyltin dialurate		0.25
Stannous octoate		0.20
Triethylene diamine		0.10
CFC–11		10.0
Water		4,57
Molded foam density (overall)	kg/m³	25
	lb/cu.ft	1.6

High–Resilient Foam (HR Foam). HR foam is characterized by high ball rebound, low hysteresis loss and high sag factor. HR foam can be produced either as slabstock or as molded foam.

In automotive applications, e.g., automotive seats, molded HR foam is the major process used. HR foam is produced by using four kinds of technologies: (a) crosslinker technology, b) graft–polyol technology, (c) PHD polyol technology, and (d) specialty isocyanate technology.

HR foam usually contains closed cells which become a cause of foam shrinkage. In general, therefore, immediately after demolding, the foams are passed through a crushing roll to break the closed–cell membranes, and are then cured in an oven. The choice of proper surfactant for HR foam is also an important solution to the problem of shrunk or collapsed foams.

Crosslinker Technology. A low–molecular–weight organic compound containing two or more active hydrogen atoms, such as N–methyldiethanolamine, diethanolamine, triethanolamine, trimethylolpropane, is added to the basic polyol to form hard segments in the resulting foams.

Graft Polyol Technology. Graft polyols (or polymer polyols) are prepared by grafting both acrylonitrile and styrene monomer or acrylonitrile alone to conventional polyether polyols. Graft polyols provide increased load–bearing ability as well as cell–opening, which prevent or minimize the formation of closed–cell foams, because closed–cell flexible foams readily shrink.

Normally a 70/30 to 50/50 blend of a 4500–6500 EO–capped polyether triol with a polymer polyol is used, together with an 80:20 blend of TDI (80/20 isomer ratio) and polymeric MDI. Recently, higher–solids–content graft polyols, e.g., 30–50% solid polyols, have become available in the market.

PHD Polyol Technology. PHD polyols are prepared by adding TDI slowly into a blend of conventional polyol and hydrazine under vigorous stirring. The hydrazine reacts instantaneously with TDI to produce a polyurea dispersion. The dispersion is stable and does not precipitate. PHD polyols can give high–load–bearing foams having improved flame retardance.

Specialty Isocyanate Technology. The polyisocyanates employed for HR foams are TDI, liquid MDI, mixtures of TDI and liquid MDI or crude MDI. In addition to these polyisocyanates, specialty isocyanates can be used alone or in combination with the above polyisocyanates.

Specialty polyisocyanates include isocyanurate–containing TDI, biuret–containing TDI, allophanate–containing TDI, and urethane prepolymers. Specialty polyisocyanates also provide some advantages, such as high resilience, short tack–free time, and higher cross–link density. In general, the formation of open–cell, closed–cell or collapsed foams is explained by Figure 12 as shown below.

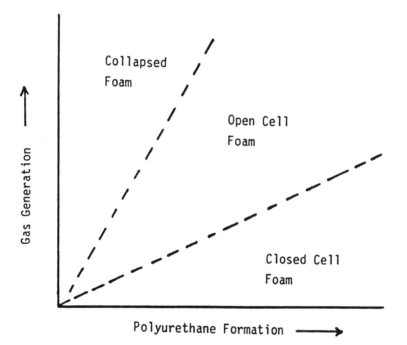

Figure 12. Balance of polymer formation and gas generation

It is clear from Figure 12 that open–cell foams can be obtained only by a good balance between two reactions, i.e., polyurethane formation and gas generation. When gas generation is too fast, the rising foam may collapse like beer bubbles, because the foam is not stable.

In contrast, when polyurethane formation is too fast, the resulting foam cells are stable, cell membranes are not broken during foam rise, and closed cells are formed. The closed cells are the cause of shrunk foams, because the internal gas pressure becomes lower than the atmospheric pressure.

Polyurethane formation is mainly accelerated by a tin catalyst, and gas generation, supplied by the water–isocyanate reaction, is mainly promoted by a tertiary amine catalyst. Therefore, if collapsed foams are obtained, increased tin catalyst can solve the foam collapse. Likewise, if closed–cell foams are obtained, increased tertiary amine catalyst and/or reduced tin catalyst concentration can solve the resulting shrinkage problems.

The effect of the balance between gas generation and polymer formation on foam–cell structure is shown in Figure 12. A formulation for high–resilience, flexible urethane foams is shown in Table 20.

Table 20: Formulation GM 25 (26)

	Weight Percent
Polyol–Glycerol–based poly(oxypropylene) poly(oxyethylene) polyol capped with primary hydroxyl groups, 6000 mol.wt.	62.4
Isocyanate – modified TDI	30.5
Blowing agent – CO_2 produced by isocyanate reaction with water	1.5
Catalyst – Tertiary amines	4.4
Processing aid – Phosphate ester	1.2

Physical Properties	US Unit	SI Unit
Density, lb/ft^3 (kg/m^3)	2.75	44
ILD		
25% R, $lb/50$ in^2 $(N/323$ $cm^3)$	26.	116.
65% R, $lb/50$ in^2 (N/m)	75.	334.
Sag factor	2.88	2.88
Tensile strength, lb/in^2 (kPa)	10.	69.
Elongation, %	100.	100.
Tear strength, lb/in (N/m)	0.8	140.
Compression set, %	3.5	3.5
Ball Rebound, %	73.	73.

A solution to the requirements is to make dual–hardness foams by one–step molding. Two kinds of processes are proposed. One process is to pour two kinds of systems through two mixing heads into a mold. Another method is to pour a single system through a mixing head into a mold and change the isocyanate index (159). This system is mainly based on MDI systems. An overview of different molding techniques for producing flexible urethane foams is shown in Table 21.

Table 21: Comparison of Molded Flexible Foams

TECHNIQUE	HOT-CURE	COLD - CURE	GRAFT POLYOL	PHD POLYOL	SPECIALTY ISOCYANATE
ISOCYANATE	TDI	MDI / TDI	TDI / MDI	TDI / MDI	Specialty isocyanate
POLYOL					
Molecular wt.	2800-3500	4500-6500	4500-6500	4500-6500	4500-6500
Terminal OH	Secondary/primary	primary	primary	primary	primary
Functionality	triol	triol	triol	triol	triol
MOLD TEMP.($^\circ$C)	35-65	25-35	35-45	35-45	35-45
POST CURING	No	No	Yes	Yes	Yes

Dual-Hardness Molded Foam: A number of formulations for molded foams have been reported. A new development in recent years is dual-hardness molded foams. Automotive seats and furniture have been labor-intensive products. These products provide good comfort by using different layers of different hardnesses. Molded foams produced by using a single formulation have relatively comfortable cushioning properties in terms of sag factor. However, this advantage is not enough to satisfy customers' requirements for comfort.

Microcellular Urethane Elastomers

Foamed urethane elastomers are also called microcellular elastomers. The densities are in the range of about 20 to 60 pcf (320 to 960 kg/m^3. Integral-skin foams having a density of ca. 700 to 1,000 kg/m^3 are sometimes referred to as microcellular elastomers.

Preparation of Microcellular Foams. The major polyols for microcellular elastomers include aliphatic polyester diols having a molecular weight of about 1,000 to 3,000, and poly-epsilon-caprolactones. Poly(oxytetramethylene) glycols (PTMEG) can also be used. The polyisocyanates to be used for microcellular elastomers are TDI-prepolymers and liquid MDI, i.e., carbodiimide-modified MDI or urethane-modified MDI. Low-molecular-weight, active-hydrogen compounds such as chain extenders (difunctional compounds) and crosslinkers (at least three functional compounds) are used with the above high–molecular–weight polyols. Examples of chain extenders are glycols, e.g., 1,4–butanediol, ethylene glycol, and aromatic diamines, such as MBCA (formerly MOCA) [4,4'–methylene–bis–(2–chloroaniline)]. MBCA results in excellent elastomers, but it is a suspected carcinogen, and therefore its use should be carefully monitored. Recently many MBCA substitutes have been commercialized.

Crosslinkers include triethanolamine and trimethylolpropane. The chain extenders and crosslinkers listed above form hard segments in the resulting elastomers and high–molecular–weight polyols form soft segments. A good balance of soft and hard segments gives superior elasticity.

The blowing agent for microcellular elastomers is water. The amount of water should be accurate, and its accuracy can be obtained by a water–containing solution, such as liquid sodium sulfonate of vegetable oils containing a small amount of water. The catalysts to be used are those used in urethane foams, e.g., tertiary amines, and tin catalysts. The above ingredients are mixed and poured into a hot mold and cured in a defined period of time. After demolding, a post cure is applied to complete the polymer–formation reactions.

Applications of Microcellular Elastomers. Microcellular elastomers are widely used for various shock absorbing materials in automotive applications such as bumper cores, shock–absorbing elements in vehicle–suspension elements, machines, electrical equipment, cameras, precision machines, shoe soles and heels, and sports shoes.

Integral–Skin Flexible Urethane Foams

Integral–skin foam is referred to as self–skinning foam or self–skinned foam. The foams have high–density skin layers and low–density cores. The overall densities vary in the range of about 200 to 1,100 kg/m^3 (12 to 70 pcf).

Integral–skin urethane foams are classified into two types, flexible and rigid. The former foams will be described here, and the latter will be described in the rigid–foam section.

Preparation of Integral–Skin Flexible Foams. The major polyisocyanates for use in making integral–skin flexible urethane foams are liquid MDI and TDI prepolymers. In order to make light–stable integral–skin foams, aliphatic diisocyanates, e.g., HDI (hexamethylene diisocyanate) and IPDI (isophorone diisocyanate) in modified forms are used.

Polyols used for the above foams include polyether polyols having a molecular weight of 2,000 to 8,000 and a functionality of two or three. Primary hydroxyl–terminated polyether polyols are preferably used. Chain extenders or crosslinkers are also used with the polyether polyols for the purpose of forming hard segments in the resulting elastomer foams.

The major blowing agents for making integral–skin foams are non-reactive solvents having low boiling points, such as trichloromonofluoro-methane (CFC–11, boiling point is 24°C). Such solvents do not evaporate on the cold mold surface and result in high–density–resin skin. The thickness of the skin is controlled by the difference between the mold temperature and the temperature of the foaming mixture inside the mold, as well as the foaming pressure generated by the overpacking of the foaming system into the mold.

Water is not suitable as a blowing agent for making solid skins, and for this reason it is used in only a few cases. Due to the ban on the use of CFCs, other methods are being studied. One method is 100% water-blown integral skin foams. Its physical properties was reported by Madaj et al. (248). The use of high molecular weight polyether polyols is proposed by Wada et al. (247). Another method is the use of inert solvents other than CFC–11, such as hydrocarbons (228, 243). The typical catalysts for making integral–skin foams are tertiary amines such as triethylene diamine, organotin salts such as dibutyltin dilaurate, and mixtures thereof. Additives, such as pigments, flame retardants, etc. can also be incorporated in the foaming systems.

Properties of Integral–Skin Flexible Urethane Foams. The foam densities of integral–skin foams of commercial products are in the range of about 300 to 950 kg/m³ and their Shore A hardnesses are in the range of about 90. Shore D hardnesses are about 40 to 90. Figure 13 shows an example of density distribution of integral–skin foams.

The scope of the applications of integral–skin foams depends mainly on the foam densities, as shown below:

Density (kg/m³)	Applications
300 (13–19 pcf)	Bicycle seats, interior car parts
600 (25–38 pcf)	Shoe soles
1,100 (44–69 pcf)	Automotive exterior body

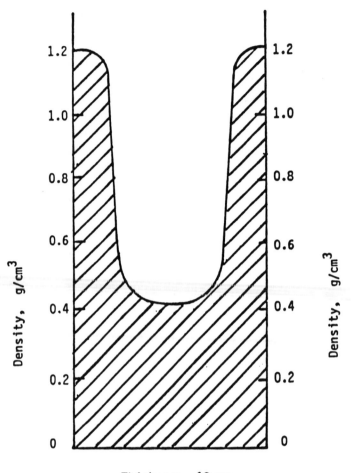

Thickness, 10 mm.

Figure 13. Density profile of integral–skin flexible polyurethane foam.

Flame Retardant Flexible Foams

The high flammability and toxic–gas generation of flexible and rigid urethane foams have been major problems in the urethane–foam industry, and accordingly considerable efforts have been focused on the production of substantially flame–retardant flexible foams.

Due to serious fire hazards caused by flexible urethane foam furniture, very strict regulations have been issued in the State of California (California Technical Bulletin 133), and in the U.K. (British Standard 5852, Part 2, Source 5), as well as in Italy. Because of this problem intensive research and development is being conducted to produce substantially flame–retardant flexible foams.

In 1983, Mobay Chemical announced its CMHR Foam (Combustion–Modified High–Resilience Foam). The foam is based on a PHD polyol and TDI and contains high loadings of hydrated alumina and a carefully chosen combination of flame–retardant additives (152). BASF Corp. proposed the use of high loadings of melamine as the flame retardant (130, 135). Recently the combined use of melamine and phosphorous–containing polyol was reported by Olin (132). The use of a blend of melamine and liquid flame retardant as well as the use of water as the sole blowing agent was reported by UCC (134).

The effect of the amount of melamine on the flame retardance of the resulting foams was studied in detail by Shell (131) and Dow Chemical (133). A combined use of ammonium sulfate and aluminum hydroxide was proposed by Toyo Rubber Chemical Industries Corp. (136). Recently, blends of ammonium sulfate and weakly basic organic compounds, e.g., sodium carboxylates, were patented (158, 165).

Halogenated phosphate esters (26) and many other types of flame retardants such as phosphorous–containing compounds including chlorine and/or bromine were extensively reviewed by Hilado (154).

Nitrogen–containing organic compounds other than melamine, such as melamine phenolphosphoric acid salt (160), methylolmelamine and alkyl ethers (161), cyanic acid derivatives, e.g., cyanamid, dicyandiamide, guanidine, and biguanidine (162), melamine–formaldehyde precondensate (163a, 163b), and aminoplast resins (164) have appeared in the patent literature.

Non–CFC–Blown Flexible Urethane Foams

The need for alternative blowing agents to replace CFCs was discussed in the section on blowing agents. It was noted that the production and use of CFCs having an ozone–depletion effect will be banned worldwide by 1995.

Methylene chloride has been used in part as an alternative for CFC–11. The sole use of methylene chloride results in problems in processing and environment, i.e., toxicity.

Water as the sole blowing agent results in some problems, including high viscosities, higher exotherm, which may result in greater fire risk, hard foam, high compression–set values, high hysteresis loss, low elongation, low tear strength and difficulty in producing integral–skin foam. The sole use of water as a blowing agent was studied by many companies: for molded foams (153b, 153f), for slabstock foams (153e), with the use of special polyols (153b, 153d) and with the use of a modifier (153c). 100% HCFC–blown flexible urethane foams have also been investigated (153a). Numerous papers regarding alternative blowing agents for both flexible and rigid foams have appeared every year in the proceedings of the Society of the Plastics Industry (SPI), Polyurethane Division.

Viscoelastic Foams and Energy–Absorbing Foams

Viscoelastic foams show a stress–relaxation phenomenon, i.e., delay of complete deformation recovery after compression. These foams usually take 2 to 30 seconds to recover after 50% compression. In contrast, HR foams and conventional flexible polyurethane foams show very short deformation–recovery times, e.g., less than 1 second. This means that these foams have low viscoelasticity or small energy absorption.

Viscoelastic foams are characterized by high magnitudes of energy absorption, hysteresis loss %, sound damping, and vibration damping. In one particular use the foams show a high magnitude of draping by means of the cantilever test.

Due to the high energy–absorbing abilities of viscoelastic foams, the foams are sometimes called energy–absorbing foams or sound–damping foams. However, the sound–absorption and impact–absorption properties are somewhat different. Therefore, optimum formulations for making them are not exactly the same.

An energy–absorbing foam has been used as a bumper core material for automobiles (195). Sound and vibration damping of viscoelastic MDI–based IPN foams was reported by klempner et al (50) and Gansen et al (194).

Other examples of viscoelastic foams include NASA's space–shuttle seats, seats for those who spend a long time sitting, e.g., truck drivers, office workers, airline pilots, and handicapped people using wheelchairs, sporting goods which require resiliency, energy absorption and comfort, such as bicycle seats, gym mats, helmet linings and boat cushions. Leg guards for ice hockey, ski boots, hiking boots and ice skates are other interesting applications. Medical uses, such as cervical pillows, prosthetic devices for orthopedic and wheelchair padding are also important applications (192).

Polyolefinic–Polyol–Based Flexible Foams

These foams are prepared by using polyolefinic polyols, liquid MDI, and water as the sole blowing agent (81). It is interesting to note that the resulting foams are characterized by very low compression–set values over a wide range of isocyanate indices, e.g., 80 to 120. This allows the use of a wider range of mixing ratios.

In addition, this system has been modified by adding dibenzyl sorbitol to give high thixotropicity to the system to prevent penetration of the system into foam cells. For these reasons the foam system is very suitable for pour–in–place application into small foam voids, e.g., 1 to 10 cm³, for repairing surface air voids of molded automotive–foam seats to reduce the rejection rate, thereby resulting in lowering production costs.

Semi–Rigid (or Semi–Flexible) Foams

The definition of semi–rigid (or semi–flexible) urethane foams is not clear, i.e., the border line between rigid foams and flexible foams is not well defined. However, it is generally recognized that semi–rigid foams have a higher load–bearing property and higher hysteresis loss than flexible foams. Due to the complete recovery of compressive deformation, these foams are usually classified as flexible urethane foams in the statistics of urethane–foam consumption. An example of the hysteresis–loss curve of semi–rigid foams is shown in Figure 14. A classification of semi–rigid foams based on polyols has been described in the previous section.

Manufacturing Process. Both prepolymer and one–shot processes are available, but the polymeric isocyanate–based one–shot process is used in preference because of the easy processing due to the low viscosity of the system, relatively low toxicity of polymeric isocyanates, and fewer environmental problems. However, a disadvantage of the one–shot process is a possible risk of shrunken–foam formation due to its higher closed–cell content. In contrast, the TDI–based prepolymer process has advantages including better in–mold flowability and higher open–cell content.

Prepolymer Process. Semi–prepolymers are prepared mostly by the reaction of TDI and a polyol to obtain a free NCO content of 5 to 10%. The polyols employed for the prepolymers are generally branched amine– or glycol–initiated polypropylene oxide–based polyether polyols having a molecular weight of ca. 600 to 4,000. Polyesters having molecular weights of about 1,000 to 2,000 can also be used.

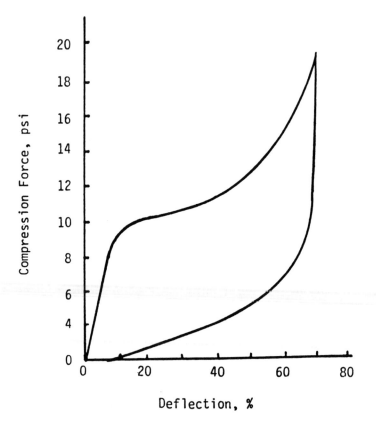

Figure 14. Hysteresis loss of semi–rigid foam.

One–Shot Process. Monomeric TDI (80/20 and 65/35 isomer ratios) was used as the polyisocyanate component in the early stages of the urethane–foam industry. But due to the toxicity problems of TDI at the present time, most of the isocyanates used for producing semi–rigid foams have been replaced by polymeric isocyanates, and the use of TDI is somewhat limited.

In most cases semi–rigid urethane foams are produced by using two–component systems, that is component A (polyisocyanate component) and component B (blend of the remaining ingredients, i.e., polyol, blowing agent, catalyst and surfactant).

As component B, a mixture consisting of 100 pbw of a polyether polyol (MW 3,000 to 6,000), 5 to 20 pbw of chain extender (e.g., ethylene glycol, 1,4–butanediol, diethanolamine, methyldiethanolamine, etc.), 1 to 3 pbw of water, 0.1 to 1.0 pbw of urethane–foam catalyst and

0.5 to 2.0 pbw of surfactant is usually employed. In rare cases, CFC–11 or alternative physical blowing agents are used as an auxiliary blowing agent. A hand–mix formulation is shown in Table 22.

Table 22: One Shot Semi–Rigid Foam Formulation

Component A:	
Polymeric isocyanate (NCO% : 31)	70 pbw
Component B:	
Polyether polyol (OH No, M.W. 3,000)	95.
Quadrol	5.0
Water	3.0
TMBDA	0.5
Dabco T–12	0.03
Surfactant, L–5310	2.0
Properties:	
Density, kg/m^3	54.
Tensile strength, kg/cm^2	1.27
Elongation, %	50.

Applications. Major applications of semi–rigid foams are in the area of automotive parts, especially in the interior of vehicles, for use as shock–absorbing pads, e.g., instrument panels, console–box lids, door trims, head restraints, arm rests, and sun visors.

Rigid Urethane Foams

Introduction. Rigid urethane foams are hard (high ratio of load bearing/density) foams having very low flexibility. They show permanent deformation, i.e., no complete recovery after compression. In other words, the compression deflection curves of rigid urethane foams exhibit yield points.

As opposed to flexible urethane foams, rigid urethane foams have a highly cross–linked chemical structure and a high percent of closed cells, e.g., over 90%. Rigid urethane foams can be classified as follows: unmodified (or pure) rigid urethane foams and modified rigid urethane foams, which include isocyanurate–modified, epoxy–modified, amide–modified and oxazolidone–modified rigid urethane foams.

Preparation. The polyisocyanates utilized in rigid urethane foams include TDI–prepolymers, crude TDI (undistilled TDI), modified MDI and polymeric isocyanates (or polymeric MDI). TDI (80/20 isomer ratio) is not widely used because its high isocyanate content results in scorched foams in large block–form preparation.

In principle, both the one–shot process and semi–prepolymer processes have been used for rigid–urethane–foam manufacturing. However, the monomeric TDI–based one–shot process was used only in the initial stage of the rigid–urethane–foam industry because of the toxicity problems of TDI and difficulties in controlling reactivity due to the high NCO percent. For these reasons TDI–prepolymers, blends of TDI prepolymers and polymeric isocyanates, and 100% polymeric isocyanate are most widely used.

In the household–refrigerator industry, however, the one–shot process employing crude TDI (which has a lower percent NCO and contains oligomeric compounds, such as isocyanurate– and carbodiimide–types of oligomers) is still widely employed because of its low cost.

In the preparation of TDI–based semi–prepolymer, TDI–80/20 and a polyol, preferably sucrose– or sorbitol–based polyol, are reacted to obtain about 30% free NCO–containing semi–prepolymers.

The major polyols used for rigid urethane foams are polyether polyols. Polyester polyols were used in the beginning of the urethane–foam industry but their use was discontinued because of their high viscosities, low functionality, low dimensional stability of the resulting foams, and high costs. The major blowing agent for rigid foams has been CFC–11.

The relationship between the amount of CFC–11 and the foam density is shown in Figure 15 (212). The relationship between the amount of water and foam density is shown in Figure 16 (212).

TDI Prepolymer Process. This process has the major advantage of better flowability in pour–in–place processes, e.g., household–refrigerator insulation, than crude TDI–based or polymeric MDI–based processes. An example of TDI–based prepolymers and a formulation for making rigid urethane foam on a small scale is shown below.

Preparation of TDI Semi–Prepolymer: The calculated amount of TDI 80/20 needed to obtain prepolymers having 30% free NCO is charged into a glass flask which is then heated with stirring to about 70° to 80°C. At the same temperature, the stoichiometric amount of polyether polyol is added slowly into the TDI with stirring. The reaction temperature is kept below 90°C by controlling the rate of addition of the polyol and by adjusting the heating mantle.

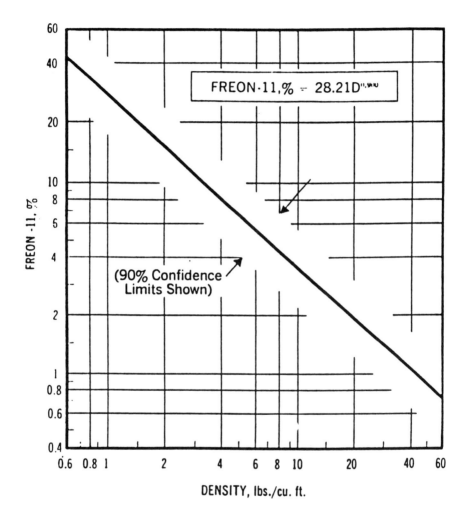

Figure 15. Effect of Freon–11 concentration on density (212.

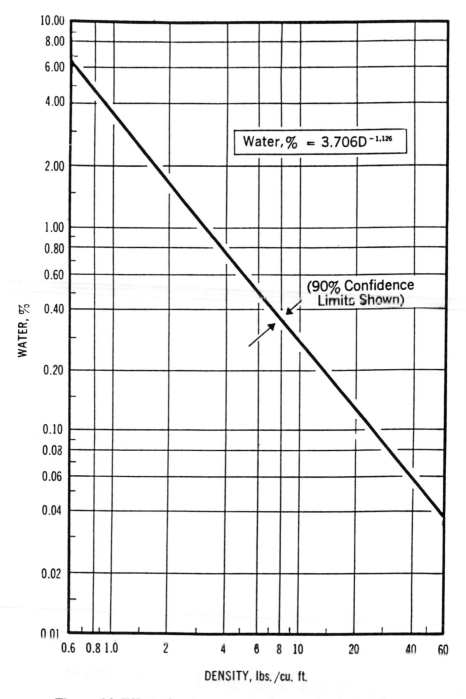

Figure 16. Effect of water concentration on density (212).

After the exothermic reaction is finished, additional heating is applied for another 0.5 to 1.0 hour at the same temperature. Then the heating mantle is removed and the reaction kettle allowed to cool. This reaction needs no catalyst. In a small–scale preparation, e.g., 1 to 2 liters, the following simple method can be used. Both TDI 80/20 and polyol are mixed at room temperature in a flask with stirring. The mixture is then heated to 70° to 80°C with stirring and maintained at that temperature for about 2 to 3 hours. The reaction product is then allowed to cool to room temperature. No catalyst is needed to prepare these polymers.

Before the prepolymer preparation, it is important to confirm that the polyols used have a low water content, e.g., less than 0.02% and a low pH value in order to prevent gelation in the prepolymer preparation. Table 23 shows rigid urethane–foam formulation by the semi–prepolymer process for obtaining a 30–kg/m^3 density foam by hand mixing.

Table 23: Formulation for Rigid Foam Made by the Semi–Prepolymer Process

Component–A: Semi–prepolymer:	
(TDI/sucrose–based polyol, % NCO:30)	100 pbw
Component–B	
Sucrose–based polyether polyol (OH No. 360)	117.
CFC–11	15.0
Silicone surfactant	1.0
Amine catalyst	1.5
Tin catalyst	0.2

Crude TDI—One-Shot Process. In place of the TDI–semi–prepolymer process, a one–shot process using crude TDI is widely used for refrigerator insulation because of its lower cost, although the flowability of this system is inferior to the TDI–prepolymer process.

Crude TDI has an average functionality of more than 2 because it contains TDI–dimer, TDI–trimer, and carbodiimide–containing compounds. The NCO percent of crude TDI is about 40%, which is lower than pure TDI but is higher than crude MDI as well as TDI–based semi–prepolymers, both of which have about 30% free NCO. Accordingly, the kinds and amount of catalysts and surfactants for the one–shot, crude–TDI process are slightly different from those of TDI–based semi–prepolymers.

Polymeric MDI—One-Shot Process. Polymeric isocyanate–based fluorocarbon–blown rigid urethane foams are the most widely used rigid urethane foams at the present time because the foams have a high thermal insulation, low toxicity, low environmental problems, ease of processing, and fast curing time (therefore, higher productivity than TDI–based foams). An example of one–shot formulations for preparing low–density rigid urethane foam is shown in Table 24.

Table 24: Formulation for One–Shot Process

	Parts by Weight
Component–A	
Polymeric MDI (Index)	105
(% NCO:30)	
Component–B	
Polyether polyol (Sucrose based, OH No.:360)	100.
Blowing agent, (CFC–11)	28.0
Silicone surfactant (L–5340)	1.5
Tin catalyst (UL–6)	0.3
Amine catalyst (Dabco 33 LV)	2.0
Reactivity, seconds	
Cream time	25
Rise time	36
Tack–free time	45
Foam Properties	
Density pcf, (core)	2.60
Compressive strength, psi	
Parallel	55.1
Perpendicular	17.7
K–factor (BTU. in/hr.ft^2 °F)	0.135

Cream time, rise time, gel time and tack–free time are influenced by environment temperature, component temperature, and catalyst level.

Non–CFC-Blown Rigid Urethane Foams. Rigid urethane foams are characterized as closed–cell foams. Their thermal conductivity is mainly affected by the blowing gases such as carbon dioxide and CFC–11. In the early days of the rigid urethane foam industry the foams were blown by water. However, the outstanding growth of rigid urethane foams is based on the high insulation (low thermal conductivity) of CFC–11–blown foams. In recent years, due to the ozone–depletion problems in the stratosphere, the production and use of CFCs have become subject to worldwide regulation.

Alternative blowing agents for CFCs are HCFCs, e.g., HCFC–141b, HCFC–123 and HCFC–22. The combined use of HCFC's and water as blowing agent was also proposed (153g). The sole use of HCFC–22 as the blowing agent for rigid urethane foams has also been studied (153h). The sole use of water as the blowing agent for these foams has been reported (153i, 153j). Methods for the use of reduced CFCs by the combined use of CFC–11 and water were studied by several companies (153k, 153n). The catalyst/surfactant systems for water–blown rigid urethane foams were studied by some companies (153l, 153m).

A blend of hydrocarbon and halogenated hydrocarbons has appeared recently in the patent literature (196). An example is a blend of pentane and methylene chloride in a weight ratio of about 50/50.

Due to the ban of the use of CFC–11, alternative blowing agents have been widely studied. The physical blowing agents include HCFCs, HFCs, FCs, HFEs, and non–fluorinated blowing agents. These blowing agents have been described in detail earlier in this section.

Flame–Retardant Rigid Urethane Foams. The major applications of rigid urethane foams are in the appliance and building industries. The latter requires strictly flame–retardant rigid foams, because serious rigid urethane foam fires have been reported in the housing, construction, and ship–building industries.

A number of flame retardants for rigid urethane foams have been developed over the past 30 years. Nevertheless, substantially flame–retardant, and fire–resistant rigid foams are not available, because the urethane linkage is thermally unstable and decomposes to produce low–molecular–weight flammable compounds.

In principle, the preparation of substantially flame–resistant rigid urethane foams is very difficult. Substantially flame–retardant and fire–resistant rigid foams can be produced only by employing more thermally stable linkages than the urethane linkage, such as the isocyanurate linkage. However, flame propagation or flame spread of rigid urethane foams can be retarded by adding flame retardants.

Flame–retardant rigid foams can be classified by the testing methods employed, but the results do not reflect actual fire situation. Fire–retardant rigid urethane foams can be prepared by using flame retardants of the additive type, reactive type, or a combination thereof. A review of flame retardants for polyurethane foams has been prepared by Hilado (154).

Another route to flame–retardant rigid foams is the use of flame–retardant polyether polyols which contain phosphorous and halogen (reactive type). In recent years, due to the fire–gas toxicity caused by halogen–, phosphorous– or nitrogen–containing flame retardants, other types of flame–retardants which do not produce toxic gases are being developed.

Production Technologies of Rigid Urethane Foam. Rigid urethane foams can be produced by different methods, as shown below.

Slabstock Production. Slabstock rigid foams are produced continuously by using a horizontal–conveyor–type machine which is similar in principle to Figure 7. In contrast, discontinuous production is carried out by using a box. The box–foaming method is not efficient, but it is suitable for small–scale production of various foams with low investment cost. Slabstock foam and box–foaming foam are used for the production of insulation boards, pipe coverings and many other insulation materials.

Foam–In–Place. The foam–in–place (or pour–in–place) method is used for the production of refrigerators, deep freezers, sandwich panels, and similar applications. This process is also used for field applications, such as indoor– and outdoor–tank insulation (79), LPG (liquefied petroleum gas) tank insulation, heavy oil–tank insulation, chemical–tank–car insulation, and pipe–covering insulation, among others (79).

Molding. This method is used for producing molded–foam products such as pipe coverings, window frames, chair shells, and picture frames.

Lamination. This method is used for producing laminated panels having flexible–facing materials, such as aluminum foil, kraft paper, and asphalt paper. An example of a machine used in lamination is shown in Figure 8. The panels produced are used as insulation board for roofs and walls. Rigid facing materials such as gypsum board can also be used for semi–continuous production of building materials.

Spraying. Two–component rigid–foam systems can be sprayed onto any surface at a temperature of about 10°C and higher. Examples of spray–insulation applications are outdoor tanks, such as heavy oil–storage tanks, wall surfaces in cold–storage warehouses, and provisions of ships.

Properties of Rigid Urethane Foams. *General.* Almost all physical properties of rigid urethane foams depend on their foam densities, as shown in the following equation:

$$P = A D^n \quad (\text{or } \log P = \log A + n \log D)$$

where P is property, D is density, and A and n are constants. Note that log (property) and log (density) form a linear relationship. The constant n is usually in the range of 1 to 2, and A relates to both foam properties and temperature.

Figures 17 through 21 show the relationships between physical properties and foam densities (29), in which Figures 17 through 19 show the effect of foam density on physical strengths, e.g., compressive, tensile and flexural strengths.

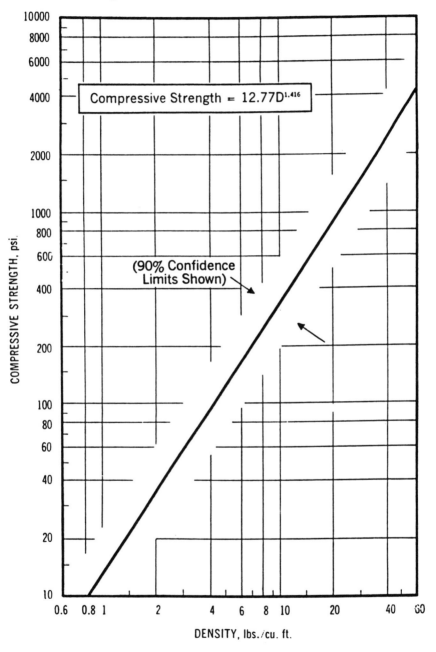

Figure 17. Effect of density on compressive strength (212)

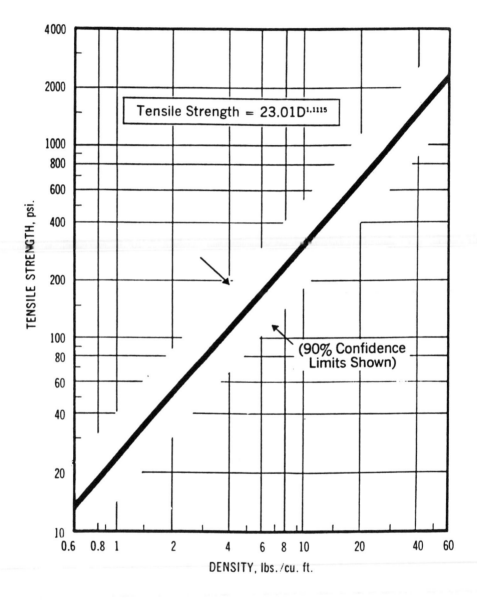

Figure 18. Effect of density on tensile strength (212).

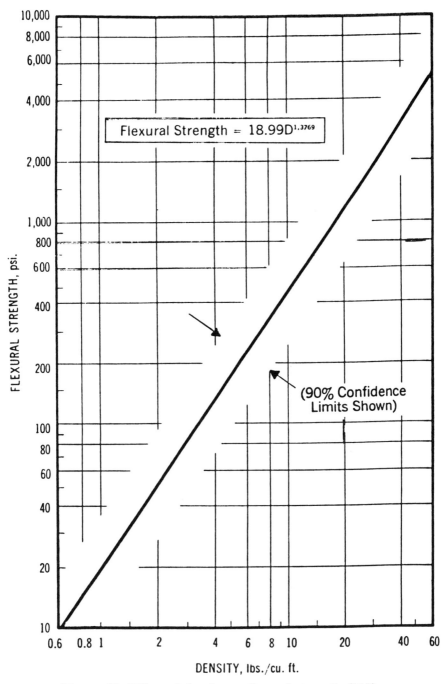

Figure 19. Effect of density on flexural strength (212).

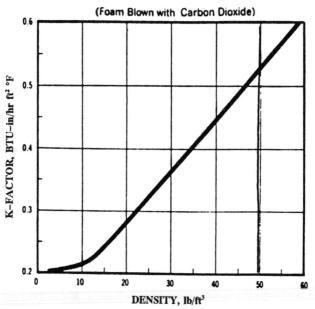

Figure 20. Effect of density on K–factor (foam blown with carbon dioxide) (212).

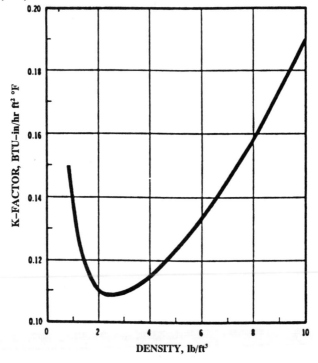

Figure 21. Effect of density on thermal conductivity (foam blown with CFC–11) (212)

Figures 20 and 21 show the effect of foam density on thermal conductivity. Water–blown (carbon dioxide blown) foam (Figure 20) shows a linear relationship between foam density and thermal conductivity (212). In contrast, CFC–11–blown foam (Figure 21) shows a minimum value of thermal conductivity at a density of about 2 lb/ft^3 (212).

Figure 22 shows aging of rigid urethane foam with cut surfaces. In the case of panel foams, however, aging is apparent only in the panel-side foam, and no aging is observed in core foams, as shown in Figure 23 (212). Figure 24 shows the effect of temperature on thermal conductivity (213).

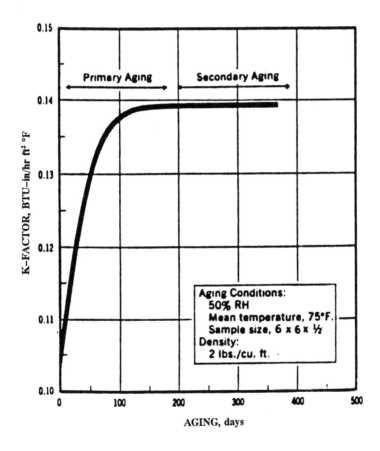

Figure 22. Aging of rigid urethane foam with cut surfaces (foam blown with CFC–11) (212).

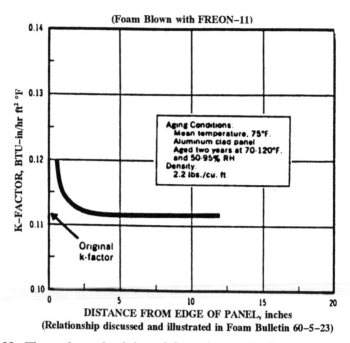

Figure 23. Thermal conductivity of foam in panels (foam blown with CFC–11) (212).

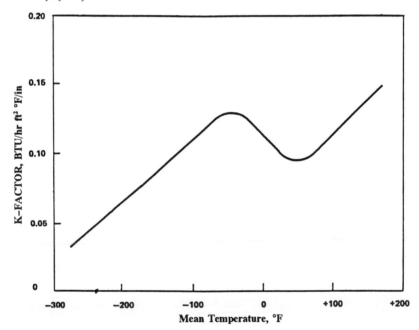

Figure 24. Effect of temperature on thermal conductivity (1 in thick specimens cut from 2 pcf rigid CFC–11 blown polyurethane foam)(213).

Miscellaneous Urethane Foams

These foams include isocyanurate–modified rigid urethane foams, isocyanurate–modified flexible urethane foams, urethane–oxazolidone foams, urethane–based IPN foams, and urethane–based hybrid foams.

Isocyanurate–Modified Rigid Urethane Foams. In the early stage of the urethane–foam industry, incorporation of isocyanurate linkages into rigid urethane foam was attempted because polyester polyols have relatively high viscosities and low functionalities which make it difficult to produce dimensionally stable, highly cross–linked rigid urethane foams. For this reason isocyanurate linkages were used to increase the crosslink density, and low–functionality and low–viscosity polyesters were used. Accordingly, the isocyanate index employed was relatively high, typically 150 (46).

This method, however, did not result in improved flame retardance because the isocyanurate content was too low. In recent years, however, high–functionality and low–viscosity polyether polyols have become available, and therefore the above method has become less important. Even so, the method was sometimes used to make foams having increased crosslink density to improve dimensional stability or chemical resistance. Therefore, a higher isocyanate index, e.g., 150 to 200, was used to incorporate isocyanurate linkages in rigid urethane foams.

It should be noted that the incorporation of isocyanurate linkages in small quantity does not improve high–temperature resistance nor flame retardance of the resulting foam. Urethane–modified isocyanurate foams which have isocyanate indices of more than 300 show outstanding high–temperature resistance and flame retardance (71). These foams will be discussed in the next section.

Isocyanurate–Modified Flexible Urethane Foams. The incorporation of isocyanurate linkages into flexible urethane foams results in lower melting points of the flexible foams and increased dripping, thereby preventing flame propagation, as tested by ASTM D 1692. Thus, this practice was applied to produce "so–called" flame–retardant high–resilience foams (49). It should be emphasized that the incorporation of isocyanurate linkages in flexible urethane foams does not increase thermal stability of the resulting foams, but decreases their melting point, and gives a low flame–spread rating by the ASTM D 1692.

Urethane–Based IPN Foams. Interpenetrating polymer networks (IPNs) are types of polymer alloys composed of the entanglement of at least two cross–linked components (112). An ideal IPN has essentially no covalent bonds between the polymers. The resulting morphology shows phase separation.

The first IPN resin was invented by Aylsworth in 1914 (51). The mixture of rubber, sulfur, phenol and formaldehyde, on heating, yields a simultaneous interpenetrating polymer network

Urethane–based IPN resins were first developed by Frisch et al (111, 113), but little literature is available on foamed IPN's. Recently, urethane–epoxy–IPN flexible foams were reported to have improved energy–absorbing and sound–absorbing properties (50). Semi–rigid IPN foams have been prepared by simultaneous polymerization of urethane–isocyanurate and epoxy–resin systems. According to Frisch et al (110), significantly higher compressive strengths were obtained with the IPN foams compared to the corresponding urethane–isocyanurate foams.

Urethane–Based Hybrid Foams.   Hydroxyl–terminated ester oligomers or vinyl esters which have terminal vinyl groups and pendant hydroxyl groups can react simultaneously with isocyanate groups and vinyl monomers in the presence of catalysts to obtain crosslinked hybrid polymers.

Unsaturated polyesters with terminal hydroxyl or carboxyl groups can also react at the same time with isocyanate groups and a vinyl monomer, such as styrene, to produce hybrid polymers. Two kinds of reactions, the NCO–OH addition reaction and radical polymerization by the vinyl groups, can occur. In parallel to the two reactions, interference between the two reactions also can occur. Detailed studies on the interference reactions were studied by Hsu et al. (190).

Examples of formulations for preparing unsaturated polyesters are shown in Table 25.

Table 25: (A) Formulations for Unsaturated Polyesters

Materials	A	B	C
Isophthalic acid, moles	1.00	1.00	1.00
Maleic anhydride, moles	1.00	1.00	1.00
Ethylene glycol, moles	0.0	1.5	1.96
Diethylene glycol, moles	2.64	2.42	0.0
Neopentyl glycol, moles	0.0	0.0	1.96
Hydroquinone, ppm of resin solids	200.	200.	200.
Styrene monomer, Wt.% of resin	35	35	35
p–Benzoquinone, ppm of resin solids	50	50	50

Table 25: (B) Properties of Polyol Diluted with Styrene

Materials	A	B	C
Chemical Properties			
hydroxyl number (actual), solid	111	151	171
equivalent weight (theoretical), solids	383	337	334
Viscosity at 25°C, Brookfield, 60% solid,Cp	248	248	385
SPI gel test(1% BPO; 82.2°C)			
gel time, min:sec.	4:00	5:05	7:00
cure time, min:sec.	7:35	7:50	10:10
peak exotherm, °C	180	211	211

Figure 25 shows model reactions of some urethane/unsaturated compound–based hybrid–resin formation.

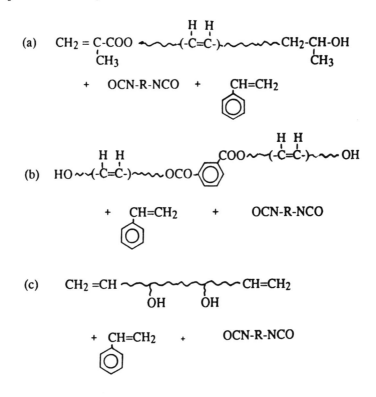

Figure 25. Urethane–based hybrid foams.

Some examples of hybrid–resin–formation reactions are shown below.

Example (1): Unsaturated monoalcohol (acryloesterol) and styrene monomer are reacted with a polyisocyanate in the presence of a urethane catalyst and a radical polymerization catalyst to form hybrid resins (52). Styrene monomer acts as a crosslinker and at the same time, acts also as a reactive diluent. The trade name of a commercial product of such systems is Arimax (Ashland Chemical) (107).

Example (2): Isophthalic acid–based unsaturated polyesters may be reacted with polyisocyanates in the presence of styrene monomer, a peroxide and a urethane–forming catalyst, as shown in the model reaction (b). Formulations for producing polyesters and the physical properties of the resulting polymers are listed in Table 25. Rigid foams having differing densities were prepared by adding a blowing agent, such as CFC–11 or water (53, 54).

Example (3): Vinyl ester–urethane hybrid foams may be prepared according to the model reaction (c) in the presence of a blowing agent such as CFC–11 (55, 127).

Urethane/Oxazolidone Foams. A urethane/oxazolidone structural foam (which is an integral–skin rigid foam) with surface area larger than 10 ft^2 has a flame–spread index of 50 or less, as required by UL–478 for large components (56). This foam will offer new markets for RIM products for aircraft, building, and military applications. The foam system is composed of component A (polyisocyanate) and component B (epoxy–terminated prepolymer prepared by the reaction of polyalkylene glycols with phthalic anhydride and a diepoxide). When the epoxy–terminated polyol reacts with polyisocyanate, the resulting foams have both urethane and oxazolidone linkages.

Polyisocyanurate Foams

Introduction. It has been recognized that CFC–11–blown rigid urethane foams are the insulation materials with the lowest thermal conductivities, in comparison with other insulation materials, such as water–blown rigid urethane foams, glass–fiber materials, or polystyrene foams.

Rigid urethane foams have been used exclusively in areas which require high thermal insulation and flame retardance, such as household refrigerators, deep freezers, and cold–storage warehouses. However due to the low flame retardance and poor fire resistance of rigid urethane foams, serious fire hazards have been reported and building applications have been restricted by building codes.

The addition of flame retardants, either additive or reactive types, can provide flame–retardant foams having low flame spread or surface flammability, but flame retardants do not improve the temperature resistance of these foams because the thermal stability or the dissociation temperature of the urethane linkage is relatively low and unchanged by the addition of flame retardants, i.e., the linkage dissociates at about 200°C to form the original components in polyol and polyisocyanate. The dissociation can result in further decomposition of polyol and polyisocyanate into low–molecular–weight compounds at elevated temperatures. For these reasons urethane foams are not temperature–resistant nor thermally stable.

In contrast, the isocyanurate linkage is thermally stable, as determined by TGA, as shown by a model compound study (58) and produces less combustible gases. Accordingly, unmodified isocyanurate–based polymers, e.g., resins and foams, are thermally stable, and therefore, temperature– and flame–resistant. In other words, the unmodified polyisocyanurates decompose at higher temperatures than the polyurethanes, and generate lower amounts of combustible gases than polyurethanes.

The isocyanurate linkage is obtained by the cyclotrimerization of isocyanate groups, as shown in the following model reaction.

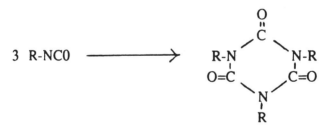

The first patent regarding isocyanurate–containing urethane foams was disclosed by Windemuth et al (46). The aim of the patent was to increase the crosslink density of urethane foams by incorporating isocyanurate linkages as crosslinkers with the co–use of low–functionality and low–viscosity polyester polyols, because the handling and the dispensing of high–functionality polyester polyols were very difficult due to their high viscosities. This type of foam can be classified as "isocyanurate"–modified urethane foam (46).

Burkus (201) and Nicholas and Gmitter (58) reported on TDI prepolymer–based isocyanurate foams prepared by trimerizing NCO–terminated prepolymers. TGA data (58) showed that there was no significant difference in thermal stability between the urethane foam and urethane–modified isocyanurate foams, probably due to cleavage of labile urethane linkages present in significant amounts. However, no discussion of flame retardance was given in this reference.

Substantially flame–retardant and high–temperature–resistant isocyanurate foams were first invented independently in 1965 by Ashida (39) and in 1966 by Haggis (199). Since 1966 a number of papers and patents regarding modified isocyanurate foams have appeared. A few reviews on urethane–modified–isocyanurate foams are also available (144, 202, 203, 204).

The above modification philosophy, i.e., lowering crosslink density, has been extended to various modified–isocyanurate foams. The methods for modifying isocyanurate foams are summarized in Table 26.

Table 26: Modified Isocyanurate Foams

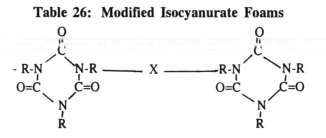

Where   X : Linkage of modifier

Modifier		Linkage
Polyol	Urethane	-NH-CO-O-
Polycarboxylic	Amide	-NH-CO-
Carboxylic anhydride	Imide	-R⟨CO,CO⟩N-
(No modifier, catalyst or heat)	Carbodiimide	-N=C=N-
Polyepoxide	Oxazolidone	-N⟨CH₂,CH-,C-O,O⟩
Diamine	Urea	—NH—CO—NH—

The modification linkages include urethane, amide, imide, carbodiimide and oxazolidone linkages. A urethane–modified isocyanurate foam (trade name: Airlite Foam SNB, Nisshinbo, Ind. Inc.) was first applied to the petrochemical industry as a seamless fire–resistant insulant in 1965 (39).

The commercialization of similar foams was followed by Hexaform (trade name: ICI) in 1968 (57). The commercialization was later followed by Upjohn Company (trade name: Kode 25) and Jim Walter Corp. (trade name: Thermax). Foaming methods of modified–isocyanurate foams include slabstock, laminate, block, pour–in–place and spraying.

Principles of Urethane Modification. The flame retardance and temperature resistance (or flame endurance) of modified isocyanurate foams are affected by the following factors:

(a)   Choice of polyisocyanate
(b)   NCO/OH equivalent ratio
(c)   Choice of polyol in molecular weight and functionality
(d)   Content of aliphatic component (aliphaticity index)
(e)   Choice of catalyst
(f)   Choice of polyol structure

Figure 26 shows a comparison of burn–through time between the two foams based on polymeric isocyanate (polymeric MDI) and a TDI–prepolymer (65). In the case of polymeric isocyanate–based foams, a remarkable increase in burn–through time resulted after increasing the NCO/OH ratio, which reached a maximum value at 3.0.

In contrast, TDI–based modified isocyanurate foams did not show any increase in burn–through time, even if the NCO/OH equivalent ratio was increased. This significant difference could be attributed to the difference in flash point of the two isocyanates. The flash points of polymeric isocyanate and liquid–modified MDI oligomers are >200°C, and that of TDI is 135°C.

When urethane linkages dissociate by dry heat or flame the dissociation of the urethane linkage starts at temperatures below 200°C, which temperature is already higher than the flash point of TDI. Therefore, in the presence of air, the dissociated TDI vapor ignites immediately.

In contrast, polymeric isocyanate does not immediately ignite, because its decomposition temperature is lower than its flash point. Accordingly, only polymeric isocyanates or liquid oligomeric MDI should be used for urethane–modified isocyanurate foams having substantial flame retardance.

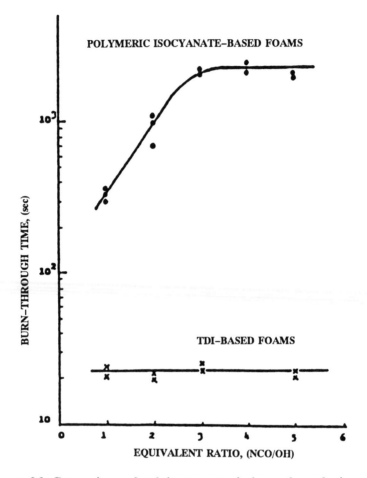

Figure 26. Comparison of polyisocyanates in burn–through time (65).

The second factor important in obtaining highly flame–retardant isocyanurate foams is the NCO/OH equivalent ratio. Fire endurance, i.e., flame retardance) and temperature resistance, can be increased with increase of the NCO/OH equivalent ratio when polymeric isocyanate is used as the polyisocyanate component.

Figure 26 shows that fire endurance in terms of burn–through time increased sharply when the equivalent ratio was increased from 1 to 3 and then the burn through times leveled off, even at higher equivalent ratios. However, at higher NCO/OH equivalent ratios (over 5.0), the resulting foams usually have high friability. Accordingly, preferable NCO/OH equivalent ratios for making commercial products are 3 to 5 (39). The fire endurance of the foams is related to the content of the isocyanurate linkage (trimer) in the foams. The percent trimer is expressed by the following equation.

$$\% \text{ Trimer } = 4200 \ \frac{E_i - E_h}{W_i + W_h}$$

where E_i is the number of isocyanate equivalent, E_h is the number of hydroxyl equivalent, W_i is weight of isocyanate and W_h is weight of polyol.

At higher levels of the trimer, friability (percent weight loss by tumbling) increases to an unacceptable level (>30%), while lower trimer contents increase the combustibility. Moss et al chose a theoretical trimer level of 18 to 22% (72).

The third factor important in obtaining high–flame–endurance foams is the choice of molecular weight, functionality of polyol and aliphaticity index (65). The polyol content, which is largely responsible for the low flammability of isocyanurate foams, is expressed by the aliphaticity index, as defined by the following equation:

$$\text{Aliphaticity Index } = \frac{\text{Weight of Polyether Polyol}}{\text{Weight of Polymeric Isocyanate}}$$

When the polyol component is varied, both factors, i.e., equivalent ratio and aliphaticity index, are changed at the same time, as shown in Figure 27.

The effects of the NCO/OH equivalent ratio and aliphaticity index on burn–through time are shown in Figures 28 and 29.

Higher NCO/OH indices give higher contents of isocyanurate linkages in the resulting foams, which result in higher–fire–endurance foams. However, if higher–molecular–weight polyols are used at the same NCO/OH index, the resulting foams have higher aliphaticity indices and result in higher flammability.

In order to clarify the relationship among the three factors, i.e., NCO/OH equivalent ratio, aliphaticity index and polyol molecular weight, a series of experiments was conducted by varying the three factors. The results obtained are shown in Figure 30. By using the results in Figure 30, the three factors governing satisfactory flame endurance are shown by a three–dimensional figure in Figure 31. This figure shows that the optimum ranges of the three factors required to obtain high flame endurance are as follows: molecular weight, 300 to 900; aliphaticity, 0.15 to 0.4; and NCO/OH equivalent ratio, 3.0 to 8.0.

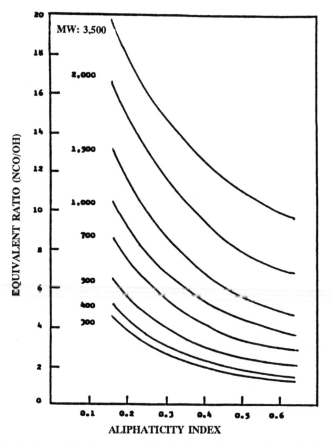

Figure 27. Relationship between Aliphaticity Index and equivalent ratio at different molecular weights of polyether triol (65).

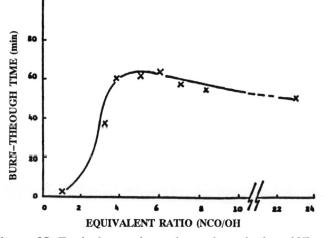

Figure 28. Equivalent ratio vs. burn–through time (65).

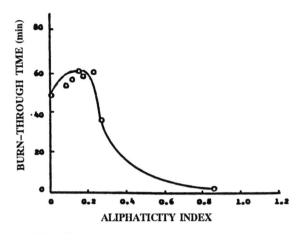

Figure 29. Aliphaticity Index vs. burn-through time (65)

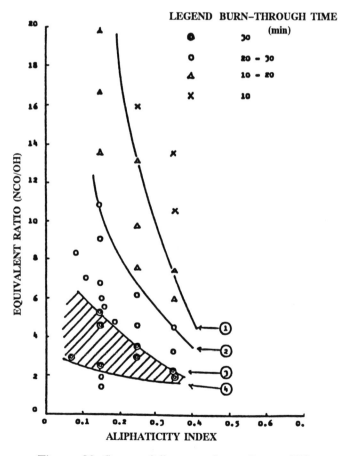

Figure 30. Scope of flame-resistant foams (65).

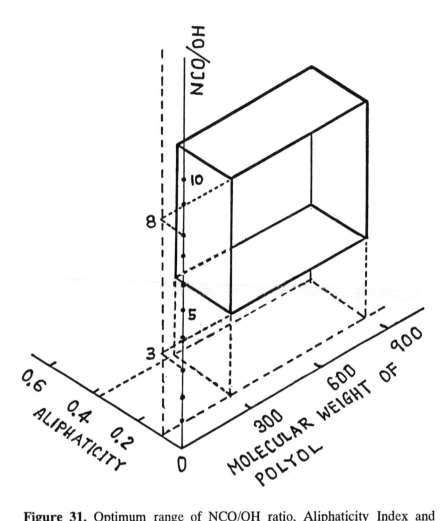

Figure 31. Optimum range of NCO/OH ratio, Aliphaticity Index and molecular weight (65).

The results shown in Figure 31 are based on experiments using glycerol–based polyether triols having differing molecular weights. In the case of other modifiers, such as aromatic polyesters, the optimum factors of the above are somewhat different. The fourth factor important in obtaining high–flame endurance foams is the choice of catalyst.

In the case of urethane–modification, two kinds of catalysts, i.e., urethane–formation catalysts and isocyanate–trimerization catalysts, are usually used. Major trimerization catalysts are listed in Table 11 and major urethane catalysts are listed in Tables 8 through 10.

There are many combinations of the two catalysts, and a detailed description is not possible in these limited pages, but it should be noted that proper choice of the two catalysts provides smooth–rise profiles and good foam properties.

The sole use of a trimerization catalyst, such as potassium 2–ethylhexanoate, shows two–step rise curves. In the initial stage the catalyst slowly accelerates urethane linkage formation, and then, after the reaction exotherm rises, trimerization begins.

The fifth factor important in obtaining high–flame–endurance foams is the choice of polyol structure. Since the beginning of the isocyanurate–foam industry, the major polyols used for modification have been polyether polyols (65, 66), which include polyols having a variety of functionalities (e.g., 2 to 8) and varying equivalent ratios.

The polyols which function as modifiers include ethylene glycols, 1,4–butanediol, polyether polyols and polyester polyols. In recent years aromatic polyesters prepared from reclaimed PET (polytetraethylene terephthalate) or the distillation residue of DMT (dimethylterephthalate) have appeared as modifiers for urethane–modified isocyanurate foams (73, 78). These aromatic polyesters are produced by the transesterification of reclaimed PET or DMT distillation residue.

The advantages of these aromatic polyesters are lower cost than conventional polyether polyols, better flame retardance, and high–temperature resistence. However, their disadvantages include compatibility problems with chlorofluorocarbons and quality deviations in viscosity and hydroxyl values. In order to improve the compatibility problems, amine–based polyether polyols have been blended.

Preparation. Urethane–modified isocyanurate foams are mostly prepared by the one–shot process based on the principle discussed previously in the urethane modification section. The semi–prepolymer process is used only in limited cases because of viscosity problems. This section describes several examples of formulations for producing block foams, slabstock foams, laminate foams, and spray foams.

Urethane–modified isocyanurate foams using polyols of functionality of at least 3 were described by Ashida et al in 1967 (71). Other types of urethane–modified isocyanurate foams are shown below.

Sorbitol Polyol–Modified Isocyanurate Foam (71). This foam was prepared according to the following procedure. Potassium 2–ethylhexanoate containing 4% water was used as the catalyst (soluble in polyether polyols). 94 g of sorbitol–based polyether polyol having a hydroxyl number of 490, 8.5 g of the potassium 2–ethylhexanoate, 120 g of tris(chloroethyl) phosphate, 5 g of silicone surfactant, and 90 g of trichloromonofluoromethane were mixed to make a solution. 630 g of polymeric isocyanate were mixed and, after vigorous stirring by means of a mechanical mixer for ten minutes, a foam was obtained.

The NCO/OH equivalent ratio was 5.0 and the aliphaticity index was 0.15. The resultant foam had a cream time of 15 sec, rise time of 20 sec, foam density of 0.041 g/ml, tensile strength in the direction of foam rise of 2.0 kg/cm^2 and a closed–cell content of 90%.

Ethylene Glycol–Modified Foams. These foams were studied in detail by Moss et al (66–69, 72). Examples of formulations are shown below. The following formulation was used for bench–scale one–shot foams.

Table 27: Ethylene Glycol–Modified Foam (66)

Component	Parts by Weight	Equivalent
Mondur MR	560.	4.2
Ethylene glycol	42.	1.4
CFCl$_3$	70.	
DC–193	12.	
2,4,6–tris(dimethylaminomethyl) phenol	4.	
N–(2–hydroxyethyl) aziridine	4.	

In order to meet the requirements for low friability and low combustibility the following formulations (Table 28) were disclosed for laminate production (72). The resultant foam has a trimer content of 20%.

Table 28: Formulation for Lamination (72)

Component	Parts by Weight	
	#1	#2
Ethylene glycol	21	0
Propylene glycol	0	24
Mondur MR (Mobay Chem.)	280	276
CFCl$_3$	35–40	35–40

Aromatic Polyester–Modified Foam. The formulations shown in Table 29 are examples of the co-use of an aromatic polyester and a compatibilizer co–polyol (Armol 401 or 402) (76). Other papers also discuss the use of aromatic polyesters as modifiers for isocyanurate foams (216–222). Such compatibilizers increase the blend ratio of conventional polyol and aromatic polyols.

Table 29: Aromatic Polyester–Modified Foams (76)

Formulation (pbw)	#1	#2
Armol 402	22	22
Terate 203	78	0
Chardol 570	0	78
DC 193	2.5	2.5
T–45	1.2	0.9
CFC–11A	51.9	55.8
Mondur MR	239	266
NCO/OH	3.5	3.5
Reaction Profile (min:sec)		
Cream time	0:19	0:22
Gel time	0:59	0:48
Tack free time	1:54	1:58
Foam Density (pcf)	1.80	1.74

Properties. The physical strengths of modified isocyanurate foams are proportional to foam densities. The values are about the same as those of urethane foams, as discussed in the previous chapter. Typical properties of modified isocyanurate foams to be discussed are flame endurance and friability. Figure 32 shows the relationship between OH/NCO equivalent ratio (inverse equivalent ratio of NCO/OH) and friability in terms of the ASTM C–421 tumbling test (16).

Acceptable percent friability in industrial applications is less than 30% weight loss, i.e., the OH/NCO equivalent ratio is more than 0.2 (NCO/OH ratio is less than 5.0). Figure 33 shows the relationship between OH/NCO ratio and oxygen index in terms of ASTM D 2863. Oxygen index relates to flame retardance (16).

Figure 34 shows a comparison of temperature stability in terms of TGA at different isocyanate indices (16). The results indicate that the higher the isocyanate index, the higher the temperature resistance.

Processing. All kinds of processing methods used for rigid urethane foams can be used by modifying processing conditions and/or formulations, including box foaming, slabstock foaming, laminates, frothing, spraying, molding and pour–in–place.

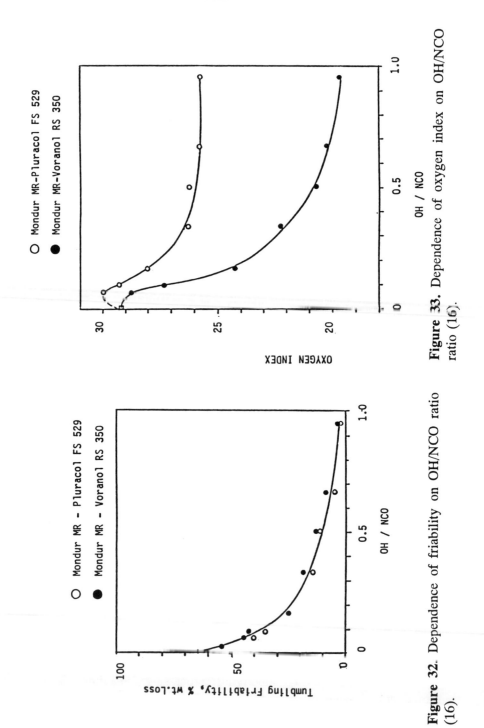

Figure 33. Dependence of oxygen index on OH/NCO ratio (16).

Figure 32. Dependence of friability on OH/NCO ratio (16).

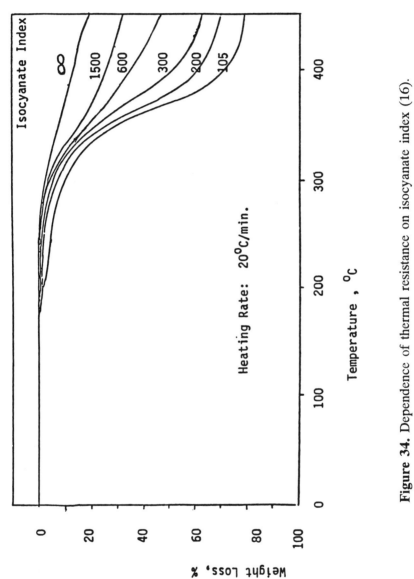

Figure 34. Dependence of thermal resistance on isocyanate index (16).

In the case of box foaming, the foam in a box must be kept in the box for post curing at room temperature. If the box foam is taken out immediately after foaming, many cracks may be formed. The same attention must be paid to bun foam (block foam). These foams must be post cured at ambient temperature, either by indoor or outdoor curing for at least a week.

The major application fields of urethane–modified isocyanurate foams lie in building applications, such as warehouses, high–rise buildings and residential houses. In the USA, laminates are one of the major applications of urethane–modified foams. Figure 35 shows an example of the laminate process (72).

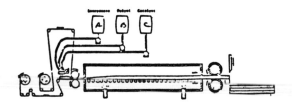

Figure 35. Free rise, continuous lamination process (72).

The pour–in–place process is applied to chemical–plant insulation and building insulation. These insulations can be made by pouring the system into the void space formed by surface materials and their reinforcing board in the presence of spacers. A number of pipes and reaction towers of petrochemical plants in Japan have been insulated by the pour–in–place process (79).

The spraying process is carried out at ambient temperatures. The spraying of urethane–modified isocyanurate foam is not as easy as urethane–foam spraying because the cyclotrimerization reaction of isocyanate groups requires relatively higher temperatures than for urethane foams. An example of the spraying of urethane–modified isocyanurate foams was reported (198). The spraying was conducted with formulations at a low–NCO/OH equivalent ratio.

The frothing process is widely used in rigid urethane foam pour–in–place applications. The frothing process of urethane–modified isocyanurate foams has been used for the insulation of petrochemical plants, e.g., spherical tanks, reaction towers, etc (79). An example of the frothing system is shown below (71).

The frothing system was prepared as follows. 96 parts of sucrose-based polyether polyol (OH No. 460), 60 parts of trichloromonofluoro-methane, 120 parts of tris(chloroethyl) phosphate, 20 parts of silicone surfactant DC 113 and 11 parts of potassium octoate dissolved in 11 parts of the polyether polyol were mixed homogeneously to make a solution.

Frothing was conducted by the co–use of dichlorodifluoromethane, in which the mixing ratio was 2.2 parts of crude MDI per 1.0 part of the solution. The equivalent ratio of NCO/OH was 5.6 and the weight ratio of crude MDI/polyol was 0.14. The frothing mixture was poured into a 100 mm thick panel. The physical properties of the resulting foam were as follows: overall density, 0.047 g/cm^3; core density, 0.042 g/cm^3; compressive strength in the direction of foam rise, 2.2 kg/cm^2; the same in the perpendicular direction, 1.6 kg/cm^2; closed–cell percent, 93.8; thermal conductivity, 0.015 kcal/mh°C (71).

Non–CFC–Blown   Urethane–Modified   Isocyanurate   Foams. Recently, methods of making non–CFC–blown urethane–modified isocyanurate foams have been reported. These methods involve the partial replacement of water for CFC–11 (97). The methods however, have the disadvantages of (a) the higher thermal conductivity of the resulting foams due to the presence of carbon dioxide in the foam cells and (b) the higher friability of foams due to increased urea and biuret linkages (197).

Water as the sole blowing agent is not practical for producing these foams, because it results in extremely high–friability foams which make it very difficult to handle in practical applications. An alternative method proposed by Ashida involves the use of specific solvent blends, e.g., an 80/20 blend of methylene chloride/pentane (196). Advantages of this solvent blend are:

(a) This blowing agent can give very low foam density, e.g., 1.0 pcf foams (Figure 36) (196), without any processing problems, i.e., foam scorching (water–blown foams have scorching problems);

(b) This blended solvent is substantially self extinguishing, i.e., an 80/20 blend in weight ratio of methylene chloride/pentane is substantially non–flammable, and isocyanurate foams blown by this method have the same flame retardance as CFC–11 blown foams in terms of burn–through time, as shown in Figure 37 (196);

(c) This blended solvent is cheaper than HCFC's as well as CFC's;

(d) The consumption of expensive polyisocyanate is lower than the water–blown foams at the same foam density, especially at lower foam densities;

(e)   Foam spraying is possible using the solvent blend as the sole blowing agent because of its low viscosity. However, this is not feasible with water as the sole blowing agent;

(f)   This solvent blend makes it possible to produce integral–skin foams, but water–blown foams do not result in integral–skin foams;

(g)   This blended solvent makes it possible to make pour–in–place foams. Water–blown foams have problems in pour–in–place foaming. The blended solvent results in higher–thermal–conductivity foams than possible with CFC–11–blown foams. However, water–blown rigid foams also have higher thermal conductivities than CFC–11–blown foams.

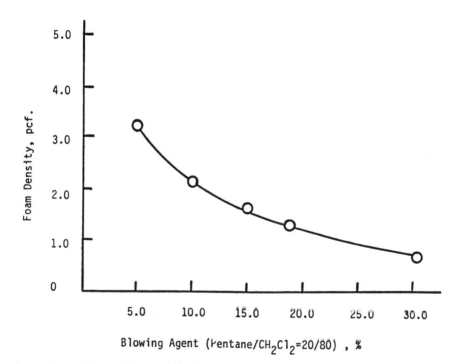

Figure 36. Percent blowing agent vs. foam density (196).

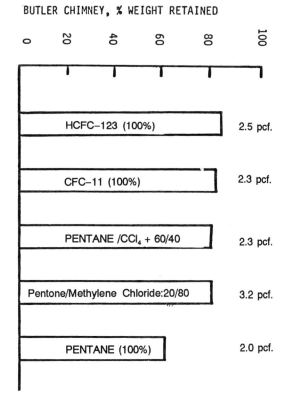

BUTLER CHIMNEY, % WEIGHT RETAINED

HCFC–123 (100%)   2.5 pcf.

CFC–11 (100%)   2.3 pcf.

PENTANE /CCl$_4$ + 60/40   2.3 pcf.

Pentone/Methylene Chloride:20/80   3.2 pcf.

PENTANE (100%)   2.0 pcf.

Figure 37. Comparison of blowing agents in Butler Chimney Test (196).

Oxazolidone–Modified Isocyanurate Foams. The 2–oxazolidone, or 2–oxazolidinone, linkage is considered to be a cyclic urethane linkage, but its thermal stability is much higher than that of a urethane linkage. Kordomenos et al (207) compared the thermal stabilities of urethane, oxazolidone and isocyanurate linkages in terms of activation energy by using model compounds. The results obtained were as follows.

Linkage	Activation Energy, Kcal
Urethane	.38.7
Oxazolidone	48.1
Isocyanurate	60.3

The reason for the thermal instability of the urethane linkage can be attributed to the existence of labile hydrogen, which results in the dissociation of the linkage as shown below:

$$R-\underset{\underset{\displaystyle O}{\|}}{\underset{\displaystyle C}{N}}-C-O-R' \rightarrow R-NCO + HO-R'$$

(with H above the N)

In contrast, the oxazolidone linkage has no labile hydrogen, thereby resulting in higher thermal stability. For this reason, oxazolidone–modified isocyanurate foams are expected to have higher thermal stability than urethane–modified isocyanurate foams. Oxazolidone–modified isocyanurate foams were first prepared by Ashida et al by a prepolymer process (60, 61, 62, 63). The foams were prepared by the two–step process, i.e, (a) preparation of NCO–terminated oxazolidone prepolymers (Figure 38); and (b) cyclotrimerization of the NCO–terminated oxazolidone prepolymers in the presence of a trimerization catalyst, a blowing agent, preferably an inert solvent such as CFC–11, and a surfactant.

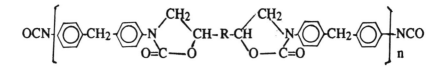

Figure 38. NCO–terminated polyoxazolidone prepolymer (63).

In this process the most important factor necessary to provide foams having high flame endurance in terms of burn–through time is the ep–oxy/NCO equivalent ratio. It has been found that the best ratio is in the range of about 0.05 to 0.10. Figure 39 shows this relationship.

An example of the surface flammability in terms of the Butler chimney test of oxazolidone–modified isocyanurate foams is shown in Figure 40. The effect of equivalent ratio on friability is shown in Figure 41. The effect of equivalent ratio on smoke density is shown in Figure 42.

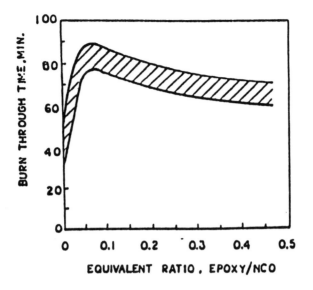

Figure 39. Effect of equivalent ratio on burn–through time (63).

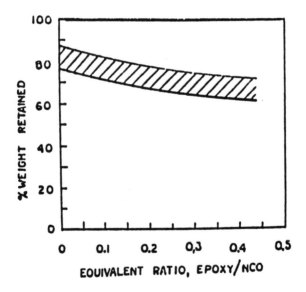

Figure 40. Effect of equivalent ratio on surface flammability (Butler Chimney Test) (63).

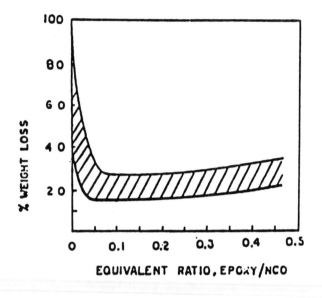

Figure 41. Effect of equivalent ratio on friability (63)

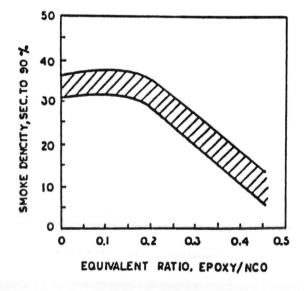

Figure 42. Effect of equivalent ratio on smoke density (63).

In view of these figures, the optimum equivalent ratio is about 0.10. A comparison of the thermal stability of modified foams employing different epoxy resins is shown in Figure 43.

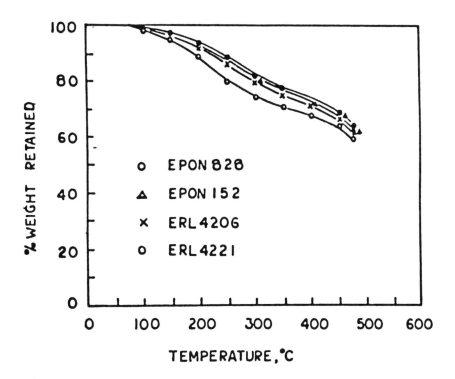

Figure 43. TGA profile in nitrogen. Comparison of epoxy resins (63).

A one–shot method for oxazolidone–modified isocyanurate foams was proposed by Hayash et al (129). The process, however, resulted in a high–exotherm reaction which caused thermal degradation of the resulting foams.

Recently, oxazolidone–urethane–modified isocyanurate foams made by the one–shot process have been reported by Fuzesi et al (128, 209, 210). This process consists of a three–component system. Unmodified oxazolidone foams will be described in the following section.

Amide–Modified Isocyanurate Foams. The amide linkage is a thermally stable, difunctional linkage. Amide–modified isocyanurate foams can be prepared by using carboxylic acids having at least two carboxylic groups as modifiers.

The first example is the use of dimer acids as modifiers (39, 166). The acid has two carboxylic groups and a small percentage of three car-boxylic groups. The carboxylic groups react with the isocyanate groups to form amide linkages by liberating carbon dioxide gas and result in isocyanurate foams having reduced crosslink density. Recently, tempera-ture-stable, flame-resistant and low friable foams were prepared by the cyclotrimerization of NCO-terminated polyamide prepolymers (253).

Carbodiimide-Modified Isocyanurate Foams. The carbodiimide linkage is a thermally stable, difunctional linkage very suitable for modifying isocyanurate foams having low friability and high flame endurance. The linkage is produced by the condensation reaction of isocyanate groups in the presence of a carbodiimide-forming catalyst, as shown below:

$$—R—NCO + OCN—R— \quad —R—N{=}C{=}N—R— \quad + \quad CO_2$$

Both the one-step and two-step processes can be used (80, 252, 253).

Two-Step Process. A mixture composed of 100 g of 4,4'-diphenylmethane diisocyanate and 1.1 g of 1-phenyl-3-methyl-1-phospholene oxide is stirred for 3 minutes at room temperature. After carbon dioxide is evolved 4.0 g of 2,4,6-tris(dimethylaminomethyl) phenol is added into the reaction mixture. The mixture generates a reaction exotherm and expands to form a foamed product. After curing at 125°C for one hour, the resultant foam is colorless and posesses low friability.

One-Step Process. Into 100 g of polymeric isocyanate, 0.8 g of 1-phenyl-3-methyl-1-phospholine oxide, 2.1 g of methanol and 2.1 g of 2,4,6-tris(dimethylaminomethyl) phenol are mixed and stirred for 15 sec. After 10 sec of mixing, foam rise starts, and the rise time is 60 sec. Low-friable foamed products result.

Another literature reference (68) also shows an example of a prepolymer process. The prepolymer was prepared by using PAPI 901, and the resulting prepolymer had 25.5% NCO and its viscosity was 7,000 cp at 25°C. The resulting foam had a density of 2.1 pcf, 9% weight loss by means of a 10-min tumbling test and its oxygen index was 25.8. The co-use of tertiary amine-containing s-triazine compounds, e.g., 2,4,6-tris(diethanolamino)-s-triazine, and methanol or furfuryl alcohol was applied for producing carbodiimide-modified isocyanurate foams (92, 93, 94).

The co-use of furfuryl alcohol with a triazine derivative or triethanolamine and a quaternary borate ester-salt catalyzed the polymerization of polyisocyanates or polyisocyanate–polyol mixture in the manufacture of polymeric foams containing carbodiimide and isocyanurate groups (92).

A modification of isocyanurate foams by incorporating urethane–, carbodiimide– and imide groups has been reported (97). An example is as follows. The amounts of ingredients used are shown in parts by weight in parenthesis.

A foam was prepared from polymeric isocyanate (134 pbw); 3,3',4,4'–benzophenonetetracarboxylic dianhydride (45 pbw); epoxy novolak resin (20 pbw); wetting agent (1 pbw); methanol (2.5 pbw); N,N',N'–tris(dimethylaminopropyl)–s–hexahydrotriazine (2.5 pbw); triethylenediamine (2.5 pbw); and 1–phenyl–3–methyl–2–phospholene 1–oxide (2.5 pbw).

The foam had a density of 0.0317 g/cm^3, closed–cell content of 96.9% compression resistance 1.8 kg/cm^2, and brittleness (ASTM C–421) 27.4% weight loss after a 10–min tumbling test.

Imide–Modified Isocyanurate Foams. The imide linkage is a thermally stable linkage, and therefore, imide–modified isocyanurate foams have higher thermal stability and flame retardance than urethane–modified isocyanurate foams. R. Grieve (114) prepared such foams in a one–shot process by reacting a polycarboxylic acid anhydride with an organic polyisocyanate in the presence of a catalytic amount of a monomeric homocyclic polyepoxide and a tertiary amine.

The use of this catalytic system permits foam formation to proceed without the need to supply external heat to the reaction mixture after the reactants are brought together. This permits pouring–in–place of the foam–forming system, for example, in the insulation of cavity walls for construction purposes, in trailer walls, and in cold–storage frameworks and the like.

Filled Isocyanurate Foams. The addition of specific inorganic powders, such as graphite and talc, to urethane–modified isocyanurate foams, has been proven to produce high temperature– and flame–retardant insulation materials (59).

After large fire tests, graphite powder–loaded, urethane–modified isocyanurate foam insulation was found to be superior to Perlite board, which is a typical insulant in the petrochemical plant insulation (79), because Perlite is a non–flammable inorganic insulant but its flame resistance is poor. Figure 44 shows that this foam insulation can protect petrochemical tanks from huge fires occurring at petrochemical complexes. The testing conditions are shown in Figure 45.

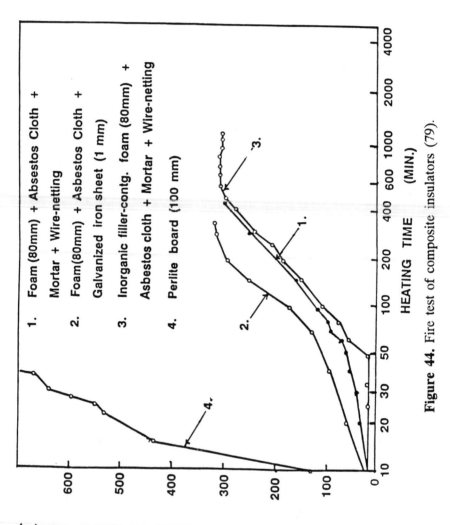

Figure 44. Fire test of composite insulators (79).

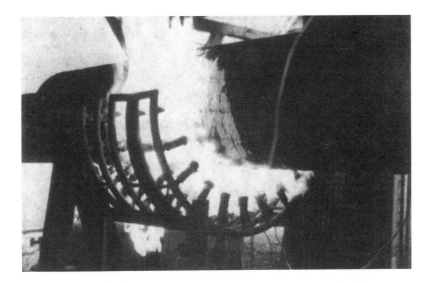

Figure 45. Fire test of a composite composed of U–modified iso–cyanurate foam and mortar (79).(Courtesy of Nisshin Spinning Co. Ltd.)

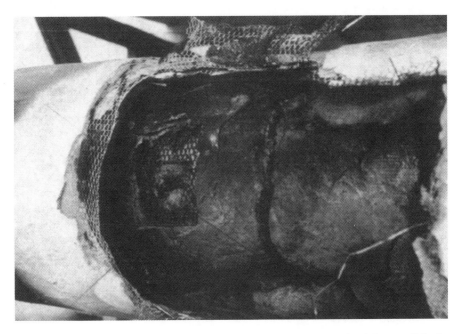

Figure 46. Results of fire test of a composite composed of U–modified isocyanurate foam and mortar (79). (Courtesy of Nisshin Spinning Co. Ltd.)

The heat input corresponds to that of a large fire which might occur at a petrochemical plant, and was estimated to be 86,000 Btu/hr. The insulation layers after the test are shown in Figure 46 which shows that graphite–loaded urethane–modified isocyanurate foams remain heavily charred, thereby protecting against flame and high temperature. The formulations employed for this test (59) are shown in Table 30.

Table 30: Formulation Employed for Large Fire Test (59)

Formulation (parts by wt.)	A	B
Polymeric isocyanate (PAPI)	64.	64.
Graphite powder	30	0
Talc powder	0	30
Trimerization catalyst	12.5	12.5
Polyether polyol (sucrose–based)	18.5	18.5
Tris(chloroethyl)phosphate	13.0	13.0
Silicone surfactant (L530)	1.0	1.0
CFC–11	13.0	13.0
Foam Properties		
Foam density, kg/m³	52.8	48.6
Foam density, lb/ft³	3.3	3.0
Flame Penetration Time, Bureau of Mines 6366(1964)	150	180
Thermal conductivity, Kcal/m/hr.°C,	0.021	0.019
Compressive strength,		
parallel to rise, kg/cm²,	1.49	1.57
perpendicular to rise, kg/cm²	0.98	1.07

Polyurea Foams

In this section isocyanate–based polyurea foams produced by the reaction of the water–isocyanate or amine–isocyanate reactions will be described. Urea–formaldehyde foams will be excluded. The isocyanate–based urea–linkage formation is shown by the following model reactions:

(a) $2R-NCO + H_2O \rightarrow R-NH-CO-NH-R + CO_2$

(b) $R-NCO + H_2N-R' \rightarrow R-NH-CO-NH-R'$

Reaction (a) was applied in preparing low-density packaging foams. Reaction (b) was recently used for preparing low-density flexible urea foams (81, 141)

Some examples of urea foams prepared by the water-isocyanate reaction are as follows. ICI disclosed the foam prepared by the water-isocyanate reaction in the presence of imidazole compounds (137). PRB NV disclosed a foam prepared in the presence of water-soluble saccharide and polyol (139). Bayer AG disclosed a foam prepared by using 1.5 to 50 parts of alkanolamine with water and 100 parts of polyisocyanate (138). Schaum Chemie disclosed foams prepared by using lower alkanols and alkylene diols (140).

The urea linkage is thermally more stable than the urethane linkage due to its higher content of hydrogen bonding and urea foams are suitable for higher-temperature-service applications.

Priester et al (141) have found a method of controlling the amine-isocyanate reaction by using substituted-amine-terminated polyethers, and they obtained high-resiliency polyurea foams by this method. Ashida et al (81) prepared low-density flexible polyurea foams by using a primary-amine-terminated polyether (Jeffamine D-2000, Texaco Chemical).

Polycarbodiimide Foams

The carbodiimide linkage is a thermally stable difunctional linkage preferably used for producing carbodiimide-modified isocyanurate foams. This linkage is formed by the condensation reaction of isocyanate groups accompanied by the generation of carbon dioxide gas. The amount of carbon dioxide generated is enough to produce low-density foams with open-cell structures.

$$2n(-R-NCO) \xrightarrow{\text{catalyst}} (-R-N{=}C{=}N-R-)_n + nCO_2$$

Therefore, the preparation of foams with a high percentage of closed cells or high-density foams is difficult. These foams are quite different from other isocyanate-based foams, e.g., urethane foams, isocyanurate foams and oxazolidone foams.

The polyisocyanates used for preparing carbodiimide foams include TDI, TDI-based prepolymers, liquid-MDI oligomers and polymeric isocyanates. Many catalysts for producing carbodiimide foams have been disclosed in the patent literature. Some of these are shown in Table 13.

Most of the catalysts are active only at elevated temperatures, but some are active at room temperature. Co-catalysts or accelerators have also been proposed. A carbodiimide-forming catalyst, 2,4,6-tris(dialkanolamino)-s-triazine was used for producing carbodiimide foams.

An example of the production of carbodiimide foams is as follows (84). 50 parts of TDI and 0.5 parts of 2,4,6-tris(diethanolamino)-2-triazine were mixed and heated with agitation at 100°C. After 30 min from the start of mixing, carbon dioxide was generated and an exotherm was observed. After 60 min the foaming reaction was completed and the maximum temperature, about 200°C, was reached.

An active co-catalyst system was proposed (83). The system is composed of 2,4,6-tris(diethanolamino)-s-triazine and 1,3,5-tris(3-dimethylamino propyl) hexahydro-s-triazine. This catalyst was used to prepare rigid carbodiimide foams from TDI without external heating.

Another type of triazine-derivative proposed for use as a carbodiimide catalyst is 2,4,6-tris(N-methylethanolamino)-s-triazine. This derivative was used alone or in combination with tris(dimethylaminomethyl) phenol or similar compounds (85).

Other catalyst combinations proposed consisted of a blend of 2,4,6-tris(dialkylamino)-s-triazine and 1,3,5-tris(dialkylaminoalkyl)-s-hexahydrotriazine and tris(dimethylaminomethyl) phenol or organotin compounds (86). This catalyst system is reported to give a higher percentage of closed cells to the resulting foams. In contrast, conventional carbodiimide foams have only small amounts of closed cells, and are not suitable for thermal-insulation materials.

Other methods used to obtain carbodiimide foams having a higher closed-cell percentage involve urethane-modified carbodiimide foams. Acrylonitrile-grafted polyether polyol was used for the modification (87). Also a combined use of a carbodiimide-forming catalyst selected from triazine derivatives having N-alkanolamine groups and a urethane-forming catalyst was used for producing urethane-modified carbodiimide foams (88).

A combination catalyst system comprising 2,4,6-tris(n-methylethanolamino)-s-triazine (I), and optionally, 1,3,5-tris(3-dimethylaminopropyl)-hexahydro-s-triazine, bis(tributyltin) oxide or bis(triphenyltin) oxide was used for preparing carbodiimide foams. For example, a mixture of 0.5 g (I) and 50 g of tolylene diisocyanate was stirred at 100°C. The foaming started after 9 min and about 1 min after the foaming, the temperature in the vessel rose to 192°C; the resulting foam occupied a volume of 800 ml (89, 90).

NCO–terminated prepolymers have been used for producing carbodiimide foams (91). Bayer AG also developed carbodiimide foams (82).

Polyoxazolidone Foams

The oxazolidone linkage is a thermally stable and difunctional linkage that has been used for modifying isocyanurate foams (61, 63). Unmodified oxazolidone foams are expected to have improved properties such as temperature stability, flame retardance, and low friability. An attempt to obtain unmodified polyoxazolidone foams was reported (98). The formulation employed is shown in Table 31.

Table 31: Formulation for Unmodified Oxazolidone Foam (98)

Bisphenol A–epichlorohydrin adduct	94.5	g (0.5 equiv.)
(Epikote 819, Shell Chem.)		
Silicone surfactant,	2.0	g
(SH–193, Torey Silicone)		
CFC–11	15.0	g
Polymeric isocyanate (PAPI, Upjohn)	67.0	g
AlCl$_3$ hexamethylphosphoric triamide	6.1	g
Processing		
Cream time,	60	sec.
Rise time,	480	sec.
Foam density,	40	kg/m^3

A premix was prepared by blending the bisphenol A–epichloro–hydrin adduct, silicone surfactant and CFC–11A. Into the premix, polymeric isocyanate and the complex catalyst were added and the mixture was immediately agitated vigorously to produce a foamed material. The resulting foam had a density of 40 kg/m^3 with fine cell structure.

Polyimide Foams

Polyimide foams have outstanding advantages of high–temperature resistance, oxidation resistance, and low smoke evolution. They can be prepared by two routes, as shown below. One method consists of the reaction of aromatic diamines with carboxylic dianhydrides (99–102), as shown in model reaction (1) Another synthetic route is the reaction of aromatic diisocyanates with carboxylic dianhydrides, as shown in model reaction (2).

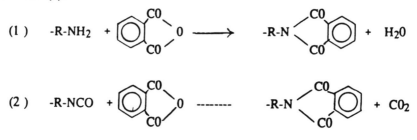

In this section, isocyanate–based imide foams, according to model reaction (2) will be described, and other imide foams prepared by model reaction (1) will be described in a separate section.

Methods for making isocyanate–based polyimide foams include a one–shot process by admixing carboxylic dianhydrides with an organic polyisocyanate at room temperature in the presence of a dipolar aprotic organic solvent (103, 106, 142, 251). The resulting foams from this method exhibited outstanding thermal resistance, as shown in Figure 47.

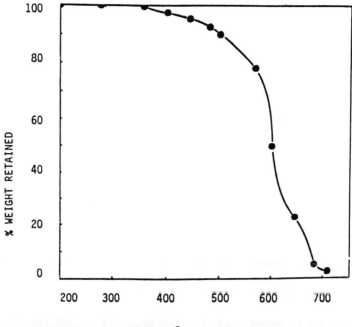

Figure 47. TGA profile of polyimide foam (103).

An example of the preparation method at room temperature is as follows (106). A mixture of 161 parts by weight (1.00 equivalent) of 3,3',4,4'-benzophenone tetracarboxylic dianhydride and 132 parts by weight (1.00 equivalent weight) of polymeric isocyanate was prepared by mechanical blending.

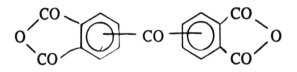

To this liquid mixture was added a mixture of 50 parts by weight (0.473 equivalent) of a polyol of equivalent weight 105.6 (the adduct of propylene oxide and a mixture of polyamines containing 50% by weight of methylenedianiline obtained by the acid condensation of aniline and formaldehyde); 50 parts by weight of dimethyl sulfoxide; and 10 parts by weight of a silicone surfactant.

These mixtures were mechanically blended for 10 seconds at room temperature and poured rapidly into a wooden mold (14" x 6" x 4") and allowed to expand freely at room temperature. After approximately 10 to 15 minutes, the resultant foam was very rigid, posessing fine cells. The residual solvent was removed from the resultant foam by placing it in a 100°C oven for 4 days. The foam had the following physical properties: density, 2.82 pcf; compressive strength (parallel to rise), 30.2 psi; and flame test (ASTM D 1692), total length burned, 0.1 inch.

Polyimide foams can be prepared with alkanolamine as catalyst and without an aprotic solvent (104). An example is as follows. Ten grams of 1,4,5,8-naphthalenetetracarboxylic dianhydride, 30 grams of polymeric isocyanate, and 0.4 gram of water were blended. Into the mixture 0.4 gram of dimethylethanolamine and silicone surfactant DC–193 (Dow Corning Corp) were mixed, and the mixture was heated at 232°C for 3 hours, resulting in a rigid foam.

Polyimide foams can be prepared at high temperatures without any catalyst. An example is shown below (105). 4,4'–Diphenylmethane diisocyanate in an amount of 44 grams was melted gently in a beaker with continuous agitation. After 16 min when the temperature of the melt was 390°C, 24 grams of trimellitic anhydride was added. After 4 min, the temperature was 485°F and shortly thereafter agitation was stopped. As a yellow foam cake developed, the beaker was covered with a metal plate in order to prevent expansion in excess of the volume of the beaker. After cooling, the foam cake was removed from the beaker and was postcured for two hours at 350°F. The density was 2.1 pcf, and the compressive strength was 11.5 psi. The modulus was 624 psi.

Polyamide Foams

Polyamides are temperature– and solvent–resistant polymers, and, therefore, these foamed products have the same advantages as the basic polymers. The first polyamide foam was prepared by the reaction of an organic diisocyanate and a carboxyl–terminated polyester (71).

The reaction of NCO groups and carboxylic acid groups resulted in the formation of amide linkages and carbon dioxide as blowing agent. This reaction has led to the invention of urethane foam preparation, and the polyurethane industry has become one of the biggest plastic industries. A model reaction of a polyamide foam formation is shown below:

$$n \ \ \text{HOOC} \sim \sim \sim \text{COOH} \ + \ n \ \text{OCN-R-NCO} \longrightarrow$$

$$-(\sim \sim \text{CO-NH-R-NH-CO} \sim \sim)_n \ + \ 2n \ CO_2$$

Recent advances in polyamide foams are based on caprolactam–based foamed products, i.e., foamed nylon 6. Epsilon caprolactam, sodium lactamate as a ring–opening polymerization catalyst, a blowing agent, a surfactant, and an aliphatic isocyanate as an activator were used as major raw materials (143). A method of preparing the foams is as follows. Into 135 g of dried epsilon caprolactam, 4.6 g of metallic sodium was added with stirring to form sodium lactamate. 20.7 g of the product and 83.3 g of epsilon caprolactam were charged into a beaker. This component is termed Component 1.

Component 2 was prepared by mixing 10 g of toluene, 1 g of silicone surfactant (Toray SH–193), and 7.6 g of hexamethylene diisocyanate. The two components were mixed in a container for 15 seconds. Rise time was 35 seconds, and the foam density was 0.042 g/cm^3. Aliphatic diisocyanates such as hydrogenated MDI and m–xylylene diisocyanate also act as effective activators, but aromatic isocyanates, such as MDI, TDI, were found not to be effective.

REFERENCES FOR ISOCYANATE–BASED FOAMS

1. DeBell, J.M., *German Plastic Practice* (1946)

2. Ferrigno, T.H, *Rigid Plastic Foams (2nd Edition)*, Reinhold Publishing Corp., New York (1967).

3. Saunders, J.H. and Frisch, K.C., *Polyurethanes, Chemistry and Technology*, Volume I and 2, Interscience Publishers, New York (1962)

4. Vieweg, R. and Hoechtlen, A., *Polyurethane*, Carl Hanser Verlag, Munchen. (1966).

5. Phyllips, L.N., and, Parker, D.B., *Polyurethanes*, Iliffe Books Ltd., London (1964).

6. Dombrow, B.A., *Polyurethanes*, Reinhold Publishing Corp., New York (1965)

7. Buist, J.M. and Gudgeon, H., *Advances in Polyurethane Technology*, Maclaren and Sons, Ltd., London (1968)

8. Bruins, P.F., *Polyurethane Technology*, Wiley–Interscience, New York (1969)

9. Benning, C.J., *Plastic Foams*, Vol. I and 2, Wiley–Interscience, New York (1969)

10. Frisch K.C. and Saunders, J.H., *Plastic Foams*, Vol. 1 and 2, Marcel Decker, Inc., New York (1972).

11. Bayer, O. Das, *Diisocyanate–Polyadditionverfahren*, Carl Hanser Verlag, Munchen (1963).

12. Frisch,K.C., and Reegen, S.L. *Advances in Urethane Science and Technology*, Vol. 1–7, Technomic Publishing Co. Inc., Lancaster, PA (1971–1979)

13. Frisch, K.C. and Klempner, D. *ibid*, Vol. 8. (1981)

14. Stewart, S.A., *A Glossary of Urethane Industry Terms*, The Martin Sweets Co, Inc. (1971).

15. Landrock, A.H., *Polyurethane Foams—Technology, Properties and Applications*, Plastics Technical Evaluation Center, Dover, New Jersey (1969)

16. Frisch K.C., and Hemandez, A., *International Progress in Urethanes*, Vol. 1, Technomic Publishing Co. Inc., Lancaster, PA (1975).

17. Ashida,K., and Frisch, K.C., *International Progress in Urethanes*, Vol. 2–5, Technomic Publishing Co. Inc., Lancaster, PA.

18. Minecke, E.A. and Clark, R.C., *Mechanical Properties of Polymeric Foams*, Technomic Publishing Co. Inc., Lancaster, PA (1973)

19. Oertel, G., *Polyurethane Handbook*, Hanser Publishers, Munich, Vienna, and New York (1985)

20. Richter, R. and Ulrich H., *The Chemistry of Cyanates and Their Thio Derivatives*, Part 2, Chapter 17, pp. 610–818, John Wiley and Sons, New York (1977)

21. Ozaki, S., *Chemical Reviews*, Vol. 72, No.5, pp. 457–496 (1972)

22. Amold, K.G., Nelson, J.A., and Verbanc, J.J., *Chemical Reviews*, Vol. 57, p. 47 (1959)

23. Saunders, J.H., and Slocombe, R.I., *Chemical Reviews*, Vol. 43, p. 203–216 (1948)

24. Bayer, O., (a) *Angew. Chem.*, Vol. 59, p. 254 (1947), (b) *ibid.* Vol. 62, p. 57, & p. 523 (1950)

25. Knox, R.E., *Chem. Eng. Progress*, Vol. 57, No. 10, p. 40 (1961)

26. Product Research Committee (National Bureau of Standards), *Materials Bank Compendium of Fire Property Data* (Feb. 1980)

27. Dow Chemical Company Form No. 109–613–84. *The Flexible Polyurethane Foam Handbook* (1984)

28. Dow Chemical Company Form No. 109–598–84. *Polyether Polyol Voranol and General Urethane Chemistry* (1984)

29. Du Pont Technical Bulletin BA–13. Freon Products Information, *Properties of Rigid Urethane Foams*.

30. Kresta, J.E., *Reaction Injection Molding and Fast Polymerization Reactions*, pp. 11–30, Plenum Press, New York (1982).

31. Chono, M., Fukuoka, S., et al, *Proceedings of the SPI–6th International Technical/Marketing Conference*, Nov. 2–4 (1983) San Diego, Calif. p. 394

32. Zengel, H.G., *Proceedings of SPI International Conference*, Strasbourg, France, June 9–13, (1980), p. 315.

33. Ashida, K., *International Progress in Urethanes*, Vol. 2, p. 153, Technomic Publishing Co., Inc., Lancaster, PA (1980).

34. Dishart, K.T. & Creazzo, J.A., *Proceedings of Polyurethanes World Congress–SPI/FSK*, September 29–October 2 (1987) Aachen, F.R.G., p. 59.

35. *Proceedings of the SPI–31st Annual Technical/Marketing Conference*, Oct. 18–21 (1988), Philadelphia, PA (a) p. 130, (b) p. 141, (c) p. 148, (d) p. 153, (e) p. 159, (f) p. 278 & (g) p. 285.

36. *Bayer–Polyurethane Handbook*, Bayer, A.G., Polyurethane Research Department (1979).

37. Hashimoto, S., *International Progress in Urethanes*, Vol. 3, Technomic Publishing Co., Inc., Lancaster, PA (1981) p.43.

38. Casati, F.M., et.al., *Proceedings of the SPI 27th Annual Technical/Marketing Conference*, Oct. 20–22, (1982), Bal Harbor, Florida, p.35.

39.   Ashida, K. and Yagi, T., (to Nisshinbo Ind. Inc.) *French Pat.* 1,511,865 (1965).

40.   Ashida, K., *Proceedings of 30th Annual Technical/Marketing Conference*, SPI, Oct. 15–17, (1986), Toronto, Canada. p.354.

41.   Speranza, G.F. & Pepper, W.J., *J. Org. Chem.* Vol. 23, 1922, Dec. (1958).

42.   Ashida, K. and Frisch, K.C., (to Mitsubishi Chemical) *U.S. Pat.* 3,817,938 (1974).

43.   (a) Iseda, Y., Kitayama, M., et.al., (to Bridgestone Co.) Japanese examined patent publication No. Sho–51–33,905 (1976).

      (b) Kitayama, M., *International Progress in Urethanes*, Vol. 5, p. 116, Edited by Ashida, K., and Frisch, K.C., Technomic publishing Co., Inc., Lancaster, PA.

44.   Kimura, S., and Samejima, H., (to Mitsubishi Petrochemial), *Japanese Examined Patent Publication* Sho–54–30,040 (1979).

45.   Ashida, K., (to Mitsubishi Chemical) *U.S. Pat.* 4,022,721 (1977).

46.   Windemuth, E., Braun, G., et.al., (to Bayer A.G.), *German Pat.* 1,112,285 (1961).

47.   Ashida, K., *Japanese Pat.* 724,247 (1973).

48.   The International Isocyanate Institute,Inc., (a) Technical Information No. 1, *Recommendation for the Handling of Toluene Diisocyante*, (b) *ibid, A report on the International Institute,Inc.*, Second Edition, September, 1981 (c) *ibid, Recommendations for the Handling of 4,4′ – Diphenylmethane Diisocyanate MDI*, momeric and Polymeric.

49.   Zacharias, G., and Schuhman, J.G., *Proceedings of the SPI 26th Annual Technical Conference*, Nov. 1–4, (1981), San Francisco, CA, p.87.

50.   Klempner, D., Wang, C.L., Ashtiani, M., and Frisch, K.C., *J. Appl. Polymer Science*, Vol. 32, 4197 (1986).

51.   Aylsworth, J.W., *U.S. Pat.* 1,111,284 (1914)

52.   Butwin, F.J., and Howes, W.C. *Preprint of 41st Annual Conference, World of Composites Focus '86*, SPI., Jan. 27–31, 1986, Session 5–C.

53.   Reed, W.N. and Purdy, J.W., *Proceedings of Advanced Composites Conference*, ASM, SAMPE ESD, and SPE, Nov. 18–20, (1986), Dearborn, Michigan, p. 245.

54.   Edwards, H.R., *Preprint of 42nd Annual Conference and Expo '87*, SPI Composites Institute, Feb. 2–6, (1987), Session 8–C.

55.   Olstowski, F., and Parrish, P.B. (to Dow Chemical), *U.S. Pat.* 4,098,733 (1978).

56.   Kraft, P., and Konkus, D., *Modern Plastics*, Oct. (1987), p. 104.

57.   Ball, G.W., Haggis, G.A., et al, *J. Cellular Plastics*, Vol. 4, 248 (1968).

58.   Nichols, L. amd Gmitter, G.T., *J. Cellular Plastics*, Vol. 1, Jan. (1965), p. 85.

59.   Ashida, K. (to Nisshinbo Ind.), *U.S. Pat.* 3,625,872 (1971).

60.   Ashida, K. and Frisch, K.C., (to Mitsubishi Chem.), *U.S. Pat.* 3,849,349 (1974).

61.   Ashida, K. and Frisch, K.C., (to Mitsubishi Chem.), *U.S. Pat.* 3,793,236 (1974).

62.   Ashida, K. and Frisch, K.C., *J. Cellular Plastics*, May/June, (1972), p. 160.

63.   Ashida, K. and Frisch, K.C., *J. Cellular Plastics*, June/August, (1972), p. 194.

64. Bemard, D.L., Backus, J.K., et al, (to Mobay Chem.), *U.S. Pat.* 3,644,232 (1972).

65. Ashida, K., *International Progress in Urethanes*, Vol. 3, p. 83, Edited by Ashida, K., and Frisch, K.C, Technomic Publishing Co., Inc., Lancaster, PA (1981).

66. Moss, E.K. and Skinner, D.L., *J. Cellular Plastics*, Nov/Dec (1976), p. 332.

67. Moss, E.K., and Skinner, D.L., *ibid*, July/August (1977), p. 276.

68. Moss, E.K., and Skinner, D.L., *ibid*, Nov./Dec. (1977), p. 399.

69. Moss, E.K., and Skinner, D.L., *ibid*, May/June (1978), p. 143.

70. Frisch K.C., Patel, K.J. and Marsh, R.D., *J. Cellular Plastics*, Sept/Oct., (1970), p. 203.

71. Ashida, K. and Yagi, T. (to Nisshinbo Ind.), *Fr. Pat.* 1,511,865 (1967), USP. 3,931,065 (1976).

72. Moss, E.K. and Skinner, D.L., *J. Cellular Plastics*, July/August (1978), p. 208.

73. Canaday, J.S. and Skowronski, M.J., *Proceedings of the SPI–29th Annual Technical/Marketing Conference*, Oct. 23–25 (1985), Reno, Nevada, p. 66.

74. White, K.B. Largent, B., et al, *Proceedings of the SPI–6th International Technical/Marketing Conference*, Nov. 2–4 (1983), San Diego, California, p. 80.

75. Guenther, F., *Proceedings of the SPI–28th Annual Technical/Marketing Conference*, Nov. 5–7 (1984), San Antonio, Texas, p. 44.

76. White, K.B., Jirka, L., Largent, B., and Bailey III, B., *ibid*, p. 47.

77. Tanabe, K., Kamemura, I. and Kozawa, S., *ibid*, p. 53.

78. Carlestrom, W.L., Stoehr, R.T., and Svoboda, G.R., *ibid*, p. 65.

79. Ashida, K, *14th Annual Technical Conference*, Cellular Plastics Division, SPI, Cobo Hall, Detroit, Michigan, Feb. 24–26 (1970), p. 168.

80. Bemard, D.L. and Doheney, A.J., *Belg. Pat.* 723,151 (1967).

81. Ashida, K., *Proceedings of the SPI–32nd Annual Technical/Marketing Conference*, Oct. 1–4 (1989), San Francisco, CA, p. 379.

82. Newman, W. and Fischer, P., *Angew. Chem. Internat. Ed.*, Vol. 1 (1962) No. 12, p. 621.

83. Kan, Peter T.Y. (to BASF Wyandotte) *Ger. Offen.* 2,102,603 (1972).

84. Kan, Peter T.Y (to BASF Wyandotte Corp.), *Japanese Examined Patent Publication*, Sho–47–41,439 (1972).

85. Kan, Peter T.Y. (to BASF Wyandotte Corp.) *Japanese Unexamined Patent Publication*, Sho–48–25,796 (1973).

86. Kan, Peter T.Y. (to BASF Wyandotte Corp.), *Japanese Unexamined Patent Publication*, Sho–48–25,795 (1973).

87. Cenker, M., Kan, P.T., and Robertson, R.J. (to BASF Wyandotte Corp.), *Japanese Unxamined Patent Publication*, Sho–48–43,798 (1973).

88. Kan, Peter T.Y. and Cenker, M., (to BASF Wyandotte Corp.), *Japanese Unexamined Patent Publication*, Sho–50–92,398 (1975).

89. Narayan, T.L., Cenker, M., Kan, R.T., and Patten, J.T. Jr. (to BASF Wyandotte Corp.) U.S. Pat. 3,806,475 (1974).

90. Kan, P.T.Y., Cenker, M., and Patton, J.T. (to BASF Wyandotte Corp.) *U.S. Pat.* 3,717,596 (1973).

91. Campbell, T.W. (to E.I. du Pont de Nemours), *U.S. Pat.* 2,941,966 (1960).

92. Narayan, T.L., and Cenker, M., (to BASF Wyandotte Corp.) *U.S. Pat.* 3,887,750 (1975).

93. Narayan, T.L., and Cenker, M., (to BASF Wyandotte Corp.) (a) *U.S. Pat.* 3,887,501 (1975). (b) *U.S. Pat.* 4,166,164 (1979).

94. Narayan, T.L., and Cenker, M., (to BASF Corp.), (a) *U.S. Pat.* 3,894,972 (1975), (b) *U. S. Pat.* 3,928,526 (1975). (c) *U.S. Pat.* 3,922,238 (1975).

95. Cenker, M., Narayan, T.L. and Kan, P.T., (to BASF Wyandotte Corp), *U.S. Pat.* 3,981,829 (1976).

96. Kan, P.T.Y., and Cenker, M., (to BASF Wyandotte Corp.), *Ger. Offen.* 2,754,011 (1978).

97. McLaughlin, A., Nadeau, H., and Rose, J.S., (to Upjohn Co.) *Ger. Offen.* 2,208,336 (1972).

98. Ashida, K., *Proceedings of the International Confernce*, Strasbourg, France, June 9–13 (1980), Sponsored by SPI and FSK., p. 349.

99. Serlin, I., Markhart, A.H., and Lavin, E., et al, *Modern Plastics*, July (1970), p. 120.

100. Fincke, J.K., and Wilson, G.R., *ibid*, April (1969), p. 108.

101. Serlin, I., Markart, A.H., and Lavin, E., *Proceedings of 25th Annual Technical Conference*, (1970) Reinforced Plastics/Composites Division, SPI., Section 19–A, Pages 1–6.

102. NASA–CP–147496, *Fire Resistant Foams Final Report*, Jul. 1975–16, Feb. (1976), (Solar).

103. Farrisey, W.J. Jr., Rose, J.S. and Carleton, P., *J. Appl. Polym. Sci.*, Vol. 14, p. 1093 (1970).

104. Rosser, R.W. (to USA), *Japanese Unexamined Patent Publication*, Sho 48–91,200 & 91973.

105. Frey, H.E., (to Standard Oil Co.) *U.S. Pat.* 3,300,420 (1967).

106. Farrissey, W.J. Jr., McLaughlin, M., and Rose, J.S., (to Upjohn Co.), *U.S. Pat.* 3,562,189 (1971).

107. Borgnaes, D., Chappell, S.F., et.al, *Proceedings of the SPI–6th International Technical/Marketing Conference*, Nov. 2–4 (1983), San Diego, CA, p. 381.

108. Farkas, A., Mills, G.A., Emer, W.E., and Maerker, J.R., *Ind. Eng. Chem.* Vol. 51, 1299 (1959).

109. Orchin, M., (to Houdry Process Corp.), *U.S. Pat.* 2,939,851 (1960).

110. Frisch, K.C., Sakhpara, D. and Frisch, H.L., *J. Polymer Science: Polymer Symposium* 72, 277–293 (1985).

111. Frisch, H.L., Klempner, D., and Frisch, K.C., *Polymer Letters*, Vol. 7, p. 775 (1969).

112. Klein, A.J., *Advanced Materials & Processes*, July (1986), p. 25.

113. Klempner, D., Frisch, H.L. and Frisch, K.C., *J. Polymer Sci.*, Part A–2, Vol. 8, p. 921.

114. Grieve, R.L., (to Upjohn Co.) *U.S. Pat.* 3,644,234 (1972).

115. Pankratov, V.A., Frenkel, Ts.M. and Fainleib, A.M., *Russian Chemical Reviews*, Vol. 52, (6), (1983), p. 576.

116. Zhitinkina, A.K., Shibanova, N.A., and Tarakanov, O.G., *Russian Chemical Reviews*, Vol. 54, (11), (1985), p. 576.

117. Singh, B., Chang, L.W., and Fergione, P.S., (to American Cyanamide Co.) *U.S. Pat.* 4,439,616 (1984).

118. Schuchardt, J.L., and Harper, S.D., *Proceedings of SPI–32nd*

Annual Technical/Marketing Conference, Oct. 1–4 (1989), San Francisco, CA, p. 360.

119. Ashida, K. *ibid*, p. 379.

120. Narayan, T.L., (to BASF Wyandotte Corp.), *U.S. Pat.* 3,806,475 (1974).

121. Hoechtlen, A., and Droste, W., (to I.G. Farbenindustrie A.G.), DRP 913,474, Apr. 20, (1941), (*Japanese Patent Publication No.* Sho–31–7541).

122. Zaunbrecher, K., and Barth, H., (to I.G.Farbenindustrie A.G.), DRP 936,113, Dec. 15, (1942) (*Japanese Patent Publication No.* Sho 33–8094).

123. Griffiths, T., and Shreeve, P., *Proceedings of the SPI–6th International Technical/Marketing Conference*, San Diego, CA, Nov. 2–4 (1983), p. 76.

124. *Freon Products Information* BA–13, E.I. Du Pont de Nemours & Co., Freon Products Division.

125. Hubert, H.H., and Gluzek, K.H., *Proceedings of the FSK/SPI Polyurethanes World Congress–1987*, Aachen, F.R. Germany, Sept. 29–Oct.2 (1987), p. 820.

126. Ramlow, G.G., Heyman, D.A., and Grace, O.M., *Proceedings of the SPI 27th Annual Technical/Marketing Conference*, Oct. 20–22, (1982), Bal Harbor, Florida, p. 279.

127. Olstowski, F.D. (to Dow Chemical Co.), *U.S. Pat.* 4,125,487 (1978).

128. Fuzesi, S., and Skiroanick, N.J., *Proceedings of the SPI 32nd Annual Technical /Marketing Conference*, Oct. 1–4 (1989), San Francisco, p. 342.

129. Hayash, E.F., Reymore, H.E., and Sayigh, A.A.R., (to Upjohn Co.) *U.S. Pat.* 3,673,128 (1972).

130. Grace, O.M., Mericle, R.E. and Taylor, J.D., *Proceedings of the SPI–29th Annual Technical/Marketing Conference*, Oct. 23–25 (1985), Reno, Nevada, p. 27.

131. Batt, A.M., and Appleyard, P., *Proceedings of the SPI–32nd Annual Technical/Marketing Confemce* Oct. 1–4 (1989), San Francisco, California, p. 433.

132. Puig, J.E., Natoli, F.S., and Baranski, J.R., *ibid* p. 218.

133. Schrock, A.K., Solis, R., Real, G.E., Skorpenske, R.G., and Parrisch, D.B., *ibid*, p. 451.

134. Sullivan, D.A., and Televantos, J., *ibid*, p. 246.

135. Grinberg, E., Lovis, W.W., and Gagnon, S.D., (to BASF Corp.) *U.S. Pat.* 4,745,133 (1988).

136. Kumasaka, S., Suzuki, S., Yoshino, T., Ibayashi, T., and Kobayashi, T., (to Toyo Rubber Chem. Ind. Corp.) *U.S. Pat.* 3,737,400 (1973).

137. Wooler, A.M., (to ICI) *U.S. Pat.* 4,234,693 (1980).

138. Creyf, H.S., (to PRB NV) *U.S. Pat.* 4,334, 944 (1982).

139. Wiederman, R., (to Bayer A.G.) *U.S. Pat.* 4,404,294 (1983).

140. Frisch, K.C and Baumann, H., (to Schaum Chemie) *U.S. Pat.* 4,454,251 (1984).

141. Priester, R.D. Jr., Peffley, R.D., Tumer, R.B., and Herrington, R.M., *Proceedings of 32nd Annual Technical/Marketing Conference*, Oct. 14 (1989), San Francisco, CA, p. 21.

142. Albefino, L.M., *4th SPI International Cellular Plastics Conference*, Montreal, Nov. 15–19 (1976), p.l.

143. Okuyama, T., (to Bridgestone Tire Co.) *Japanese Unexamined Patent Publication*, Sho 50–77, 497, (1975).

144. Ashida, K., "Polyisocyanurate Foams" in *Handbook of Polymeric Foams and Foam Technology*, eds., Klempner, D., and Frisch, K.C., Hanser, New York (1992).

145. Zajacek, J.G., McCoy, J.J., et.al. (to Atlantic Richfield), *U.S. Pat.* 3,895,054 (1975).

146. Hirai, Y., Miyata, K., and Hasegawa, S. (to.Mitsui Toatsu) *U.S. Pat.* 4,170,708 (1979).

147. Becker, R., Grolig, J., Rasp, C., and Scharfe, G., (to Bayer A.G.) *U.S. Pat.* 4,219,661 (1980).

148. Onoda, T., Tano, K. and Fujii, S., (to Mitsubishi Chem.) *Japanese Unexamined Patent Publication*, Sho 54–22,339 (1979).

149. Singh, B., Chang, L.W.K., and Henderson, W.A. Jr., (to American Cyanamid Co.), *U.S. Pat.* 4,879,410 (1989).

150. Mascioli, R.L., *Proceedings of the SPI–32nd Annual Technical/Marketing Conference*, Oct. 1–4 (1989), San Francisco, CA, p. 139.

151. Spitter, K.G., and Lindsay, J.J., *Proceedings of International Conference*, Strasbourg, France, June 9–13 (1980), sponsored by SPI/FSK, p. 719.

152. Szabat, J.F. and Gaetano, J.A., *Proceedings of the SPI–6th International Technical/Marketing Conference*, Nov. 2–4 (1983), San Diego, CA, p. 326.

153. *Proceedings of the SPI–32nd Annual Technical/Marketing Conference*, Oct. 1-4, (1989), San Francisco, CA (a) p. 232, (b), p.9, (c), p. 484, (d) p. 2, (e), p. 510, (f), p. 239, (g), p. 56, (h), p. 52 (i), p. 69, (i) p. 531, (k), p. 527, (1), p. 538, (m), p. 610, (n), p 522.

154. Hilado, C., *Flammability Handbook for Plastics*, 2nd Edition, Technomic Publishing Co., Inc., Lancaster, PA, (1974), p. 155.

155. Frost, C.B., (to General Tire & Rubber) *U.S. Pat.* 3,072,582 (1963).

156. Ashida, K. (to Fischer, H.A.) *U.S. Pat.* 4,780,485 (1988).

157. Private communication.

158. Ashida, K. (to Fischer,H.A.) *U.S. Pat.* 4,908,161 (1990).

159. Kleiner, G.A., Tham, T. and Tenhagen, R.J., *Proceedings of the SPI-6th International Technical/Marketing Conference*, Nov. 2-4, (1983), San Diego, CA, p. 188.

160. Simon, E., *U.S. Pat.* 4,061,605 (1977).

161. Rudner, B. & Noone, T.M. (to Tenneco Chem.) *U.S. Pat.* 4,139,501 (1979).

162. Jaffe, W., Mueller, G., Zinn, O., and Ropte, E., (to BASF) *U.S. Pat.* 4,258,141 (1981).

163. (a) Mahnke, H. and Kreibiehl, G. (to BASF) *U.S. Pat.* 4,334,971 (1982) and (b) Hahn, K., Hors, P., Marx, M., et al, (to BASF) 4,367,294 (1983).

164. Reichel, C.J., Pattern, J. T., Narayan, T., (to BASF), *U.S. Pat.* 4,454,254 (1984).

165. Ashida, K. (to Fischer,H.F.), (a) *European Pat. Appl.* 0-347-497 A2, (1988) (b) *U.S. Pat.* 4,908,161 (1990).

166. Bradley, T.F., and Johnston, W.B., *Ind. Eng. Chem.*, Vol.32, 802 (1940).

167. France, H., Lister,A., (to ICI), *Brit. Pat.* 809,809 (1959).

168. Private communication.

169. Dabi, S., and Zilkha, A., *European Polymer J.*, (1980), 16(1), 95.

170. Dabi, S., and Zilkha, A., *ibid*, (1980), 16(1), 471.

171. Herbstman, S., *J. Org. Chem.* Vol.30, 1259 (1965).

172. *Union Carbide Corporation's Technical Bulletin* "Intermediates for Flexible Foams p. 54.

173. Inoue, S. and Nishimura, T., (to Nippon Polyurethane Ind.Co.) *Japanese Examined Patent Publication*, Sho–46–28777 (1971).

174. Inoue, S., and Nishimura, T., (to Nippon Polyurethane Ind.Co.) *ibid*, Sho–47–3267 (1972).

175. Inoue, S., and Nishimura, T., (to Nippon Polyurethane Ind. Co.), *ibid.*, Sho 46–28776 (1971).

176. Matsui, H., Yasuda, K., and Goto, G., (to Takeda Chem.), *ibid*, Sho–4615298 (1971).

177. Diel, H.C. Martin, R., Ulrich, K., Piekata, H., and Weber, C., (to Bayer A.G.) *ibid*, Sho. 46–41610 (1971).

178. Balon, W.J. (to E.I. du Pont), *U.S. Pat.* 2,801,244 (1957).

179. Bock, P.J., and Slawyk, W., (to Bayer A.G.), *U.S. Pat.* 4,288,586 (1981).

180. Markiwitz, K.H., (to Atlas Chem.) *Japanese Examined Patent Publication*, Sho–46–13257 (1971).

181. Beitchman, B.D. (a) *I & EC, Product Research and Development*, Vol.5, No.1, March, (1966), p. 35 (b), *Rubber Age*, Vol. 98, Feb. (1966) p. 65.

182. Tsuzuki, R., Ichikawa, K., and Kase, M., *J. Org. Chem.*, Vol. 25, June (1960), p. 1009.

183. Kogon, I.C., *J. Am. Chem. Soc.*, Vol.78, 4911 (1956).

184. Jones, J.J. and Savill, N.G., *J. Chem. Soc.* (1957) p. 4392.

185. Kresta, J.E., Chang, R.J., Kathiriya, S. and Frisch, K.C., *Makromol.*

Chem., Vol. 180, 1081 (1979).

186. Pitts, J.J. and Babiec, Jr., J.S., (to Olin Corp.), *U.S. Pat.* 3,759,916 (1973).

187. Bachara, I.S., *J. Cell. Plast.*, Mar/Apr. (1979) p. 102.

188. Bachara, I.S., *ibid*, Nov./Dec. (1979) p. 321.

189. Imai, Y., Hattori, F., and Teramoto, T., *Proceedings of SPI/FSK International Conference on Polyurethanes*, Strasbourg, France, June 9–14 (1980), p. 91.

190. (a) Hsu, C.F., *Interference Reaction in the Catalysis of Polyurethane–Unsaturated Polyester Hybrid IPN's.*, Ph.D Dissertation, April (1990), University of Detroit.

    (b) Frisch, K.C., Ashida, K., and Hsu, C.F., *Preoceedings of Advanced Composite Materials Composites Conference*, Detroit, Michigan, U.S.A., Spet. 30–Oct. 3 (1991), p. 301.

191. Iwata, K., (Editor), *Polyurethane Resin Handbook*, p. 157, Japanese version Nikkan Kogyo Shimbun–Sha. Tokyo, Japan (1987).

192. Specialty Composites Corp., *Catalogue of Confor Foam.*, Newark, Delaware, 19713 (1986).

193. Lickei, D.L., *Proceedings of the SPI-32nd Annual Technical/Marketing Conference*, Oct. 1–4 (1989), San Francisco, CA, p. 183.

194. Gansen, P. and McCullough, D., *Preprint of 1990 SAE Int'l Congress and Exposition*, Feb. 26 – Mar. 2, (1990), Detroit, MI. # 900093.

195. Kath, H., Avan, G. and Thompson–Colon, J.A., *Proceedings of the SPI-32nd Annual Technical/Marketing Conference*, Oct. 1–4 (1989), San Francisco, CA, p. 587.

196 Ashida, K., (to H.A.Fischer), *U.S. Pat.* 4,898,893 (1990).

197. Hickey, L., *Proceedings of the SPI–32nd Annual Technical/Marketing Conference*, Oct. 1–4 (1989), San Francisco, CA, p. 64.

198. Ohmura, Y., Kakuhari, T., and Sawachika, Y., *ibid*, p. 335.

199. Haggis, G.A., (to I.C.I.) *Belg. Patent* 680,380 (1966).

200. Hoshino, T. and Iwakura, Y., *Preprint of Research Conference, Rikagaku Kenkyusho*, Tokyo, Japan, Dec. (1940).

201. Burkus, J. (to United States Rubber Co.) *U.S. Pat.* 2,993,870 (1961).

202. Reymore, H.E., Carleton, P.S., Kolakowski, R.A., and Sayigh, A.A.R., *J. Cellular Plastics*, Nov./Dec. (1975), p. 328.

203. Sayigh, A.A.R., *Advances in Urethane Science and Technology*, Vol. 3, (1974), p. 141, Edited by Frisch, K.C. and Reegen, S.L.

204. Klempner, D. and Frisch, K.C., *Handbook of Polymeric Foams and Foam Technology*, Hanser, New York, (1992).

205. (a) Herold, R.J. and Livigni, R.A., *Advances in Chemistry-- Polymerization Kinetics and Technology*, No. 128, p. 208 (1973) (b) Livigni, R.A., Herold, R.J., Elmer O.C.,and Aggarwal, S.L., *ACS Symposium Series–Polyethers*, No.6, P. 20 (1975).

206. *PB Report* 103 & 373 Feb. 29, 1952, and 1122, Sept. 1945.

207. Kordomenos, P., Kresta, J.E., and Frisch, K.C., *ACS Organic Coatings and Plastic Chemistry*, Vol. 38, p. 450 (1978).

208. Scott, P.H., *Advances in Urethane Science and Technology*, Vol. 6, p. 89, Edited by Frisch, K.C., and Reegen, S.L., (1978).

209 Fuzesi, S., and Brown, R.W. (to Olin Corp.), *U.S. Pat.* 4,699,931 (1987).

210. Fuzesi, S., and Brown, R.W. (to Olin Corp.), *U.S. Pat.* 4,766,158

(1988).

211. David, D.J., and Staley, H.B., *Analytical Chemistry of the Polyurethanes*, Wiley–Interscience, New York (1969).

212. Du Pont de Nemours & Co.(Inc.)., *Properties of Rigid Urethane Foams*, June 21, (1963), Revised Sept. 1966, In a booklet "New Information About Urethane Foam made with Du Pont Hylene".

213. Harding, R.H., *J. Cellular Plastics*, Vol. 1, No. 3, p. 385, July (1965).

214. Iwakura, Y., Nabeya, A., Hayano, F., and Kurita, K., *J. Polymer Sci.*, Part A–11 Vol. 5, 1865 (1965).

215. Dileone, R.R., *J. Polymer Sci.*, Part A–1, Vol. 8, 609 (1970).

216. Guenther, F., *Proceedings of the SPI–28th Annual Technical/Marketing Conference*, Nov. 5–7 (1984) San Antonio, Texas, p. 44.

217 Tanabe, K., Kamemura, I., and Kozawa, S., *ibid*, p. 53.

218. Murphy, J.A., and Wilbur, B.C., *ibid*, p. 58.

219. Carleton, W.L., Stoehr, R.T., and Svoboda, G.R., *ibid*, p. 65.

220. Canaday, J.S., and Skowronski, M.J., *Proceedings of the SPI–29th Annual Technical/Marketing Conference*, Oct. 23–25 (1985) Reno, Nevada, p. 66.

221. White, K.B., Largent, B., Jirka, L., and Bailey, B. III, *ibid*, p. 72.

222. De Guiseppi, D.T., *ibid*, p. 87.

223. Private communication.

224. *AMOCO Chem. Corp. Technical Bulletin*, RO 585.

225. Marks, M.J. and Plepys, R.A. (to Dow Chem.), *U.S. patent*

4,658,007 (1987)

226. Ashida, K., Ohtani, M., Yokoyama, T. and Ohkubo, S., *J. Cellular Plastics,* Sept./Oct. 1978, p. 255.

227. Wood, G., (Editor), *The ICI Polyurethane Book* (2nd Edition) John Wiley and Sons, New York (1990)

228. Ashida K., Morimoto, K. and Yofu, A., *Proceedings of SPI Urethane Conference,* Boston, p. 123 (1994).

229. Herrington, R. and Hock, K., *Flexible Polyurethane Foams*, Dow Chemical Company (1991)

230. Rosbotham, D., Deschaght, J., and Thomas, A.K., *Proceedings of the SPI–34th Annual Technical/Marketing Conference*, Oct. 21–24 (1992) New Orleans, Louisiana, p. 17

231. Lamberts, W.M., *Proceedings of the SPI–/ISOPA, Polyurethanes World Congress 1991*, Sept. 24–26 (1991) Nice, France, p. 734.

232. Volkert, O., *ibid.*, p. 740

233. Fishback, T.L., and Reichel, C.J., *Proceedings of the SPI–34th Annual Technical/Marketing Conference*, Oct. 21–24 (1992) New Orleans, Louisiana, p. 23

234. Speranza, G.P. (to Texaco Chem.) *U.S. Pat.* 5,093,382 (1992)

235. Gills, H.R. (to I.C.I.) *U.S. Pat.* 5,079,271 (1992)

236. Heilig, G., Prager, F., Walter, R., Wiedermann, R., and Wittbecker, F–W., *Kunststoff German Plastics*, September (1991) p. 28.

237. Decaire, B.R., Pham, H.T., Richard, R.G. Shankland, *Proceedings of the SPI–34th Annual Technical/Marketing Conference*, Oct. 21–24 (1992) New Orleans, Louisiana, p. 2

238. Doerge, H.P., and Nichola, W.J., (to Mobay Corp), *Proceedings of the SPI 33rd Annual Technical/Marketing Conference*, Sept. 30–Oct. 3, (1990) Orlando, Florida, p. 82

239. Doerge, H.P., Spitler, K.G., and Mortiner, C.e., (to Mobay Chem.) *U.S. Patent* 5,102,920 (1992)

240. a) Wallaeys, B., De Schryver, P., and Cop, P., *Proceedings of SPI World Congress*, (1991) Nice, France, Sept. 24–26 (1991) p. 63

    b) Monstrey, J., and Wallaeys, B., (to Recticell) *U.S. Pat.* 5,064,872

241. Leppkes, R., and Vorspohl, K., *Proceedings of International PU Forum*, Nagoya, Japan, June 4–5 (1990) p. 54

242. Doyle, E.N., *U.S. Pat.* 5,120,770, June 9 (1992)

243. Leitner, H., and Shaper, H. (to Phenix Gummiwerk) *German Patent No.* DE-4012030

244. Watson, S.L., Jr., Graham, J.R., et al. (to E.R. Carpenter, Inc.) *U.S. Pat.* 5,100,925 (1992)

245. Lin, C.Y. (to Dow Chem.) *U.S. Pat.* 5,114,986 (1992)

246. Volkert, O., *U.S. Pat.* 5,096,933 (1992)

247. Wada, H., Hasegawa, N., Fukuda, H., and Takeyasu, H., *Proceedings of the SPI–34th Annual Technical/Marketing Conference*, Oct. 21–24 (1992) New Orleans, Louisiana, p. 449

248. Madaju, E.J., and Jasenak, J.R., *ibid.* p. 668.

249. Ashida, K., Ohtani, M., Yokoyama, and Ohkubo, S., *J. Cellular Plastics*, September/October (1978) p. 255.

250. *European Plastic News*, July (1990)

251. Kashiwame, J., and Ashida, K., *J. Appl. Polym. Sci.* Vol. 54, 477 (1994).

252. Saiki, K., Sasaki, K. and Ashida, K., *Proceedings of the Urethane Conference* (1993)

253. Goto, J., Sasaki, K., and Ashida, K., *ibid,* (1994) p. 435.

PYRANYL FOAMS (by Kaneyoshi Ashida)

Introduction

Pyranyl monomers have at least two double bonds, producing polymeric foams that are crosslinked, thermosetting and temperature-resistant. These foams were developed by Canadian Industries, Ltd. (1–5), and in part by ICI America, Inc. (7), but the foams have not been commercialized. The reason for this is not clear, but may be due to the fact that pyranyl monomers were not cost effective in comparison with polyurethane foams. However, the technology developed for the foams could be applied for preparing other types of foams from different vinyl monomers.

Chemistry of Pyranyl Foams

Pyranyl foams are prepared by the cationic polymerization of pyranyl monomers in the presence of a catalyst and a surfactant. The pyranyl monomers are prepared via acrolein and acrolein dimer.

Raw Materials

Pyranyl Monomer. Pyranyl monomers are prepared according to the following reactions (1–7). Acrolein is converted to acrolein dimer, e.g., by means of a Shell patent (8), which then leads to pyranyl monomers by different reactions including the Tischenko reaction and the Diels–Alder reaction. Some of the reactions are shown below.

Catalyst. Any cationic polymerization catalyst can be used, but preferable catalysts are Lewis acid catalysts, such as BF_3–etherate and p-toluenesulfonic acid, a combination of it and a Lewis acid also can be used. Inorganic catalysts, such as ferric chloride, stannic chloride, phosphoric pentachloride, and trichloroacetic acid can also be used. The proper amount of the catalyst is in a range of about 0.05 to 2.0 wt % of pyranyl monomers.

Blowing Agents. A literature reference (1) describes the use of CFC–11 (trichlorofluoromethane, CCl_3F) as blowing agent. Due to the recent worldwide regulations in the use of CFC's, alternative blowing agents, such as HCFC's (HCFC–123 and HCFC–141b) may also be used.

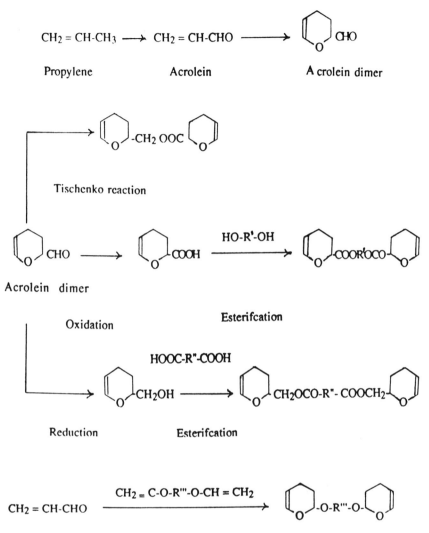

These blowing agents are inert solvents and act as blowing agents by the reaction exotherm of the cationic polymerization.

Surfactants. Silicone surfactants, which are used in rigid urethane foams, can be used as surfactants for pyranyl foam preparation. The silicone surfactants are block copolymers of polydimethylsiloxane–polyoxyalkylene ether in either linear or pendant structures.

Reaction Controllers. The cationic polymerization generates high exotherm and results in very fast foam rise, which may cause problems in pour–in–place foaming. Therefore, reaction controllers, such as organic bases, like tertiary amines derived from fatty acids of coconut oil, are added. Phenol compounds are also effective in retarding the foaming reactions.

Other Additives. The addition of alcohols was found to give fine–cell foams. Flame retardants may also be added if combustion–modified foams are required.

Foam Preparation

The foam preparation process is very similar to that of urethane foam preparation. A two–component system is conveniently used, as shown below (1):

Component A: pyranyl monomer + blowing agent (CFC–11)
Component B: catalyst + surfactant

The following are the foaming conditions: the mixing ratio of A/B is 10/1; the foaming machine is a low–pressure, high–shear mixing machine or high–pressure, low–shear mixing machine; the material temperature (A and B) is 77°±4°F; and the jig temperature is 105°±6°F. The reaction profile is as follows: cream time, 18 seconds; rise time, 35 seconds; tack–free time, 38 seconds; and jig–dwell time, 8 minutes.

The reaction times are considerably shorter than those encountered with most urethane systems, but the system remains fluid during the rise and the flow is even and controlled.

According to I.G. Morrison (1), little evidence of striations or flow lines has been observed and from poured foam 8–foot vertical rises in 3" x 15" cavities have been obtained without shear or flow lines.

Comparative tests between free–rise urethane and pyranyl foams indicate that the total pressures obtained are similar, but that the rate of pressure build–up and pressure dissipation are faster with the pyranyl systems. This property allows short jig–dwell times. Freezer cabinets varying in storage capacity from 7 to 30 ft^3, have been insulated completely satisfactorily with pyranyl foam (1).

Properties of Pyranyl Foams

Mechanical properties, thermal conductivity, thermal stability, flame retardance and chemical resistance are shown below.

Mechanical Properties. Some physical strengths of 2.2 pcf overall–density. Foam core density: 1.8 pcf are shown in Table 32 (1).

Table 32: Mechanical Properties (1)

Overall Density 2.2 Lbs./Cu. Ft.	P.S.I.
Compressive strength (10%) ⊥	18
Yield (6%) ⊥	22
Yield (10%) ‖	14
Yield (6%) ‖	16
Tensile strength (10%) ⊥	20
‖	19
Elongation at break 4%	

⊥ Perpendicular to direction of flow

‖ Parallel to direction of flow

Compressive–strength values are generally of the same order as the tensile strengths in the 2.0– to 3.0–pcf overall–density region. More important is the isotropic nature of the foam, which indicates rounded cells which have almost the same compressive strengths in both parallel and perpendicular directions to foam rise. This is very important in providing dimensionally stable foams.

In rigid urethane foams, the cell shapes are elliptical like eggs, and, therefore, the compressive strengths in the direction perpendicular to foam rise is smaller than the direction parallel to foam rise. Therefore, if urethane foams are required to have the same compressive strength as pyranyl foams in the direction perpendicular to foam rise (i.e., compressive strength in the direction vertical to the panel substrate) urethane foams must have foam densities greater than those of pyranyl foams. Adhesion of pyranyl foams to various substrates, e.g., steel, phosphated steel, stainless steel and aluminum, as well as paper and wood, is very good.

Thermal Conductivity. The thermal conductivity of pyranyl foams is almost equal to that of rigid urethane foams. Figure 48 shows the thermal conductivity of a pyranyl foam having an overall density of 2.2 pcf. The foam was cut from a freezer mock–up, and the K–factors measured were 0.117 and 0.153 Btu–in/hr/ft²/°F respectively. Of the total aging at 70°C over a six–month period, 80% of the change occurred during the first month (1).

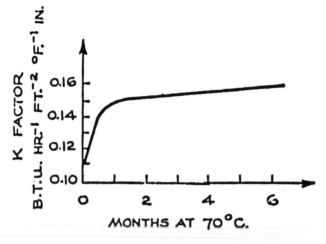

Figure 48. Change in K with time (1).

The surfactant level affects K–factors. Figure 49 shows the optimum amount of surfactant.

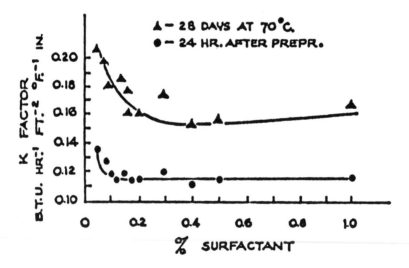

Figure 49. Effect of surfactant on the K–factor of cut foams (1).

Cell Structure and Permeability. Table 33 shows the values of water absorption and water–vapor transmission (1).

Table 33: Cell Structure and Permeability (1)

Water absorption (24 hr.)	14.4 g/1000 cm^3
	16.2 g/1000 cm^3
Water–vapor transmission	2.6 perms
Closed–cell content (corrected)	98 %
Cell size (freezer–wall sample)	2 x 10^{-4} cc.

Dimensional Stability. The dimensional changes of a pyranyl foam after a 28–day test period are shown in Table 34.

Table 34: Dimensional Stability (1)

		% Change After 28 Days (O.D. 2.2 lbs./cu.ft.)	
		Freezer Mock-up	Panel (4'x2' x 2")
-25°C	Length	+ 0.2	0.00
	Volume	+ 0.4	+ 0.1
70°C	Length	+ 2.2	+1.7
	Volume	+5.3	+ 3.0 to 5.1
70°C/100% R.H.	Length	+ 2.3	+ 0.9 to 1.9
	Volume	+ 5.0	+ 2.7 to 4.9

The data show the excellent stability of the foam over a wide range of conditions.

Thermal Stability. Pyranyl foams are crosslinked aromatic polymers, and, therefore, their thermal stability is good in comparison with polyurethane foams. The maximum service temperature for low–density pyranyl foams is 135°C (275°F), but higher temperatures are possible for short periods. The minimum temperature to which the foam has been subjected is −78°C (−108°F) (1).

Flame Retardance. The addition of flame retardants can result in self–extinguishing or non–burning according to ASTM D–1692.

Chemical Resistance. Table 35 shows some of the chemical resistance properties of pyranyl foams. The table shows that pyranyl foams are stable in polar and non–polar solvents.

Table 35: Chemical Resistance of Pyranyl Foam (1)

Sulphuric acid (5%) Hydrochloric acid (10%) Acetic acid (10%) Sodium hydroxide (40%) Gasoline Cyclohexane	Foam stable for > 1 month in each medium
Benzene Isopropanol	Foam stable for > 1 month in each medium. Although solvent-logged, foam retains shape on drying.

Possible Applications

It has been proved that pyranyl foams can be used for pour–in–place applications in household refrigerators and deep freezers. Furthermore, because of their high thermal stability the foams can withstand immersion in molten asphalt at 205°C (400°C). This property makes possible applications to roof insulation with molten asphalt as a water–protecting layer. Another possible application is the baking of the enamel finish on freezer cabinets after the foam insulation has been applied. The foaming reaction is very fast, and, therefore, spraying of the foams onto out–door tanks and pipes is possible.

Advantages of Pyranyl Foams

Properties superior to those of rigid urethane foams are as follows:

(1) Isotropic cell structures, and therefore same mechanical strengths in directions both parallel and perpendicular to foam rise

(2) Higher service temperatures, e.g., 130°C

(3) No toxicity or allergenic problems

(4) Better dimensional stability

(5) High–rise pouring, e.g., 37 feet, is possible.

Disadvantages of Pyranyl Foams

(1) Pyranyl monomers are not available as commercial products
(2) The raw–material cost is high in small–scale production
(3) Mixing ratio of 10/1 for the pour–in–place systems is not convenient for industrial production
(4) Only rigid foam can be prepared; flexible foams are not available

REFERENCES FOR PYRANYL FOAMS

1. Morrison, I.G., *J. Cellular Plastics,* August (1967), p 364.

2. Graham, N.B. and Murdock, J.D. (to Canadian Industries, Ltd.), USP.   3,311,573 (1967).

3. Bowering, W.D. and Graham, N.B., (to Canadian Industries, Ltd.), USP.   3,311,574 (1967).

4. Graham, N.B. (to Canadian Industries, Ltd.), USP. 3,311,575 (1967).

5. Graham, N.B. (to Canadian Industries, Ltd.), USP. 3,318,824 (1967).

6. Nickles, E. and Rose, F. (to Ciba Ltd.), Swiss Patent 11,818 (1961).

7. Chapman, J.F., and Ibbotson, A. (to ICI), USP. 1,069,773 (1967).

8. Whetstone, R.R. (to Shell Development Co.) USP. 2,479,284 (1949).

SYNTACTIC FOAMS *(by Kaneyoshi Ashida)*

Introduction

Syntactic foams can be defined as composites consisting of hollow microspheres (minute hollow bubbles, microbubbles or microballons) and a resinous matrix. ("Syntactic" is derived from the Greek "syntaxis" meaning "orderly arrangement" and it means absolutely isotropic physical properties).

The materials employed for making hollow microspheres include inorganic materials such as glass and silica, and polymeric materials such as epoxy resin, unsaturated polyester resin, silicone resin, phenolics, polyvinyl alcohol, polyvinyl chloride, polypropylene and polystyrene, among others, commercial products available are glass, silica, phenolics, epoxy resin, silicones, etc. Table 36 shows low–density hollow spheres, Table 37 shows physical properties of glass microspheres, and Table 38 shows comparison of some fillers on the physical properties of resulting foams (10).

The matrix is considered to be the binder for the microspheres. Typical matrix materials include (a) thermosetting resins such as epoxy resins, unsaturated polyesters, vinyl esters, phenolics, polyurethanes, and silicones; (b) thermoplastic resins such as polyethylene, polystyrene, polyvinyl chloride; (c) asphalt; and (d) gypsum and cement.

Schematic diagrams of the structure of syntactic foams are shown in Figure 50. Two–phase syntactic foam consists of microspheres and a matrix resin. Three–phase syntactic foam consists of microspheres, matrix resin and air voids.

Due to the extremely low density of the microspheres, the resultant syntactic foam is low in density. It also has high physical strength, especially compressive strength. Other features of the syntactic foam include isotropic physical properties and low water absorption. The latter has led to application in deep sea submersible vehicles. Recently applications have expanded to other areas, including aircraft.

Preparation of Hollow Microspheres

Hollow microspheres were developed by SOHIO Chem. Co. (1–5), and are produced by Emerson & Cuming, Inc. Microballon is the trade name of a phenolic resin–based hollow microsphere produced by Emerson & Cuming, Inc. and used to prevent evaporation in crude oil tanks.

The raw materials for making hollow microspheres include glass, phenolic resin, epoxy resin, polystyrene, silicone rubber, polyvinyl chloride, polyvinyl alcohol. SOHIO Chemical Co. developed two methods, as shown below.

Table 36. Low-Density Hollow Spheres [10]

Name	Description	Particle Size	Bulk Density, G/ml
Kanamite grade 200	Unicellular hollow clay spheres, primarily aluminum silicate	90% 35-60 mesh 10% 20-200 mesh	0.9
Colfoam micro-ballons	Urea-formaldehyde spheres	90% 20-100 mesh 10% through 200 mesh	0.2-0.5
Microballons, BJOA-0840	Hollow phenolic spheres	10% 40-100 mesh 74% 40-200 mesh 18% through 200 mesh	0.3-0.4
Glass micro-ballons CPR 2077	Hollow glass micropheres	89% 60%-325 mesh 11% through 325 mesh	0.26

Table 37: Properties of Glass Microballoons (10)

	Eccospheres R	Eccospheres SI	Eccospheres VT
Bulk density, pcf (g/cm²)	14 (0.22)	11 (0.18)	11 (0.18)
True particle density, pcf (g/cm²)	26 (0.42)	17 (0.28)	17 (0.28)
Particle size, μ	30-300	30-125	30-125
Wall thickness, μ	About 2	About 2	About 2
Composition	Borosilicate glass	Over 95%	SiO, plus coating
Temperature capability, °F	1000	2500	600
Thermal conductivity, loosely packed material, Btu/(hr) (ft²) (°F/ft)	0.04	0.03	0.03
Dielectric constant (dry),1 Mc to 8.6 Gc (approx)	1.3	1.2	1.2
Dissipation factor (dry), 1 Mc to 8.6 Gc (approx)	0.001	0.0005	0.0005

*Trademark, Emerson and Cuming, Inc.

(continued)

Table 37:   (continued)

Bulk density, pcf	12-15
True density (liquid displacement), pcf (g/cm^3)	21.0 (0.34)
Particle size range, μ	
>250	3
175-250	3
124-175	12
89-124	15
61-80	19
44-61	18
<44	30
Average particle size, μ	65
Melting point, °F	1400
Thermal conductivity, Btu/(hr)(ft^2 x °F/in)	0.4
Oil absorption, g oil per 100 cm^3 GMBs	40
Water absorption, % of total weight	
1 hr	0.68
3 hr	0.68
5 hr	0.68
24 hr	1.40

Table 38: Physical and Electrical Properties of Epoxy Syntactic Foam vs Various Fillers (10)

Filler	Viscosity at 250°C, cP	Shrinkage Linear, %	Hardness, Shore D	Density at 21°C g/cc	Tensile Strength, psi	Thermal Conductivity Btu/(hr)(ft²)(°F/in)	Thermal Expansion 25-100°C in/(in)(°C) x 10⁵	Weight Reduction Over Silica-filled compounds, %
None	13,500-19,500	0.12	80-85	1.17	8,000	2.68	8.7	26.5
Silica, 325 mesh, 100, phr	43,000-48,000	0.08	80-85	1.59	5,500	6.38	5.6	0
Phenolic spheres, 15 phr	34,000-38,500	0.14	80-84	0.86	3,300	1.91	8.2	46.0
Kanamite, 34 phr	34,000-39,000	0.06	75-80	1.01	2,000	2.47	6.7	36.5
Colfoam, 4 phr	45,000-48,000	0.17	80-85	1.01	4,050	4.15	8.6	36.5
Glass spheres, 14 phr	44,000-47,000	0.25	80-85	0.95	4,200	4.56	8.2	41.9

Electrical Properties

Filler	Dielectric constant at 25°C		Power factor at 25°C		Volume resistivity, ohm-cm			Dielectric strength at 25°C volts/mil
	1 kc	1 Mc	1 kc	1 Mc	25°C	65°C	100°C	
None	3.8	3.7	0.0035	0.015	8.7×10^{14}	-	5×10^{11}	400-500
Silica, 325 mesh, 100 phr	3.4	3.4	0.003	0.012	1.3×10^{14}	6.2×10^{13}	-	>330
Phenolic spheres, 15 phr	3.2	2.7	0.003	0.014	1.0×10^{14}	5.3×10^{13}	-	>330

Two Phase Syntactic Foam   Three Phase Syntactic Foam

Figure 50. Schematic diagrams of syntactic foams.

(a)   A water solution consisting of phenolic resin, PVA or water glass
       and a blowing agent is sprayed into a high–temperature oven.
(b)   Raw–material powders for glass, e.g., alkali metals, silicates and
       metal oxides, and a blowing agent are mixed and heated in a high–
       temperature oven.

Blowing agents for producing hollow microspheres include carbon–
ates, nitrates, formates, dinitrosopentamethylene tetramine, urea, and
glycerol.

Another method for producing hollow microspheres is that of the
3M Co. (6). This invention relates to minute glass bubbles formed by a
process involving direct conversion of glass particles into glass bubbles
by heating. This method consists of a two–step process, preparation of
expandable glass powder, and expansion of the powder in a high–
temperature oxidative atmosphere.

On the other hand, thermoplastic–based hollow microspheres can
be prepared by heating thermoplastics containing low–boiling–point
solvents. One example is polystyrene hollow microspheres. In the first
stage, expandable polystyrene powder is prepared, e.g., polystyrene
powder containing propane, butane or pentane is prepared by emulsion
polymerization. The powder is then exposed to steam for expansion to
form hollow microspheres.

Another example of the use of expandable powder involves the use
of polystyrene powder. The powder is mixed with a liquid–matrix–resin
system, such as epoxy. The exotherm of the curing epoxy resin expands
the expandable polystyrene powder to form a syntactic foam.

An interesting method for producing plastic microspheres is disclosed by Bayer AG (13). The microspheres consist of vinyl chloride-ethylene copolymers. This method involves the use of the difference of monomer reactivity between vinyl chloride and ethylene. Upon radical polymerization of the two monomers at 50°C, vinyl chloride reacts faster than does ethylene, and unreacted monomeric ethylene then remains in the resulting copolymer produced by the pearl polymerization. The monomer can then act as the blowing agent.

Matrix Resins

Matrix resins to be used for syntactic foams include thermosetting resins and thermoplastic resins, as shown below. Epoxy resin, unsaturated polyester and phenolic resin have been the resins of choice for industrial applications because the resulting foams have remarkably high compressive strengths. Examples of the resins used are given below.

Thermosetting Resins. Epoxy resin, phenolic resin, unsaturated resin, vinyl ester resin, silicone resin, polyurethane resin and polyisocyanurate resin.

These thermosetting resins are composed of two-component liquid systems. By mixing the liquid systems, the thermosetting resins fill the voids in the microspheres to form syntactic foams.

Thermoplastic Resins. Polyethylene, polypropylene, polyvinyl chloride, polystyrene and polyimide (9).

These thermoplastic resins are used by mixing in the melt with thermosetting microspheres.

Preparation of Syntactic Foams

Thermosetting-matrix resins consist of two-component liquid systems which can be easily blended with hollow microspheres at room temperature. In contrast, thermoplastic-matrix resins must be melted for blending with thermosetting hollow microspheres. The following are examples of syntactic foams.

Epoxy Resin-Hollow Glass Microsphere Syntactic Foam. The epoxy resin (matrix resin) and a curing agent are thoroughly mixed. Into the mixture, glass microspheres are added stepwise. The resulting blend is a thixotropic putty-like product which is then transferred into a mold and cured in an oven at an elevated temperature for a certain period of time and then demolded. Winter (14) gave detailed formulations of syntactic foams consisting of epoxy resin and glass microspheres for use in deep–submergence applications.

Phenolic Resin–Based Syntactic Foam. Powdered phenolic resin and phenolic resin microspheres are blended in a blender to make a flowable mixture, which is then poured into a mold and cured at 120°C for 2 hours. The resulting syntactic foam has a density of 0.16 to 0.24 g/cm^3.

Polyimide–Based Syntactic Foam (9). Three–phase syntactic foams were made using a polyimide solution (22% PI–2080 in DMF) and hollow glass microspheres (Type B–30–B, 3M Co.) which have a particle density of 0.25 to 0.30 g/cm^3 and a bulk density of 0.182 g/cm^3. The solution and glass spheres were hand mixed and packed into a 5" x 5" mold and compacted under pressure. Variations of foam density were obtained by molding specific quantities of blend into different volumes ranging from ½" to 1" in thickness. Greater densities required higher pressures with the maximum density obtained at a pressure of about 100 psi.

The wet moldings, after removal from the mold, were placed in an oven at 100° to 120°C. After 3 to 4 hours the temperature was raised to 240°C over a one–hour interval and held for one hour. The resulting foam was essentially solvent–free. For the maximum thermal stability, however, the foams were cured for two hours in an oven at 270° to 280°C.

The compressive strength of a three–phase syntactic foam is primarily dependent upon the properties of the microspheres, the degree to which they are packed into a volume, i.e., density, and the strength of the bond holding the spheres together.

In contrast, in a two–phase syntactic foam the resin matrix completely encapsulates the glass microspheres, and, therefore, the compressive strength of these composites is primarily a function of glass content and resin properties.

The main advantage of this type of syntactic foam lies in its thermal properties. The compressive strength retention of the foams at 288°C is in the range of 60 to 85%, depending upon the foam density.

Another measure of thermal stability is the limiting oxygen index (LOI). Polyimide 2080 has an LOI of 44% and a syntactic foam using the resin as the matrix has an LOI greater than 55%. The flame penetration test of the US Bureau of Mines is a severe test of thermal stability. A syntactic foam having a density of 0.27 g/cm^3 exhibited a flame penetration of 1.1 hr/in. Thermal conductivity of the foam is in the range of 0.38 to 0.46 Btu/hr-ft^2–°F/in over a density range of 0.22 to 0.30 g/cm^3.

Syntactic–Foam Prepregs. Syntactic foam prepregs are composed of glass microspheres supported in a resin matrix on a glass fabric, and is a strong, low–density laminating material (8). Syntactic–foam prepregs have outstandingly high physical strengths, e.g., impact and compressive strengths, and are significantly easier to fabricate into curved parts than are equivalent RP and honeycomb combinations. Combining glass, aluminum, boron, and carbon–fiber skins with syntactic–foam prepregs produces versatile design possibilities with tailored properties (8). Syntactic–foam prepregs are structural materials with properties between lightweight foams and solid laminates. A 720 kg/m³ (45 lb/ft³) syntactic–foam prepreg has ten times the compressive strength of rigid urethane foam at the same density.

Polystyrene–Epoxy Syntactic Foam. Hollow polystyrene micros–pheres are produced by heating expandable polystyrene (in other words gas–filled spheres, e.g., propane or butane–filled polystyrene) of microscopic size. The expandable polystyrene microspheres may be added to the epoxy–resin formulation, and the exothermic heat (or the heat during oven cure) can be employed for the expansion. In this manner, foams having densities as low as 80 kg/m³ (5 lb/ft³) may be developed.

Close temperature control is required to obtain reproducible results, because the expansion of the polystyrene is a function of temperature. The polystyrene microspheres may be pre–expanded by steam or radiant heat and then added to the mold, heated in advance to a convenient temperature, such as 60°C. The mixed epoxy resin is then poured over the pre–expanded beads and allowed to cure. In this case, a limited amount of further expansion will occur as a result of exotherm or cure temperature to permit complete mold fill (10). A typical formulation is shown below:

Parts

Diglycidyl ether of bisphenol A	100
Polystyrene beads	175
cured with	
60/40 Isopropyl alcohol/water	10
Diethylaminopropylamine	6

The resultant cured product will have a density of approximately 112 kg/m³ (7 lb/ft³), with a compressive strength of 0.69 MPa (100 psi). The polystyrene microspheres have about the same physical and electrical properties as those obtained with other organic syntactic foam fillers.

However, maximum use temperature is limited to about 70°C (10).

Effect of Matrix Resins on Physical Properties. Table 39 shows the effect of type of matrix resin on the physical properties of syntactic foams in which the glass microspheres employed were Emerson & Cuming Company's product.

Properties of Syntactic Foams

The structure of syntactic foams is composed of closed–cell microspheres and matrix resins, and the resultant foam has the advantages of (a) isotropic properties not available in other types of foams; (b) very low water absorption due to highly closed–cell structures; and (c) very high compressive strength/weight ratio. Table 40 shows properties of syntactic foams composed of epoxy resin and glass microspheres.

The 42 lb/ft^3 foam in the table showed a hydraulic crush point of 17,000 psi (1,190 kg/cm^2). This figure means that the foam can resist the hydraulic pressure at about 12,000 m (36,000 ft) of ocean depth.

Table 41 shows comparison of matrixes between epoxy resin and wax as matrix. The glass bubbles apparently improved the strength of the wax matrix while lowering its density, whereas the highest–strength epoxy suffered the greatest strength loss.

Phenolic microbubbles generally yield lower–strength syntactic foams than do glass microbubbles of equal density. As bubble density decreases, relative strengths also decrease (7). The most important factor for use as deep–submergence buoys is the rate of water absorption under a given hydrostatic load.

The hydraulic crush point of a foam is also important in determining to what maximum hydrostatic pressure it can be subjected without rapid failure via high water absorption. Water absorption is another important factor. A high–quality syntactic foam displayed less than 3% water absorption after six weeks of exposure to its ultimate hydrostatic strength for 1–in–diameter by 2–in–long test specimens. The test pressure employed should be not greater than 75 to 80% of the crush point.

Uniaxial compressive properties are important to the design engineer who can utilize the foams' inherent high compressive strength in reinforcing other structural members. Sandwich construction is a typical example of such a use, as in submarine–hull construction. Syntactic–foam prepregs have been developed for this application (7).

Table 42 shows properties of epoxy syntactic foam prepregs. Table 43 shows the strength of 10–ply laminates of epoxy prepregs in three orientations (8). The table indicates that the products have satisfactory isotropicity.

Table 39: Typical Properties of Syntactic Foams (Microballoons and Various Resin Binders) (12)

Binder	Two-component RT cured	Epoxy One-component	Crosslinked Polystyrene	Silicone
Density, pcf	20	23	32	25
Dielectric constant (10^9-10^{10} cycles)	1.45	1.55	1.67	1.6
Dissipation factor (10^9-10^{10})	0.01	0.01	0.001	0.002
Compressive strength, psi	1010	1500	5000	750
Operating temperature range, °F	-70 to + 300	-70 to + 500	-70 to + 350	-70 to + 800

Table 40: Properties of Syntactic Foams of Various Densities[a] (7)

Property	42 lb/cu. ft.	40 lb./ cu. ft.	38 lb./ cu. ft.	36 lb./ cu. ft.
Net buoyancy (norminal, in sea water), lb/cu. ft.	22	24	26	28
Compressive strength, uniaxial, ultimate, p.s.i.	13,400	11,000	10,200	9,600
Compressive yield, 0.2% effect, uniaxial, p.s.i.	10,400	9,000	8,500	8,100
Compressive modulus, uniaxial, p.s.i.	480,000	458,000	383,000	373,000
Hydrostatic crush point, p.s.i.	17,000	14,000	13,400	12,600
Tensile strength, p.s.i.	4,600	3,600	-	3,300
Flexural strength, p.s.i.	6,000	6,100	-	3,800
Shear strength, p.s.i.	4,400	4,100	-	3,800
Bulk modulus, p.s.i.	582,000	538,000	353,000	308,000

[a] - Data obtained from production castings of Scotchply XP-241 syntactic foam utilizing epoxy resins and standard quality 3M glass microbubbles.

Table 41: Uniaxial Compressive Strength Retention Characteristics of Epoxy and Wax Matrices Filled with Glass Microbubbles[a] (7)

Property	Epoxy resin "A"			Epoxy resins "B"			Modified wax		
	Unfilled	Filled	Strength	Unfilled	Filled	Strength retained	Unfilled	Filled	Strength retained
Density, g/cc	1.2	0.64	-	1.1	0.605	-	0.936	0.54	-
Ultimate uniaxial compressive strength, p.s.i.	21,000	12,000	56.6%	10,800	9,100	84.2%	1,230	1,235	100.3%
0.2% offset compressive yield strength, p.s.i.	16,600	10,300	62.0%	8,900	5,500	61.8%	736	1,075	146.0%
Uniaxial compressive modulus, p.s.i.	373,000	458,000	80.0%	344.00	389,000	113.0%	43,000	84,000	195.5%

[a] Data is obtained on experimental laboratory castings of 3M formulated resins and foams, utilizing standard 3M microbubbles.

Table 42: Properties of Epoxy Syntactic Foam Prepregs[a](8)

Properties	Synpreg 7801 (for use at ocean depth to 20,000 ft.)	Synpreg 7802 (for use at ocean depths to 10,000 ft.)	Synpreg 7803 (for use at ocean depths to 1,000 ft.)	Synpreg 7804 (compatible 181-glass fabric prepreg for surface finishing)	Synpreg 7201 (fire-resistant syntactic prepreg primarily for surface use)
Specific gravity	0.7-0.9	0.7-0.9	0.7-0.9	1.7-1.9	0.7-0.9
Thickness, in.	0.365	0.400	0.420	0.100	0.400
Tensile properties					
Strength, p.s.i.	13,000	13,000	13,000	60,000	16,000
Modulus, 10^6 p.s.i.	1.0	1.0	1.0	3.6	1.3
Flexural properties					
Strength, p.s.i.	20,000	20,000	19,000	92,000	-
Modulus, 10^6 p.s.i.	1.0	1.0	1.0	3.7	-
Compression, flatwise					
Yield strength, p.s.i.	19,000	17,000	15,000	-	6,000
Ultimate strength, p.s.i.	22,000	20,000	16,000	-	-
Compression, edgewise					
Strength, p.s.i.	23,000	23,000	23,000	55,000	12,000
Modulus, 10^5 p.s.i.	1.0	1.0	1.0	3.8	1.1
Water absorption, %	<3[b]	-	<1[c]	0.5	-
Izod impact, ft.-lb./in.	7	7	8	16	6
Dielectric constant					
at 9.375 GHz	2.3	2.3	2.3	4.4	2.5
Loss tangent					
at 9.375 GHz	0.014	0.014	0.010	0.020	0.017
Thermal conductivity, 10^{-4} cal. cm/cm^2 sec. °C	2.6	2.6	2.6	-	2.6
Thermal expansion 10^{-4} in./in./°F	9.1	9.1	9.1	5	9.1

Test panels 10×10 in. were fabricated from 10 layers of syntactic foam prepreg (7804 laminated 12-ply).
The laminates were bagged with top and edge bleeder on caul sheets and cured 9 min./ply at 300°F, and 50 p.s.i.
The specimens were tested per Federal Test Methods Standard 408.
After cycling 1,500 times to a hydrostatic pressure of 13,500 p.s.i.
After cycling 1,500 to a hydrostatic pressure of 300 p.s.i.

Table 43: Strength of 10-ply Laminates of Epoxy Synpreg 7801 in Three Orientations[a](8)

Property	0 deg.	48 deg.	80 deg.	Property	0 deg.	45 deg.	90 deg.
Tensile strength,				Flexural modulus,			
Dry, p.s.i.	12,150	9,000	11,825	Dry, 10^6, p.s.i.	1.27	0.90	1.34
Wet, p.s.i.	11,810	8,492	10,526	Wet, 10^6, p.s.i.	1.38	0.87	1.29
Tensile modulus,				Interlaminar shear,			
Dry, 10^6 p.s.i.	1.28	0.83	1.24	Dry, p.s.i.	1,789	1,414	1,481
Wet, 10^6 p.s.i.	1.18	0.81	1.13	Wet, p.s.i.	1,316	1,117	1,187
Compressive strength,				Bearing strength, D/t = 1 in.			
Dry, p.s.i.	22,140	14,026	21,870	Dry, p.s.i.	15,280	-	14,490
Wet, p.s.i.	21,940	14,400	21,380	Wet, p.s.i.	15,030	-	13,900
Compressive modulus,				Bearing strength, D/t = 0.5 in.			
Dry, 10^6 p.s.i.	1.27	0.82	1.30	Dry, p.s.i.	13,210	-	13,390
Wet, 10^6 p.s.i.	1.27	0.85	1.20	Wet, p.s.i.	14,030	-	12,050
Flexural strength,							
Dry, p.s.i.	22,600	14,980	20,810				
Wet, p.s.i.	22,040	13,600	20,380				

[a] -10-ply test panels 36 x 100 in. with no surface skins were cured at 300°F for 90 min. under 22-28 in. Hg. vacuum (approximately 15 p.s.i.). Compression load at 0.1 in. deflection 13,800 p.s.i.

Table 44: Effect of Cure Pressure on 10-ply Epoxy Synpreg 7803 Sandwich Panels

Skins, glass fabric	7581	7581	7581	7581	120	120	120	120
Cure pressure, p.s.i.	Vac.	15	50	100	Vac	15	50	100
Panel thickness, in.	0.398	0.364	0.370	0.344	0.375	0.341	0.327	0.316
Specific gravity	0.76	0.78	0.87	0.94	0.79	0.79	0.86	0.88
See water absorption at 700 ft. (310 p.s.i.), %	0.3	0.3	0.2	0.1	0.3	0.3	0.2	0.1

Table 44 shows the effect of cure pressure on a 10–ply epoxy prepreg (8).

Applications

The major advantage of syntactic foams is the high strength–to–weight ratios. This advantage has led to applications in deep–submergence vehicles for hydrospace use (7), aerospace applications such as interior floor panels of aircraft (8), nose cones, fins, and bodies of rockets, sonar windows (some acoustic properties of the foam are similar to those of sea water, radomes, etc. (11).

Table 45 shows that epoxy syntactic foam prepreg can be used for deep–sea applications. For example, an outer hull of Synpreg 7801 can be applied to deep–submergence vehicles to be used at a depth of 20,000 ft.

Table 45: Properties of Epoxy Syntactic–Foam Prepreg (8)

Vehicle	Use Depth (ft.)	Use Pressure (p.s.i.)
Surface support vehicle hull made from Synpreg 7201	Sea level	0
Wet swimmer, outer hull made from Synpreg 7803	– 1,000.	2,225
Deep–submergence vehicle, outer hull made from Synpreg 7801	–20,000.	9,000

(Synpreg is a trade name of Whittaker Corp.)

REFERENCES FOR SYNTACTIC FOAMS

1.   USP. 2,797,201 (1957) (to SOHIO Chem.).

2.   USP. 2,978,339 (1961) (to SOHIO Chem.).

3.   USP. 2,978,340 (1961) (to SOHIO Chem.).

4.   USP. 3,030,215 (1962) (to SOHIO Chem.).

5.   USP. 3,133,821 (1964) (to SOHIO Chem.).

6.   USP. 3,365,315 (1968) (to 3M Co.).

7.   Davis, J.W. and Johnson, R.W., *Modern Plastics,* Vol. 45, [1], 215 (1967).

8.   Kausen, R.C., and Corse, F.E., *Modern Plastics,* Vol. 46, [12], p 146 (1969).

9.   Farrissey, W.J., and Rausch, K.W. Jr., *31st Annual Technical Conference, (1976),* Reinforced Plastics/Composites Institute, The Society of the Plastic Industry, Inc., Section 21–F.

10.   Lee, H. and Neville, K., *Handbook of Epoxy Resins,* McGraw–Hill, New York (1967) pp. 19–11 to 19–17.

11.   Benning, C.J., *Plastic Foams, Vol. 1* and *Vol. 2.* Wiley–Interscience, New York (1969).

12.   Ferrigno T.H., *Rigid Plastic Foams,* Reinhold Publishing Co., New York (1967).

13.   Bayerk, A.G., *Kunststoffe,* Vol. 60, 19 (1970).

14.   Winter, A.J., *Cellular Plastics,* May (1966), p. 157.

FOAMED COMPOSITES *(by Kaneyoshi Ashida)*

Introduction

Foamed composites are lightweight materials reinforced by fibers. Some of the advantages of these materials are excellent strength/weight ratios, workability, excellent corrosion resistance, and design flexibility of molded products.

The matrix resins for foamed composites include rigid polyurethanes, unsaturated polyesters, vinyl esters, and their hybrid resins, such as, unsaturated polyester–urethane hybrid resins and vinyl ester–urethane hybrid resins. The reinforcing fibers include glass fibers, carbon fibers, and organic fibers such as polyamide fiber (Kevlar, DuPont), polyamide–

coated polyester fiber (Colback, BASF) etc.

Foamed composites appearing on the market include Thermax (Celotex Corp) (35), Elsen Neo–Lumber FFU (Sekisui Chemical Co., Ltd., Japan) (1,3,8,22), Airlite FRU (Nisshinbo Industries, Inc., Japan) (2), and Polywood (Polymetrics, Inc., Michigan, USA) (5). NPC Inc., New Hampshire, USA, has developed foamed composites, a continuous production process, and equipment for the composites (19, 21). Their product is named Synwood.

Possible applications of foamed composites include building insulation, wood substitutes for applications which require corrosion resistance, high physical strength and light weight. Very recently, a large market for foamed composites has appeared in the automotive industry. One example is a door–trim panel (7).

Raw Materials

Matrix Plastic Foams. The matrix for plastic foams includes rigid polyurethane foam, urethane–modified isocyanurate foam, unsaturated polyester–polyurethane hybrid foam, and vinyl ester–polyurethane hybrid foam.

Polyurethane Foams. Rigid polyurethane foam can be prepared by the reaction of a polyisocyanate, a polyol, a blowing agent, a catalyst and a surfactant. Detailed explanation of these foams are described in the sections on Rigid Urethane Foams and Miscellaneous Urethane Foams earlier in this chapter.

Urethane–Modified Isocyanurate Foams. Urethane–modified isocyanurate foams are prepared by the trimerization of a polyisocyanate in the presence of a polyol, a trimerization catalyst, a blowing agent, and a surfactant. The foams have high flame and temperature resistance. The combined use of an isocyanurate foam and glass fiber not only improves the physical properties, e.g., flexural strength, friability, etc. but it also improves the flame resistance because the char formed from the foam acts as thermal barrier and protects it from flame and heat. This type of composite, therefore, is widely used for building applications in the U.S.A. Urethane–modified isocyanurate foam systems have also been used in the SRIM process (26, 36, 37).

Hydroxyl–Containing Unsaturated Compounds/Polyurethane Hybrid Foams. Unsaturated polyesters have terminal hydroxyl groups and unsaturated linkages in the polymer skeleton (34). Vinyl esters have terminal vinyl groups and pendant hydroxyl groups. These model structures are shown in Equations (1) and (2) respectively.

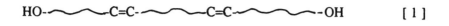

$$[1]$$

$$[2]$$

These compounds can be copolymerized with styrene to form crosslinked copolymers. When polyisocyanates are incorporated in these reaction systems, the resulting polymers are hybrid polymers containing polyurethane linkages. Rigid foams can be obtained by using the above reaction in the presence of a blowing agent.

Unsaturated polyesters having hydroxyl groups can be prepared by the reaction of glycols, maleic anhydride and dicarboxylic acids in more than the stoichiometric amount of glycols/carboxylic acids (13, 34). Edwards discussed the application of isophthalic unsaturated polyester urethane hybrids in conventional molding techniques (13). He also applied the hybrids to foamed products (21). Vinyl esters can be prepared by the reaction of bisphenol A with glycidyl methacrylate (21).

Mazzola et al used polyesters for foamed composites (6). Narkis et al (27) described foamed polyester composites made using random glass mat. Saidla et al (28) reported making foamed polyester composites using ¼-inch glass fibers. Vinyl ester/styrene copolymer foams were developed by Olstowski and Perrish (10, 11). Vinyl ester-based hybrid-foam composites were developed by Frisch and Ashida (19).

Methacrylate-modified unsaturated polyester/polyurethane hybrid resins (as matrix resins, not foamed products) were developed by Ashland Chemical (33, 40).

Styrene–Aromatic Dimethacrylate as Matrix Resin. Gonzalez and Macosko (30) prepared composites using 50% styrene and 50% aromatic dimethacrylate as the matrix resins and a chopped-strand glass fiber as reinforcement.

Epoxy Resins as Matrix Resin. Burton and Handlovits used conventional epoxy resins as the matrix resin, and fiber glass, wollastonite and inorganic fillers as the reinforcement (29).

Nylon Block Copolymers as Matrix Resin. Monsanto and DSM (Dutch State Mines) have been investigating caprolactam RIM with and without glass reinforcement (31).

Dicyclopentadiene Polymers as Matrix Resin. Klosiewicz (32) described dicyclopentadiene (DCPD) polymers with and without fiber reinforcement.

Reinforcing Materials. Glass fibers are the major reinforcing materials at the present time. E–glass fibers are conveniently used because of the low cost. S–glass fibers are used only in specific applications which require higher strength. The types of glass fiber include continuous–fiber–strand mat, long chopped–strand mat, glass rovings, and woven rovings, as shown in Figure 51.

Figure 51. Different types of glass fibers. (1) Continuous strand mat, (2) Surfacing mat, (3) Chopped–strand mat, (4) Woven roving). (Courtesy of Ownes Corning Fiberglass Corp.)

It has been pointed out that surface treatment of fibers significantly improves the physical strength of the composite (14). The reinforcing fibers should be surface–treated before use. Examples of the treatment include chemical treatment, such as silane compounds and titanates (20), and physical treatment such as corona and ultraviolet plasma (14).

Most fiber reinforcements on the market have been surface–treated for the convenience of composite production. Organic fibers, such as polyamide (Kevlar, DuPont), polyamide coated polyester fiber (Colback, BASF (26) can also be used.

Blowing Agents

The same blowing agents used for polyurethane foams can be used for foamed composites. Preferable blowing agents are chlorofluorocarbons (CFC's) and hydrochlorofluorocarbons (HCFC's). Water can be used for partial replacement of CFC's and HCFC's in the case of isocyanate–based foams, such as urethanes and their hybrid foams.

Surfactants

The surfactants employed for polyurethane foams can be also used for preparing foamed composites. The surfactants include silicone surfactants, which consist of polysiloxane–polyoxyalkylene block–copolymers.

Preparation of Foamed Composites

Both continuous and batch processes can be used for foamed–composite preparation, depending upon the products desired and market demand. The structure of foamed composites can be classified as shown in Figure 52, that is, unidirectional (or monoaxial), two–dimensional, and three–dimensional reinforcements can also be employed. In addition, combinations of these structures can be used (2).

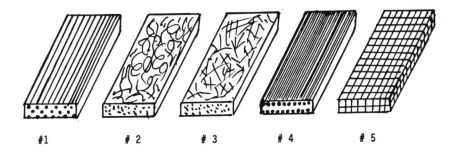

Figure 52. Schematic diagram of reinforced structure of foamed composites. (1) Unidirectional, (2) Continuous–strand mat, (3) Chopped–strand mat, (4) Uni–directional skin layer, (5) Three–dimensional.

A typical example of a monoaxially reinforced foamed composite is that developed by Sekisui Chem. Co., Ltd., Japan. Eslon Nro–Lumber FFU has this structure (22). Its production process has not been disclosed. An example of a two–dimensionally reinforced foamed composite was developed by Nisshinbo Ind. Inc., Japan. The production process employed was the compression–molding process (2). The structure of the products is shown in #2 and #3 in Figure 52. Test specimens were prepared by the following methods (2).

FRU–L: after a sheet of glass–fiber continuous–strand mat was placed in an open mold, the prescribed amount of PUF solution, prepared by hand mixing was poured into the mold and the contents were pressed to the fixed thickness (3 mm).

FRU–M: the prescribed amount of chopped glass strand (13 mm in length) was mixed with the prescribed amount of isocyanate–component and thoroughly dispersed before the addition of the polyol component. Then the following operations were carried out according to the procedure of FRU–L without strand mat.

FRU–S: Every operation was carried out according to the FRU–M using chopped –glass strand 3 mm in length instead of 13 mm. The physical properties of these foams will be discussed later.

Another example of the two–dimensionally reinforced foamed composite was developed by Celotex Corp., USA. The production process consists of the use of continuous–glass–strand mat and a horizontal conveyer process. Continuous–strand mat is supplied on a moving horizontal conveyer. The foaming mixture is then applied onto the glass mat through the mixing head of a dispensing machine (35).

In contrast, Ashida's patent (4) uses a vertical process, as shown in Figure 53. Vertically supplied glass mat is continuously fed int a two-sided release sheet. The mixed foam formulation is then applied onto the V–shaped release sheets. The mixture is immediately pressed by a squeezer which facilitates the impregnation of the foaming mixture into the strand mat, while at the same time, the squeezer controls the thickness of the resulting foamed composites.

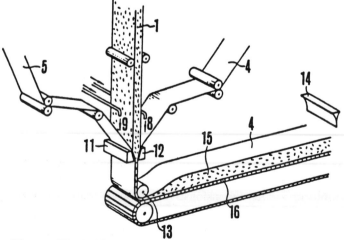

Figure 53. Ashida's continuous lamination process (4).

No. 4 structure in Figure 52 shows another composite structure consisting of unidirectional reinforcement with continuous roving. This structure was employed by Polymetrics Corp., and NPC, Inc. (25).

Three-dimensional glass fiber, as shown as #5 of Figure 52, can be used for batch and continuous processes (23, 24). This foamed composite is proposed for use in cryogenic insulation, such as liquid natural-gas tank insulation. However, its manufacturing process seems to be very complex and production costs would be high.

The combined use of continuous-glass-strand mat as a core material and uniformly and monoaxially aligned continuous-glass fiber as the surface material has also been proposed (5, 25). The incorporation of expanded polystyrene copolymers into composites was proposed for shock-absorbing composites, such as bumper beam cores, knee bolsters, etc. (17, 18).

The latest trend in foamed composites is automotive structural parts produced by the SRIM (structural reaction-injection-molding) process. The SRIM process is a method of the reactive liquid injection molding (i.e., LIM). LIM includes RIM (reaction injection molding), RRIM (reinforced RIM) and RTM (resin transfer molding). LCM (Liquid Composite Molding) is a relatively new technical term which includes SRIM, RRIM and RTM (Resin Transfer Molding).

The first step of the SRIM process is to place a fiber mat in a closed mold. Then, a low-viscosity foaming mixture is injected into the mold. The mixture penetrates into the mat and then expands to form a foamed composite. (SRIM is also called Mat-Molding RIM or MMRIM). SRIM is preferably used for making thin, large panels, such as door-trim panels, spare-tire covers, etc.

Examples of the matrix resins employed for SRIM include rigid polyurethane (7, 15, 16), or urethane-modified isocyanurate foam (9, 26, 35, 36). The matrix resins for foamed composites are required to have low viscosity and relatively slow cream time for better impregnation into fiber mat and fast-cure cycle time for higher productivity.

Before the introduction of the SRIM process, an open-mold process was employed for making fiber-reinforced door-trim panels. The open-mold process is composed of the following steps:

(1)   A random oriented continuous strand mat, precut to panel dimensions, is placed over the vacuum-formed vinyl skin prepared in a mold.

(2)   A rigid urethane foaming mixture is then dispensed over the glass to provide an even

distribution of materials.

(3) The mold is closed before the foam begins to rise, and after the two minutes cure time, the part is demolded.

The product has a molded density of 0.5 g/cm³; the part thickness is usually 4 mm.

An improvement in the production of foamed composite–door panels was reported by Schumacher et al (7). The process consists of the use of the SRIM process and a low–viscosity urethane foam system. In addition, vinyl– and glass–mat preform technology and IMR (internal mold release) were employed. The demold time at 75°C was 45 seconds, and molded densities were 400 to 800 g/cm³.

Hanak et al (9) presented a paper regarding the use of propoxylated Mannich adducts for the production of continuous laminates and fiberglass–reinforced molded rigid foam characterized by CFC reduction or elimination. Table 46 shows an example of the formulation and process conditions employed (9).

Table 46: Formulation and Process Conditions for ARCOL X6015 Based Foam (9)

ARCOL X 6015	100	pbw
Water	1.5	pbw
Glycerine	2	pbw
Tegostab B 8404 (R1)	0.5	pbw
Toyocat TF (R2)	0.9	pbw
Polymeric MDI	180	pbw
Index	115	
Reactivity Bench Mix		
Cream time	28	sec.
Gel time	39	sec.
Tack free time	47	sec.
Reactivity Machine		
Cream time	18	sec.
Gel time	30	sec.
Tack free time	38	sec.
Demold time	150	sec.
Mold temperature	50	°C

R1 Trade mark of Goldschmidt
R2 Trade mark of Tosoh Corporation

Howell (15) announced a method of improving the processing characteristics of a structural RIM system to obtain high glass loading, e.g., 40 to 50%..

Nelson (16, 39) reported a method of making polyurethane–foam composites by means of the SRIM process. Two different types of polyurethane materials were used: an amine–modified polyurethane suitable for static elastomer uses and a polyurethane material with high crosslink density designed for use as a microcellular structural material.

The former material was evaluated with and without Owens/Corning Fiberglas 737 milled glass while the latter resin was always shot neat, but evaluated with and without Owens/Corning Fiberglas M–8610 random mat in the tool. The structural polyol blend had a viscosity of about 1,780 centistokes at 77°F, and an extended gel time of about 15 seconds at the mold temperature of 140°F and component temperature of 100°F. The isocyanate had a viscosity of 200 cps at the same temperature. The A/B ratio was about 1:1.

Nelson investigated the relationship between density and physical properties, e.g., flexural modulus, Gardner impact, heat distortion, tensile/flexural strength, coefficient of linear thermal expansion, dynamic mechanical testing, and creep testing. The specific gravity of the SRIM obtained was changed from about 0.3 to 1.2.

Kuyzin et al discussed a low–density SRIM made by using glass–fiber reinforcement and a urethane–modified isocyanurate foam system (26), or by using polyamide–coated polyester fibers and a urethane–modified isocyanurate–foam system (36).

Table 47 shows typical material and process conditions for Elastolit SR low–density SRIM systems (36), and Table 48 shows the general mechanical properties for an Elastolit SR low–density SRIM system for interior trim applications. Elastolit is a trade mark for BASF's SRIM system (36).

Physical Properties

Foamed composites are lightweight materials having high strength, corrosion resistance and workability. Figure 54 shows the modulus vs density for various materials. In the figure, it is shown that natural wood has a higher modulus at relatively lower densities than other materials (1).

Table 47: Typical Material and Process for Elastolit SR
Low Density SRIM Systems (36)

Chemistry	Polyurethane Polyisocyanurate
Dispensing Machine	Elastogran Puromat 80, typ.
Component Temperature	
A-Comp. Polyol resin	23°C
B-Comp. Isocyanates	23°C
Component Impingement Mixing Pressures	
A-Comp. polyol resin	150-170 atm.
B-Comp. isocyanate	150-170 atm.
Mold Temperature	60°C
Resin viscosity at room temperature	300-1250 cp
Isocyanate viscosity at room temperature	60-200 cp
Resin/iso ratio	100/100 to 170 typ.
Blowing agent	Typically CFC-11 or carbon dioxide generated from reaction with water
Demold time	2 minutes typ.
Cream time	20 sec. typ.
Free rise density	2.8 lb./cu./ft. typ.
Shot rate	0.5 lb./sec. typ.
Shot size	2.0 lb. typ.

Table 48: Approximate Mechanical Properties for an Elastolit SR–Based Low Density SRIM System for Interior Trim Applications (with 13 wt % OCF Glass Mat Reinforcement of 1.5 oz/ft²) (36)

	Value	Test Method
Density, lb./cu. ft.	24	SAE J 315
Flex modulus, psi	150,000	ASTM D790
Flex strength, psi	4,000	ASTM D790
Moisture content, %	2.6	SAE J 315
Water absorption 2.5 hrs., %	1.0	SAE J 315
Water absorption 14 hrs., %	1.9	SAE J 315
Water swell, %	0.5	SAE J 315
Dimensional Stability, %		
Expan. in water, 70°F, 24 hrs	0.2	SAE J 315
Contrac. in water, 190°F, 24 hrs.	0.2	SAE J 315
Warpage original	0	SAE J 315
Wet	0	-
Dry	0	-
Flammability (MVSS301), inches/min.	1.5	FLTM BN24-2
Coeff. of therm. expan., in./in./°F,		ASTM D 696
- 20 to 175°F	3×10^{-6}	ASTM D 696

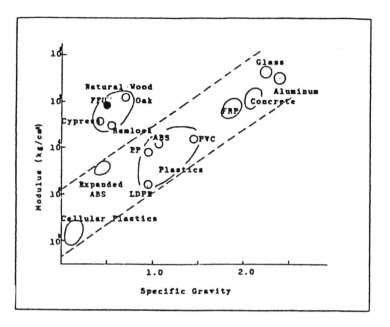

Figure 54. Comparison of strength of various materials (1).

Properties of Unidirectional Type Composites. A synthetic wood was produced using rigid polyurethane foam and continuous–glass fiber strand by Sekisui Chem. Ind. Ltd. Its trade name is Elson Neo–Lumber FFU (1, 3, 8). The product has a coarse, dense structure like natural wood, and the product has the advantages of light weight, corrosion resistance, and a high bending strength, in spite of its relatively low compressive strength (3). Table 49 shows general properties of the Eslon Neo–Lumber FFU (trade mark of Sekisui Chem) (1).

Two–dimensionally reinforced foamed composites were developed by Nisshinbo Ind., Inc. The foamed composites consist of different types of glass fibers, e.g., continuous strand mat, or chopped strand. Table 50 shows typical properties of Airlite FRU (trade mark of Nisshinbo Ind.) (2).

Morimoto and Suzuki (12) studied the flexural properties of continuous–glass–fiber–strand mat–reinforced rigid polyurethane foam, and found that both the flexural modulus and the flexural strength increased and the temperature dependence decreased when longer fibers were used as reinforcement. The density of the matrix foam also enhanced these tendencies.

Table 49: General Properties of Foamed Composite, Eslon Neo Lumber FFU (1)

Item	Test Method	Unit	Results	
			FFU-A	FFU-E
Physical Properties				
Specific gravity	JIS Z-2102	-	0.48-0.51	0.38-0.41
Brinell hardness	JIS Z-2117	-	0.85	0.55
Water absorption	JIS Z-2104	%	0.2	0.2
Mechanical Properties				
Tensile strength (MD)	JIS Z-2112	kg/cm^2	800-1,000	550-700
Bending strength (MD)	JIS Z-2113	kg/cm^2	500-700	400-500
Compression strength (MD)	JIS Z-2111	kg/cm^2	400-600	145-220
Compression strength (PD)	JIS Z-2111	kg/cm^2	35-50	20-30
Impact strength (Charpy)	JIS K-7111	$kg\text{-}cm/cm^2$	Min. 30	Min. 30
Bending modulus	JIS Z-2113	kg/cm^2	$5.0 - 5.4 \times 10^4$	$4.3\text{-}4.8 \times 10^4$
Thermal Properties				
Linear thermal expansion				
Coefficient	JIS K-6911	$1/°C$	0.8×10^{-5}	0.9×10^4
Heat transfer coefficient	DIN - 8061	Kcal/°C m.h.	0.07	0.07
Heat deformation temperature	ASTM D 648	°C	87	80
Electrical Properties				
Volume resistivity	ASTM D 250	Ω cm	$10^{13} \sim 10^{14}$	$10^{13} \sim 10^{14}$
Surface resistivity	JIS K-6911	Ω	2×10^{13}	2×10^{13}
Dielectric strength			Min. 30	Min. 30

The reinforcing effect of glass fiber was analyzed quantitatively from the fiber efficiency factor, K_e, with the simple rule of matrix (38) expressed by the following equation:

$$E_c = E_m V_m + K_e E_f V_f$$

It was concluded that the actual values of interlaminar shear strength obtained by a short–beam method agreed well with the theoretical values.

Yosomiya and Morimoto (38) studied the compressive properties of continuous–glass–fiber–strand mat with different fiber lengths, different fiber–volume fractions, and different densities of the matrix foam.

The compressive strengths of the composites obtained increased and their temperature dependencies decreased with increasing fiber length, fiber–volume fraction, and density of the matrix foam. More specifically, the compressive strength of the composite was found to be proportional to that of the matrix and increased linearly with increased fiber–volume fraction in the experimental range employed (below 2% by volume). This result could be explained by Swift's sinusoidal model, assuming that the adhesion between fiber and matrix foam is perfect.

K. Morimoto and T. Suzuki (2) showed typical properties of foamed composites in Table 50.

Physical strengths of foamed composites are significantly affected by the type of reinforcements. Table 51 shows the effect of various reinforcements on the physical strengths of foamed composites (2).

Celotex Corp. (35) has commercialized foamed composites consisting of a urethane–modified isocyanurate foam and continuous–glass–strand mat. The trade name of the composite is Thermax. Some physical properties of Thermax are shown in Table 52. The physical properties of a low–density SRIM system of urethane–isocyanurate foam for interior–trim applications are shown in Table 48.

Moss and Skinner of Jim Walter Research Corp. (41) discussed a basic formulation study for production of continuous lamination of urethane–isocyanurate foams.

Schumacher and Slocum (7) presented papers regarding low–density urethane–foam composites to be used for automotive door trims. The physical properties of glass–fiber–reinforced urethane–foam composites are shown in Table 53.

Table 50: Typical Properties of Foamed Composite, Airlite FRU (2)

Properties[*]			FRU-A20	FRU-A20	FRU-A10	FRU-A10
Specific Gravity		(g/cm³)	0.3	0.5	0.8	1.2
Forming Thickness		(mm)	20	20	10	10
Flexural	Strength ∥	(kg/cm²)	300	700	1100	1950
	Modulus ∥	(kg/cm²)	1.3×10^4	2.2×10^4	3.3×10^4	6.5×10^4
Compressive Strength	⊥	(kg/cm²)	15	35	130	600
	∥	(kg/cm²)	78	200	540	1300
Tensile Strength	∥	(kg/cm²)	200	450	750	1400
Linear Expansion Coef.	∥	(°C⁻¹)	2.0×10^{-5}	1.1×10^{-5}	1.5×10^{-5}	1.9×10^{-5}
Thermal Conductivity	⊥	(Kcal/mhr °C)	0.024	0.036	0.065	0.13
Volume Resistivity		(Ω-cm)	2×10^{16}	1.0×10^{16}	0.6×10^{16}	0.4×10^{16}
Voltage Resistance		(KV/mm)	>25	>25	>25	>25

∥ : flatwise properties.
⊥ : verticalwise properties.
* variable to some extent with its foaming size especially with thickness.

Table 51: Effect of Reinforcements on Mechanical Properties of Foamed Composite, Airlite FRU Classified by Fiber Efficiency Factor (2)

Properties	Reinforcements	Continuous Fiber (FRU-L)	Medium Fiber (13mm) (FRU-M)	Short Fiber (3mm) (FRU-S)
Flexure	Strength	O	△	×
	Modulus	O	O	O
Tension	Strength	△	×	×
	Modulus	O	△	△
Compression	Strength	O	△*	×
	Modulus	O	△*	△
Impact Strength		O	O	△
Thermal Stability		O	△	△

Effects of reinforcements: O: great (k≥0.2); △: medium (0.2>k≥0.1); ×: small (k<0.1) (k: maximum value of fiber efficiency factor).
* Estimated.

Table 52: Comparison of Physical Properties of Thermax and Polystyrene Foams (35)

Property	Thermax Insulation Board	Extruded Polystyrene	Expanded Polystyrene
Density, pcf.	2.0	2.0	2.0
Aged R value, 75°F.			
1"	7.2	5.0	3.9
2"	14.4	10.0	7.8
Compression strength, psi, ASTM D 1621	25	25–40	10–14
Water vapor transmission	< 0.03 perms	1.0 perms	1.2–3.0 perm in.
	ASTM E 96	ASTM E 96	ASTM C 355
Water absorption % by volume ASTM C 272 (24 hr.)	0.3 max	0.3	< 2.5
Maximum operating temp., °F.	250	165	167
Coefficient of linear thermal expansion, in/in °F ASTM D 696	1.3	3.5	3.5
Flexural strength, psi ASTM C 203	> 40	–	25–30

Table 53: Physical Property Comparison of Various Interior Panel Substrate Materials (7)

	Wood fiber	Wood stock	Low-Viscosity Structural Foam 0%	Low-Viscosity Structural Foam 5% Vol. Glass	Low-Viscosity Structural Foam 10% Vol. Glass
Density - total (g/cc)	1.04	1.11	0.34	0.58	0.81
Density - foam (g/cc)	-	-	0.34	0.48	0.62
Flexural modulus (MPa)	6710	1450	290	1480	2960
Flexural strength (MPa)	62	20	-	46	82
High speed impact - energy to crack (J/M)	7.0	10	18	220	340

Applications

Foamed composites have the advantages of high strength/density ratio, corrosion resistance and light weight. Possible applications include corrosion–resistant structural materials, structural–insulation materials, corrosion–resistant insulation materials and light–weight structural materials. Some application examples of foamed composites are as follows.

(a) Transportation. Foamed composites, such as low–density SRIM, have high strength/weight ratios, and therefore, foamed SRIM is used for making interior door panel substrates for U.S. cars and trucks and a sunroof sunshade for U.S. cars (26). Truck beds, floors and walls of cargo containers and freight cars are other possible application areas. These areas require sturdiness, anti–corrosion and light weight.

It is interesting to note that foamed composites are suitable sandwich core materials for FRP boats (8). Cross–ties (or sleepers) for railroads, especially for use in tunnels and subways, are very promising applications. Japanese National railways carried out a long–term test in a tunnel, and the results were excellent.

(b) Sewage Facilities. Sewage facilities require resistance to corrosion and biological attack. Natural wood is not durable for sewage uses. Foamed urethane composites are actually used in Japan for such applications (8).

(c) Electrical Equipment. Cable racks, cable crates, battery boxes and the third–rail protection plates in subways and tunnels are suitable applications for foamed composites because of their high electrical and corrosion resistance (8).

(d) Chemical Tanks and Other Anti–Corrosive Materials

REFERENCES FOR FOAMED COMPOSITES

1. Okagawa, F., and Morimoto, T., *International Progress in Urethanes, Vol. 2,* p. 85, eds. Ashida, K. and Frisch, K.C., Technomic Publishing, Lancaster, PA (1980).

2. Morimoto, K. and Suzuki, T., *International Progress in Urethanes, Vol. 5,* p. 82, eds. Ashida, K. and Frisch, K.C., Technomic Publishing, Lancaster, PA (1980).

3. Inukai, Y., and Okagawa, F., *ibid.* p. 202.

4. Ashida, K., Ohtani, M., Yokoyama, T. and Ohkubo, S., (to Mitsubishi Chemical Ind. Ltd.) *Canadian Patent 1,090,521* (1980).

5. Frisch, K.C., and Ashida, K., (to Polymetrics Corp.) U.S. Patent 4,680,214 (1987).

6. Mazzola, M., Masi, P., Nicolais, L., and Narkis, M., Fiber Reinforced Polyester Foams, *J. Cellular Plastics,* Sept./Oct. (1982), p. 321.

7. Schumacher, D.W. and Slocum, G.H., *Proceedings of the Third Annual Conference on Advanced Composites* 15–17, Sept. (1987), Detroit, Michigan, p. 151.

8. Okagawa, F., Iwata, T., and Morinoto, T., *Proceedings of the International Conference on Cellular and Non-Cellular Polyurethanes,* Strasbourg, France, June 9–13, (1980), p. 453.

9. Hanak, P. and Den Bore, J., *Proceedings of the SPI–32nd Annual Technical/Marketing Conference,* Oct. 1–4, (1989), San Francisco, California, p. 69.

10. Olstowski, F., and Perrish, D. (to Dow Chemical), *U.S. Patent 4,098,733* (1978).

11. Olstowski, R. (to Dow Chemical), *U.S. Patent 4,125,487* (1978).

12. Morimoto, K., and Suzuki, T., *Industrial and Engineering Chemistry, Product Research and Development, Vol. 23,* No. 1, (1984) p. 81.

13. Edwards, H.R., *Proceedings of 142nd Annual Conference,* Composites Institute, The Society of the Plastic Industry, Inc., Feb. 2–6, (1987), Section 8–C.

14. Yosomiya, R., Morinoto, K., Nakajima, A., Ikeda, Y., and Suzuki, T., *Adhesion and Bonding in Composites,* Marcel Dekker, New York (1990).

15. Howell, T.B., Camargo, R.E., and Bityk, D.A., *Proceedings of the Fifth Annual ASM/ESD Advanced Composites Conference,* Sept. 25–28, 1989, Dearborn, Michigan, p. 69.

16. Nelson, D., *J. Cellular Plastics,* Vol. 22, March 1986, p. 104.

17. Younes, U.E., *Proceedings of the SPI–33rd Annual Technical/Marketing Conference,* Sept. 30 to Oct. 3, 1990, Orlando, Florida, p. 610.

18. Younes, U.E., *Proceedings of the Sixth Annual ASM/ESD Advanced Composites Conference,* Detroit, Michigan, USA, Oct. 8–11, (1990), p. 657.

19. Frisch, K.C. and Ashida, K. (to NPC, Inc.), *U.S. Patent 5,091,436; Feb. 25, 1992.*

20. Monte, S.J., Sugerman, G., Damusis, A., and Patel, P., *Proceedings of the S.P.I. 26th Annual Technical Conference,* Nov. 1–4, (1981), San Francisco, California, p. 187.

21. Private communication.

22. Sagane, N., Morimoto, T., and Okagawa, F. (to Sekisui Chemical Co.), Japanese examined patent publication, *Sho– 48–30137* (1973).

23. McDonell Douglas Corp., *Modern Plastics,* Vol. 46 (11), p. 66 (1969).

24. Goldsworthy, W.B. (to McDonell Douglas Corp.), Japanese unexamined patent publication, *Sho–52–57262* (1977).

25. Private communication from NPC, Inc.

26. Kuyzin, G.S., and Schotterbeck, D.G., *Proceedings of SAE International Congress & Exposition,* #900096 (Feb. 1990).

27. Narkis, M. et al, *Proceedings of 38th Annual Conference,* Reinforced Plastics Institute, SPI, Inc., Session 23–c (1983).

28. Saidla, W., et al, *ibid,* Session 23–A (1983).

29. Burton, B.L. and Handlovitz, C.E., *ibid,* Session 23–E (1983).

30. Gonzalez, V.M. and Macosko, C.W., *Polymer Composites,* Vol. 4, (3), 190 (1983).

31. Monsanto *Chemical Technical Bulletin* 6510A.

32. Klosiewicz, D.W. (to Hercules), *U.S. Patent 4,400,340* (1983).

33. Ashland Chemical's *Technical Bulletin,* V–110–2.

34. AMOCO Chemical's *Technical Bulletin,* 1P–77.

35. Celotex *Technical Bulletin,* 07200/CEE, BuyLine 1224.

36. Kuyzin, G.S., Schlotterbeck, D.G., and Kent, G.M., *Proceedings of the 32nd Annual Polyurethane Technical/Marketing Conference,* Oct. 1–4, (1989), San Francisco, California, p. 280.

37. Carleton, P.S., Waszeciak, D.P., and Alberino, L.M., *Proceedings of the SPI–29th Annual Technical/Marketing Conference,* Reno, Nevada, Oct. 23–25, (1985), p. 154.

38. Yosomiya, R. and Morimoto, K., *I & EC Product Research & Development,* Vol. 23 (1984), p. 605.

39. Nelson, *Proceedings of the SPI–29th Annual Technical/Marketing Conference,* Oct. 23–25, 1985, Reno, Nevada, p. 146.

40. Borgnaes, D., Chappell, S.F. and Wilkinson, T.C., *Proceedings of the SPI–6th International Technical/Marketing Conference,* Nov. 2–4, (1983), San Diego, California, p. 381.

41.   Moss, E.K. and Skinner, D.L., *J. Cellular Plastics,* July/August (1978), p. 208.

PHENOLIC FOAMS *(by Kadzuo Iwasaki)*

Introduction

History. Phenolic resin is a condensation–type high polymer formed by the reaction of phenols and aldehydes. In 1910, research and industrialization of this resin were promoted in the U.S. by L. Backeland (1). In Japan and Europe research on this resin was also started at about the same time.

Phenolic foams were first used in Germany in the early 1940's to replace Balsa wood for use in aircraft (2). Notwithstanding the fact that both phenolic resin and phenolic foam itself have long histories, newer applications have undergone slow development. Quite recently, however, the heat resistance as well as fire resistance qualities of this foam have focused attention to its wide uses.

Classification. Phenolic foam is divided broadly into two categories. One is the novolac type and the other is the resol type (see Figure 55). In the novolac type, phenol and formaldehyde react in the presence of an acid catalyst and link the linear condensation product by a methylene bond. The basic curable crosslinking agent, such as hexamethylene tetramine and blowing agent are added to the linear condensation product and the resultant product is molded at elevated temperatures and high pressure.

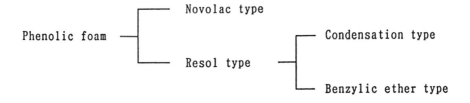

Figure 55. Classification of phenolic foams.

In the resol type, phenol and formaldehyde react in the presence of a basic catalyst and provide a liquid type resol to which an acid catalyst and blowing agent are added and the foam is then molded. In forming the resol a method is available for obtaining a benzylic ether–type liquid oligomer by reacting phenol and formaldehyde in nonaqueous phase.

Chemistry

Material Chemistry. *Phenols.* Materials to be used in making phenolic foam include phenol, cresol, and xylenol. Crude products of these materials can also be used as raw materials. The positions where the phenols can react with aldehydes are 2,4, and 6 against the OH–group. Accordingly, if substitute groups such as alkyl groups are located at the 2,4 and 6 positions, aldehydes can no longer react. In other words the existence of the substituted groups governs the functionality of phenols.

The properties of typical phenols are shown in Table 54. Generally, when using formaldehyde as the aldehyde, difunctional phenols should be used for producing linear polymers because formaldehyde is difunctional. For producing crosslinked polymers, trifunctional phenols are used. Thus, the selection of the phenols to be used is very important.

Table 54: Properties of Phenols

Name	B.p. (°C)	M.p. (°C)	Functionalities
Phenol	182	41	3
o-cresol	191	30	2
m-cresol	203	10	3
p-cresol	202	35	2
2.4-xylenol	211	26	1
2.3-xylenol	218	75	2
3.5-xylenol	222	64	3
2.5-xylenol	212	75	2
2.6-xylenol	203	45	1
3.6 xylenol	227	65	2

B.p.   : Boiling Point
M.p.   : Melting Point

Because the substituted group in trifunctional phenols is large enough to become a steric hindrance against the attack of aldehyde, it is inappropriate to decide the actual functionality by depending only on the chemical formula.

Aldehydes. Formaldehyde, paraformaldehyde, furfural, acrolein, alkyl aldehydes, and aryl aldehydes can be used as aldehydes, but formaldehyde is popularly used. An aqueous solution of formaldehyde is called formalin, and in almost all cases formalin is used industrially. When 1 mol of formaldehyde is dissolved in water, about 15 kcal of heat is generated. This heat generation results because methylene glycol is produced by solvation.

(1) $HCHO + H_2O \rightarrow CH_2(OH)_2$

However, the formalin (37% more or less) procurable on the market is already changed into trimethylene glycol.

(2) $3HCO + H_2O \rightarrow HOCH_2O\ CH_2O\ CH_2OH$

The composition of formalin sold on the market is as shown in Table 55. Methanol is used in formalin to prevent the formation of precipitation as a result of the formation of high–molecular–weight polymethylene glycol, $HO-(CH_2O)_n-H$. As formaldehyde is added to the reaction mixture, the following reactions are possible: polymerization (causing precipitate formation), methylol formation, formic acid formation (via oxidation), Cannizzaro reactions, saccharide formations, etc. are generated in formalin.

Table 55: Composition of Commercial Formalin

Ingredients	wt %
HCHO	37 to 40
CH₃OH	6 to 15
HCOOH	0.02 to 0.04
Others	trace

Resol–Type Foam Chemistry. Resol is obtained as a result of the reaction between phenols (P) and formaldehyde (F) in the presence of basic catalyst. Generally, the reaction is made at a temperature below

100°C (e.g., at 90°C), followed by neutralization and dehydration. Reaction progress is followed by monitoring viscosity changes, rate of P consumption (quantity of free P), change in average molecular weight, white precipitation (solubility with water), etc. The reaction mechanism for synthesizing resol was studied in 1894 by L. Lederer and O. Manasser, and accordingly, this mechanism is called Lederer–Manasser's reaction (2, 3).

As described above, formaldehyde exists as trimethylene glycol in aqueous solution. Phenol reacts quickly with the alkali–hydroxyl group and produces resonance structural phenoxide ion, and trimethylene glycol is added to the O and P positions in the phenoxide ion. This quinoid–transition–state is stabilized by the movement of proton. The mono–methylene–derivative produced in this way reacts further with formaldehyde and produces two types of dimethylol derivative and one type of trimethyl derivative. These reactions are expressed as second–order reactions:

$$(3) \qquad \frac{dx}{dt} = k[p] \cdot [HOCH_2OH]$$

The methyl derivative thus produced forms a multi–nucleus structure by the dehydration condensation. Generally, the numbers of benzene nuclei contained in resol for making phenolic foam are less than 10. The addition and condensation reactions used in synthesizing resol, and the reaction during foaming and curing are summarized as follows:

(1) Resol–producing reaction mechanism (Lederer–Manasser's reaction)

(4)

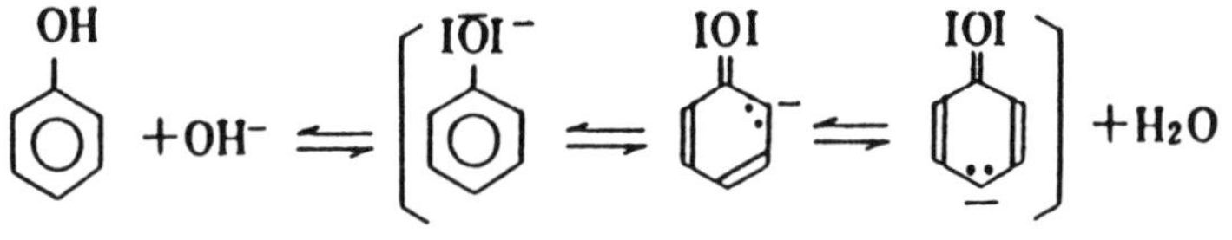

(5)

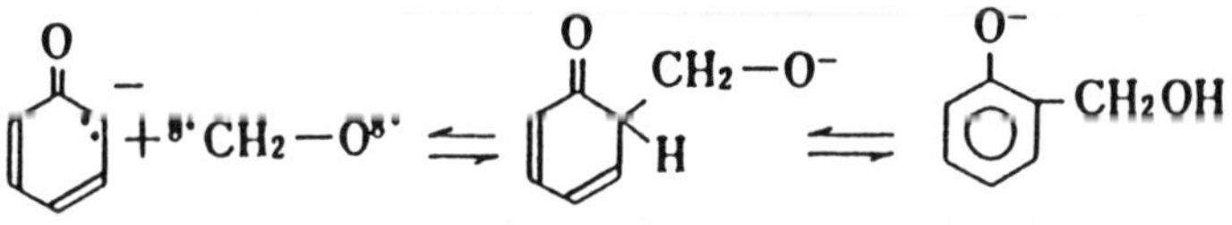

(6)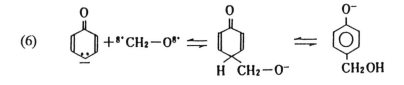

(2)   Resol synthesizing, foaming and curing reaction—

when resol synthesizing (addition, condensation)

(7)

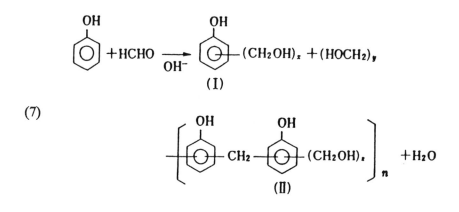

when foaming and curing (condensation)

(8)

where x = 1 ~ 3, y = 0 ~ 2, z = 0 ~ 2, n = ~ 5.

In manufacturing phenolic foams, a blowing agent such as R–113 (trichlorotrifluoroethane) is evaporated by using an exothermic reaction (Equation 7), and the blowing agent is included in the polymer. The reaction involved in forming the benzylic ether–type foam is shown below. Similarly, this foam will finally form a network polymer as do resol–type foams.

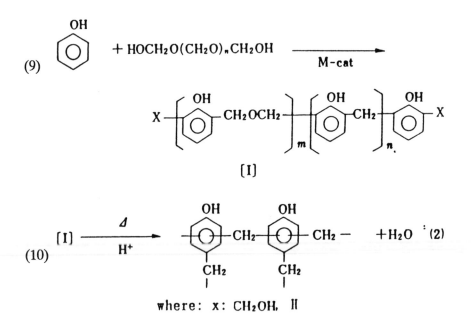

(9)

(10)

where: x: CH_2OH, H

$$m/n \geq 2$$

Novolac-Type Foam Chemistry. In this type excess P reacts with F in the presence of acid catalyst and forms a linear-condensation product to which the basic crosslinking agent, such as hexamethylene tetramine, is added to form the foamed product.

In the presence of an acid catalyst, F is changed to a hydroxymethylene carbonium-ion, reacts as the hydroxy-alkylating agent with P, and produces methylol derivatives. These derivatives are unstable under the acid condition, and immediately form methylene bisphenol. Crosslinking agents such as hexamethylene tetramine form aminomethylol compounds and undergo a Mannich reaction with P and its condensation product, finally producing a high polymer. The reactions at this stage are as follows:

(1)   Formation of Novolac

(11)
$$HO-CH_2-OH \xrightleftharpoons{H^+} {}^+CH_2-OH + H_2O$$

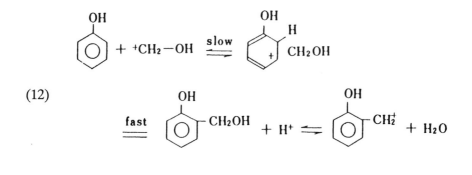

(12)

(13)

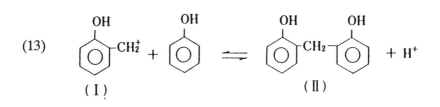

(I)   (II)

(14)   (I) + (II) ⟶

(III)

(2) Crosslinking by hexamethylene tetramine

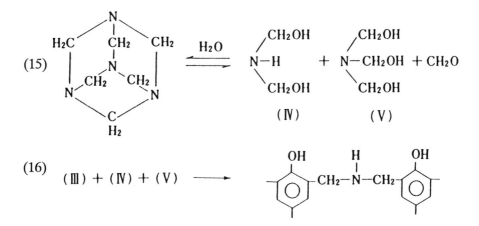

(15)

(IV)   (V)

(16)   (III) + (IV) + (V) ⟶

Foaming Mechanism. The process for producing resol–type phenolic foams is similar to that of polyurethane foam. The foaming process in phenolic foam may be divided into five steps. This process does not progress by step–by–step, but the several phenomena progress spontaneously.

(1)   Ingredient compatibilization
(2)   Bubble formation
(3)   Bubble growth
(4)   Bubble stabilization
(5)   Cell opening or cell stabilization as closed cell

The factors involved in each step are broadly divided into two factors, one chemical and the other mechanical. The former involves resol, the surfactant, and curing catalyst; and the latter involves the mixing head of the foaming machine. Among these factors, the influence of the surfactant is quite important. The surfactant has three functions. The first is to improve the compatibility of important ingredients. The surfactant has both hydrophilic as well as hydrophobic properties, and accordingly, it acts as a go–between in the compatibility of resol and curing catalyst (both are hydrophilic) to the hydrophobic blowing agent.

The second function of the surfactant is to lower the surface tension of the system, thereby forming finer bubbles. The third function is to prevent the cell wall from becoming thin and unstable during the period of growth. This is called the Marangoni effect (3, 4).

In the latter part of the bubble–forming process, namely in steps 4 and 5, the influence of the surfactant is rather small, and it is necessary to consider the material system in general. The functions of resol (molecular weight, crosslinking density, non–volatile content, water content, etc.) and curing agent are very important. The primary factor controlling the cell opening of phenolic foam is the formation of free formaldehyde and evaporation of water (16) in the latter stage of the curing process.

When evenly mixing the non–miscible ingredients with varying viscosities, the effect of the foaming machine's mixing head is considerable. The mixing head should be designed that it can maintain a high shear force and prevent the generation of Joule heat.

In general it can be said that in the initial stage of the foaming process, the function of the surfactant and the mixing function of the foaming machine are considerable, but in the latter stage, the reaction behavior of resol and curing agent are important. These relations are shown in Table 56.

Table 56: Effect of Chemical and Mechanical Factors
on Foaming Processes

	Chemical Factors			Mechanical Factor
	Surfactant	Resol	Catalyst	Mixing Head
1) Ingredients Compatibilization	O	O	O	O
2) Bubble Formation	O	O		O
3) Bubble Growth	O			
4) Bubble Stabilization	O			
5) Cell Opening/Stabilization as Closed Cell		O	O	

Note:
　　O : Effective

Raw Materials

Materials for Resol–Type Foams. *Resol (Resin)*. As described above, resol is formed as a result of the reaction of phenol (P) with aldehyde (F) in the presence of a basic catalyst. A representative preparation process is shown in Figure 56. Representative manufacturing conditions and the properties of resol are shown in Table 57 (5).

The P/F mol ratio is usually 1/1 to 1/3, but preferably 1/1.5 to 1/2.0. The basic catalysts to be used then should be hydroxides of barium, sodium, potassium and ammonium; carbonates of sodium and potassium; alkylamines, etc. The quantity of these catalysts to be used is 0.005 to 0.1 mol against 1 mol of P. The preferable quantity of these catalysts is 0.01 to 0.05 mol. After these materials, in aqueous solution, react at a temperature below 100°C they are neutralized with an appropriate acid and then dehydrated to provide a liquid–type resol. Representative properties of resol are shown in Table 57.

The manufacturing conditions and the properties of resol are the most important factors in controlling the final properties f phenolic foams. For this reason all manufacturers have their own highly developed procedures regarding manufacturing conditions.

Curing Catalysts. If resol is heated to a high temperature, it can be cured without catalyst; but in order to make curing it at an ambient temperature (room temperature), an acidic catalyst must be used. Both organic acid and inorganic acid catalysts can be used. Inorganic acids used include hydrochloric acid, sulfuric acid, and phosphoric acid, while benzene sulfonic acid, toluene sulfonic acid, and phenol sulfonic acid, are organic acids used.

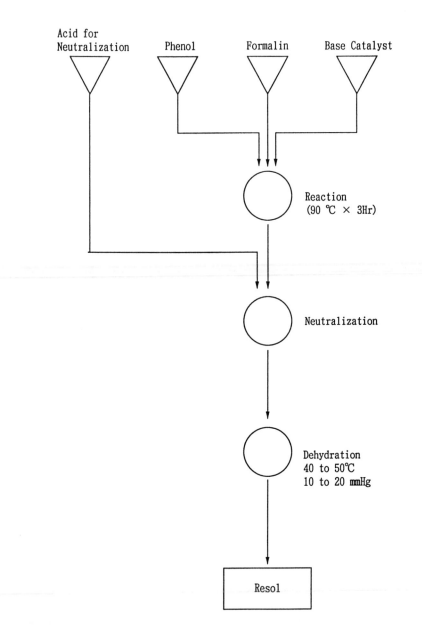

Figure 56. Typical preparation process for resol (5).

Table 57: Preparation Conditions and Properties of Resol (15)

Items		Value
Conditions		
Recipe	┌— Phenol	1
(mole)	├— Formalin (37%)	1 to 3 (Pref. 1.5 to 2.0)
	└— Alkaline catalyst	0.005 to 0.1 (pref. 0.01 to 0.05)
Temperature		Lower than 100 °C
Time		Approx. 3 hours
Properties		
Nonvolatile content (%)		70 to 90
Viscosity (Cp/25 °C)		3,000 to 10,000
pH		6.8 to 7.2
Free Phenol (%)		Less than 5
Free Formaldehyde (%)		Less than 5

In some cases, inorganic acid (e.g., phosphoric acid) and organic acid (e.g., phenolsulfonic acid) are mixed with each other and used as a catalyst. Acid remaining in the final foam causes acidity of the foam. To minimize corrosion of the metal in contact with the foam the use of organic acid catalyst is preferable.

Surfactants. Various types of surfactant can be used for forming phenolic foams. There are two types of surfactant, one is the silicone–oil type and the other is the non–silicone–oil type. The latter is used most commonly.

The silicone–oil type (the surfactant used for making rigid polyurethane foam) is also used for making phenolic foam. Examples are L–5340 (Union Carbide) and SH–193 (Dow Corning).

Non–silicone oils are nonionic surfactants, which are polymers of castor oil, lanolinic acid and various kinds of alkylene oxide. These products are made by Daiichi Kogyo Seiyaku, Ltd. and are sold on the market under the trade names of Resinol F–140 and F–520 (6). Polyoxyethylene sorbitan fatty acid ester, such as the Tween series (Atlas Chemical Industries), can also be used.

Blowing Agents. Blowing agents are divided into two types, one volatile and the other decomposable. The volatile type is mainly used for compositions of the resol type. This type is evaporated by exothermic reaction and then forms a blowing gas. On the other hand, the decomposable type is used for novolac–type compositions. This type is decomposed into carbon dioxide, ammonia gas and the like by exothermic reaction heat, heat supplied from the outside, or by acid catalyst, and then forms a blowing gas.

As for volatile types, there are Freons®, saturated hydrocarbons, etc. Examples of Freons® are R–11 (monofluorotrichloromethane) and R–113 (trifluorotrichloroethane), and examples of saturated hydrocarbons are n–hexane, n–heptane, and pentane, etc. Methylene chloride can also be used. Of course, it is possible to use combinations of these blowing agents. When Freons® are used as blowing agents, the foam exhibits high thermal insulating properties. Recently, however, it has been reported that Freons® cause destruction of the atmospheric layer of ozone, and accordingly, other materials are used as substitutes.

As for decomposable matter, examples are sodium carbonate, ammonium carbonate, sodium nitrite, sodium sulfite, and sodium bi-carbonate. Other well–known examples are di–N–nitrosopentamethyl-enetetramine, N,N–disubstituted 5–amino–1,2,3,4–thiatriazole, diesters of azodiformic acid, diazonium salts, sulfonhydrazides, and N–alkyl N–nitrosodiacetoneamine (7). The water produced during the exothermic resol condensation reaction is used as a blowing agent for producing high–density foamed composites.

Modifiers. Phenolic polymers are known to have high rigidity, and this property extends to phenolic foams, which are highly friable. In order to reduce friability and permit some flexibility and toughness, various kinds of modifiers are sometimes used. Reactive modifiers are used in the course of resol–resin preparation, and they become integral parts of the polymer structure. Examples include PVA (polyvinyl alcohol), PVA–PVC (polyvinyl alcohol–polyvinyl chloride–copolymer), resorcinol, o–cresol, furfuryl alcohol, and other various types of polyols.

Other types of resin modifiers compatible with resols may also be used. Examples include UF (urea–aldehyde condensation product) and epoxide. Japanese workers have found that the most effective and economical ingredients as modifiers for phenolic foam are resorcinol, o–cresol, and fulfuryl alcohol.

Flame Retardants. Even when phenolic foam is exposed to an open flame, it still maintains its high flame retardance and nonignition properties. Accordingly, flame retardants are not necessarily required. However, for the purpose of improving the LOI (limiting oxygen index) and meeting various types of fire–test requirements, flame retardants become necessary. Generally, boric acids (boric acid and the salts) have high fire retardance and low smoke–generation properties. Boric acids are excellent fire retardants, but if they are mixed with resol the viscosity of the resol is raised, and gelation induced in due course. It is, therefore, necessary to investigate the most practicable method of use. The use of boric acid in a hand–mixing and filler–mixing method is recommended. No other satisfactory method for using boric acid has been discovered.

PAP (polyammonium phosphate) is a practical fire retardant in the form of white powder at normal temperatures. The fire retarding efficacy of the PAP is very high, but smoke generation is liable to increase somewhat with its use. The fire–retardant action of the PAP is a result of its behavior as a Lewis acid in the char–formation reaction of phenolic polymer viz. carbonium–reaction (called SN–1 reaction).

Other fire retardants used include aluminum hydrate, antimony oxide, and molybdenum compound. Halogenated phosphate esters used in polyurethane foams and bromine compounds used in polyolefin foams are not used in phenolic foams. Flame retardants are used mostly in powder form, and accordingly, their distribution conditions are dependent on their particle size and shape.

Others. Acid–neutralizing agents are also used as additives. The acid is used as a curing catalyst for resol–type resins and remains in the foam. This agent is used for neutralizing any residual acid and for reducing the possibility that the acid might corrode any metal in contact with the foam. Examples of acid–neutralizing agents are metal powders (zinc and aluminum) and metal oxides (calcium carbonate and magnesium carbonate).

Materials for Benzylic Ether–Type Foams. *Resin (Benzylic Ether–Type Resin).* Benzylic ether–type resins can be prepared by reacting phenols and aldehydes in nonaqueous phase in the presence of metallic catalysts. For representative production conditions and for resultant resin properties, refer to Table 58 (15).

Modifiers. The foam–producing method using benzylic ether–type resin is very similar to that using resol–type resins. Accordingly, the same curing agents, surfactants, and foaming agents are commonly used for both methods. In order to improve the foaming characteristics and mechanical properties of benzylic ether–type foams, polyisocyanates can be used (8). Polyisocyanates used include polyphenylene polymethylene polyisocyanate (MDI), toluene diisocyanate (TDI) and their crude products. MDI is the preferred polyisocyanate.

Materials for Novolac–Type Foam.

Novolac (Resin). The raw materials used for novolac–type resins are phenols and aldehydes just as for resol–type resins. These materials are reacted in the presence of an acidic catalyst and a novolac–type resin is produced. Typical production conditions and properties are shown in Table 59 (15).

Table 58: Preparation Conditions and Properties of Benzylic Ether Resin (15)

Items		Value
Conditions		
Recipe	Phenol	94 (parts by weight)
	Paraformaldehdye	33 (parts by weight)
	Lead naphthenate	1.2 (parts by weight)
Temperature		100 to 130 °C (pref. 110 °C)
Time		3 to 5 hours
Properties		
Nonvolatile (%)		More than 95
Viscosity (Cp/25 °C)		30,000
Water content (%)		Less than 1

Table 59: Preparation Conditions and Properties of Novolac Resin (15)

Items		Value
Conditions		
Recipe	Phenol	1 (mole)
	Formalin (37%)	0.8 (mole)
	Acid catalyst	0.02 (mole)
Temperature		90 to 130 (°C)
Time		1 to 2 (hours)
Dehydration		150 °C , 10 to 20 mm Hg
Properties		
Nonvolatile (%)		More than 95
Water content (%)		Less than 5
Free Phenol (%)		Less than 5
Free Formaldehyde (%)		Less than 1

Basic Curable Crosslinking Agents. Hexamethylene tetramine (HMTA) is used as a curable crosslinking agent for novolac resin. The amounts used average between 5 to 20% of the novolac resin. The HMTA is decomposed into formaldehyde and ammonia during foaming and acts in crosslinking the resin.

Blowing Agents. In novolac–type foam decomposing–type materials are used. Typical organic compounds can be used as described in the section on blowing agents. Dinitroso–compounds, especially di–N–nitrosopentamethylenetetramine, is commonly used.

Auxiliary Materials. Glass fibers, polypropylene fibers and similar fibers may sometimes be used as reinforcing materials for phenolic foams. Mechanical properties and heat resistance are improved by reinforcing phenolic foams with these fibers.

It is easy to use these reinforcing materials for novolic–type foams, but they are difficult to use for resol–type foams because the viscosity of the latter foams is high and it is hard to distribute the reinforcing fibers evenly in the foam. However, if even distribution becomes technically possible, a bright future is expected for these reinforcing materials in such applications.

Facing materials for use on the laminate board and sandwich panel, etc. include metal plates (steel, iron, and aluminum), metal foils, non–woven cloth and various kinds of paper (glass–fiber, ceramic–fiber, and flame–retardant), and the complex of these materials. Because the adhesive strength of phenolic foam is weak, special measures must be taken on the side of the facing material.

Foaming Processes and Facilities

Foaming Process of Resol–Type Foam. The foaming processes for phenolic foam of both the resol type and benzylic–ether type are the same as those for rigid polyurethane foams. The block–foaming process (slab foaming), pouring process, continuous–laminate process, and spray process are used. See Figure 57.

In any process, metering, mixing, and discharging of foamable compositions are common. The differences become clear on the following stage, viz. in what space, and by what method foaming is made. In order to facilitate an understanding of this foaming principle, an explanation is made based on the hand–mixing process, which is carried out in the laboratory.

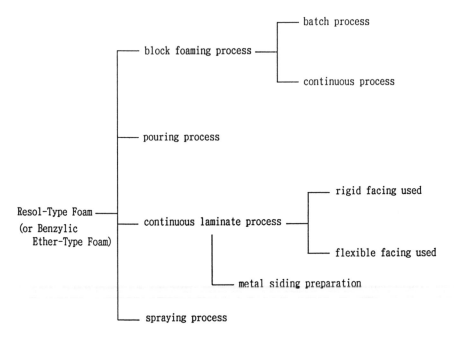

Figure 57. Foaming processes for resol–type and benzylic ether–type foam.

Hand–Mixing Process. Connect a mixer (stirrer) to the tip of an electric drill capable of maintaining 2,500 to 3,000 rpm, and use this as a mixing device. Use the formulation shown in Table 60. Adjust the temperature of the components A, B, and C at 20°±2°C each and maintain them. Weigh components A and B and put them in a beaker and mix with the above mixer for 10 to 15 seconds. Next, weigh component C and put it into the beaker and mix vigorously for about 20 seconds. Then, throw the mixture into a wooden case, the inside of which is lined with silicone paper. Foaming is completed within 5 to 6 minutes after which curing is carrier out in an air oven at 70°C for 15 minutes. In this manner a free–rise foam is produced. This method is called the hand–mixing process. This process is used in the laboratory for research and development of resin formulations.

Block–Foaming Process. By this process a large–size foam (block–shaped) can be manufactured. Continuous or batchwise processes are both used to manufacture large–size foams. With a very large–capacity foaming machine it is possible to make a very large block foam by using a formulation like the one specified in Table 60. By means of the continuous process it is possible to use large equipment for making rigid polyurethane foam (block process), for instance, conveyor. By splitting and cutting the large block foam it is possible to make a pipe cover and a board. Water–soaked foam to be used in the floral field can also be produced by the block–foaming process using the formulations specified in Table 61.

Table 60: Typical Formulation of Resol–Type Foam (5)

Pre-blend Conponents	Ingredients	Parts by Weight
A	Resol [a]	100
	Surfactant	2
	Acid neutralizing agent [b]	2
	Flame-retardant [c]	3
B	R-113	11
	R-11	3
C	Acid catalyst [d]	15

Notes: (a) Nonvolatile: 80%, viscosity: 5000 Cp at 25 °C
       (b) Zinc powder/Dioctyl phthalate: 50/50
       (c) Polyammonium phosphate
       (d) 65% Phenolsulfonic acid

Table 61: Typical Formulation of Floral Foam (15)

Preblend Conponents	Ingredients	Parts by Weight
A	Resol [a]	100
	Surfactant [b]	1
	Cell-opening agent [c]	5
	Colouring	2
B	R-113	10
C	Acid catalyst [d]	13

Notes: (a) Nonvolatile: 80%, Viscosity: 4000 Cp (25 °C)
       (b) Silicone surfactant L-5420 (Union Carbide Corp.)
       (c) Anionic surfactant Neogen AS-20 (Diichi Kogo Seiyaku Co.)
       (d) 65% Phenolsulfonic acid

Pouring Process. This process is employed for making sandwich panels with a multistage hot press. With rigid polyurethane foams the closed–mold method can be used, but for phenolic foams foaming by the closed–mold method is difficult and the open–mold method is used. That is to say the upper facing material is removed in advance and after pouring this facing material, is put and then set on the hot press. However, the author of this paper has been successful in developing an improved closed–mold method, and as a result it has now become possible to foam by pouring exactly as in making rigid polyurethane foam (9).

Continuous–Laminate Process. This process is to foam continuously between two sheets of facing material, employing a laminator (double–belt conveyor). The productivity of this process is the highest, i.e., the yield of raw material is very good, and a high–quality product can be produced. The selection of material, foaming machine and laminator design are very important factors in this process. Excellent quality products have been manufactured by this process in the United States, Western Europe, and Japan.

In Japan very large quantities of metal–siding board (consisting of metal sheet/phenolic foam/flexible–facing material) are manufactured as exterior materials in building construction. This material is also manufactured using the same principle as that in the laminate process.

Spraying Process. This process has been theoretically possible, but practical application has not been realized. In Japan, however, Bridgestone Ltd. has developed (10) a method which makes spraying–in–place possible. By this method three components of resol–type composition can be foamed at a temperature 10°C and higher (preferably at 20°C) with a spray–gun. Detailed technical data are unavailable, but it is believed that the material as well as foaming machine should have been improved very much.

Manufacturing Facilities *Foaming Machine.* Resol–type and benzylic ether–type phenolic foam–producing–machine designing is very important. The makers of foaming machines design and construct manufacturing facilities to satisfy the customers requirements as much as possible. As to useful hints when designing a foaming machine, refer to Table 62 (5). Important points to consider in designing are how to plan the metering pump and mixing head. Metering pumps for three components (3 streams) are for factory use, and pumps for four components (4 streams) are for laboratory use.

**Table 62: Important Points to Consider When Designing
Foaming Machine for Phenolic Foam**

Items	Important Points to Consider
Tank, Piping	To prevent corrosion by acid catalyst (Selection of material)
Metering Pump	(A) Component of resol ••• High viscosity, large volume, prevention of precipitation (B) Component of foaming agent ••• Low viscosity, high-precision discharging (C) Component of catalyst ••• Measure against corrosion, high-precision discharging
Mixing Head	(1) To evenly mix materials having large difference in viscosity and in mixing ratio (2) To prevent generation of heat (3) To prevent air entrainment
Temperature Adjustment	(1) Designing heating and cooling capacity (2) Selection of heat exchanger (3) Measures against Joule's heat by recycle

The greatest problem in phenolic foam production machines is how to protect the metal parts used in the foaming machine from corrosion by acid catalysts. Usually corrosion–resistant materials are used for this purpose. However, the cost of such materials is very high and machining is rather difficult. For these reasons the cost of phenolic foam–producing machines are 1.5 to 2 times as high as that of the polyurethane–producing machines. An example of a foaming machine–flow sheet is shown in Figure 58. The exterior appearance of such a machine is shown in Figure 59.

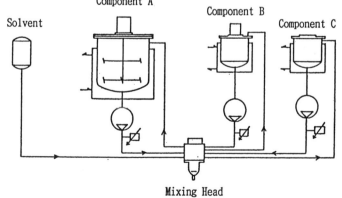

Figure 58. Standard Type of Flow Sheet for Phenolic Machine (11).

Figure 59. Phenolic foam machine (11). (Courtesy of Toho Machinery Co.)

Continuous Laminator. A continuous laminator for polyurethane foam may fundamentally be used as phenolic foam laminator. Distinguished points of difference between two foams are the viscosity of dispensed material, foam characteristics, and cure conditions. The points which are particularly important in designing are as follows:

(1)   To evenly distribute widthwise the high–viscosity dispensing liquid
(2)   To set up the platen heater properly for each foam formed
(3)   To decide setting conditions of the top facing with double conveyor
(4)   To select the proper materials for the conveyor construction
(5)   To provide sufficient curing of the foam
(6)   To remove water

It is important to specify a laminator which will provide the required thickness of the product, kinds of facing material, and production speed. Examples of a typical phenolic foam board laminator are shown in Figure 60 (11). This laminator is the fixed–gap type which employs a slat conveyor.

Figure 60. Continuous laminator for phenolic foam board (11). (Courtesy of Toho Machinery Co.)

Foaming Process of Novolac–Type Foam. In the initial process, novolac, crosslinking agent, blowing agent, etc. are mixed to provide a foamable compound. A formulation for a typical novolac foam is shown in Table 63.

Table 63: Typical Formulation of Novolac–Type Foam (15)

Ingredients	Parts by Weight
Novolac (Solid)	100
Surfactant	1
Hexamethylenetetramine	15
Blowing agent [a]	5
Urea [b]	1

Notes: (a) Di-N-nitrosopentamethylenetetramine
(b) Auxiliary agent to adjust decomposing speed of blowing agent

After thorough mixing of the prescribed components, grind well (to 100 mesh) with a ball mill, add other fillers if required, and make up into a compound. Fill up the metal mold with this compound or put this compound on the hot press and heat it in steps from 100° to 200°C and foam, then cure it after the foam is formed. Hot presses commonly used in the compression–molding process can be used in the manufacturing facility for novolac–type foams.

Processing Facilities. The phenolic–foam–producing process is similar to that used with rigid polyurethane foams. Production of composite sandwich panels, board–type products, pipe covers, are possible.

When splitting or cutting phenolic foam, band saws and/or circular saws are used, but these tools produce a large volume of sawdust. Therefore, it is necessary to install a sawdust–eliminating device. If such a device is not attached, or if its eliminating capacity is insufficient, the sawdust adhering to the cut–out face may cause the cutting line to be irregularly curved, or the cut–out face to be damaged by the saw. For this reason sawing must be done carefully. Also, if the sawdust is not removed immediately after sawing, it will become very difficult to remove it later because of static electricity.

For manufacturing composite sandwich panels by laminating, a coating apparatus for the adhesive (chloroprene–base and polyurethane–base adhesives are preferable) and a press is needed.

Properties

Foaming Characteristics. When producing resol–type and benzylic ether–type foams, the index expresses apparent reactivity. Following are the foaming times:

CT   *Cream Time*—the time required for the foamable mixture to become creamy after the reaction is started

GT   *Gel Time*—the time required for the inside of the mixture to become hard during foaming, while the outer surface begins to have spinability

RT   *Rise Time*—the time required for the rise of the mixture to be almost completed

TFT   *Tack–Free Time*—the time required for the mixture to become tack–free while it is rising or has finished rising

These designations are used in polyurethane–foaming technology but they are also applicable to phenolic–foam–producing technology. These time designations do not express the chemical reaction rates directly, but these are the factors used to compare reactivity and accordingly, they are widely utilized. The factors which influence these foaming times are as follows:

(1) Kind of resin
(2) Kind and quantity of catalyst
(3) Kind and quantity of foaming agent
(4) Temperature (liquid temperature, mold temperature and ambient temperature)
(5) Mixing conditions

Examples of foaming characteristics of resol–type phenolic foams (free–rise) are shown in Table 64 in comparison with rigid urethane foam (3). Rise–profile curves, foaming–pressure curves and temperature curves for phenolic and polyurethane foams are shown in Figure 61.

Table 64: Foaming Characteristics of Free–Rise Foams (3)

Items	Types	Phenolic Foam	Polyurethane Foam
Formulation		Resol: 100 Catalyst: 30 Foaming Agent: 15	Component A (MDI)100 Component B: 100
Material Temperature (°C)		Resol: 20 Catalyst: 20 Foaming Agent: 20	Component A: 20 Component B: 20
Room Temperature (°C)		20	20
Foaming Time CT (seconds) GT (seconds) RT (seconds)		20 150 200	20 85 110
Density (kg/m³)		36	28
Maximum Temperature (°C)		96	167
Maximum Foaming Pressure (kg/cm²)		0.06	0.05

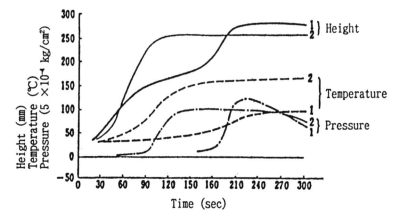

Figure 61. Foaming characteristics (3). (1) Phenolic foam, (2) Poly-urethane foam).

General Properties. Resol–type foams are classified roughly into four types, as follows:

Type A: High closed–cell content, low thermal conductivity, high fire resistance

Type B: High closed–cell content, low thermal conductivity, but low fire resistance

Type C: Open cell, high strength

Type D: Open cell, low strength

General properties of these four types are shown in Table 65.

In the past, types C and D were common in resol–type foam. Type D is low in density and is an open–cell–type foam used widely. Type C is high in density and mechanical strength. Type B was introduced by the American manufacturers, Koppers, Monsanto and others. This type has a high closed–cell content and has attracted special interest recently. Since Type B is a high closed–cell it is excellent for thermal insulation. This type has been used in construction in the U.S. and Western European countries.

However, when Type B comes in contact with flames, popping and punking phenomena occur, and this type has been rejected for construction materials as a result of various fire–resistance tests. The author of this section and his co–workers have undertaken studies to overcome the popping and punking phenomena and have been successful in developing Type A (3).

Table 65: General Properties of Typical Phenolic Foams (3)

Items \ Material	Phenolic Foam				(cf) PIR Foam
	Closed Cell, Type A	Closed Cell, Type B	Open Cell, Type C	Open Cell, Type D	
Density (kg/m³)	40	40	50	25	35
Thermal Conductivity (kcal/mh °C)	0.020	0.020	0.035	0.035	0.020
Closed-Cell Content (%)	90	90	0	0	90
Water Absorption (g/100 cm²)	4	4	12	12	3
Limiting Oxygen Index (%) [a]	50	33	33	33	26
Surface Fire Test [b]					
C_A	5	25	10	5	50
T θ (min, °C)	50	170	50	70	80
After-glowing Time (sec)	0	0	0	0	0
Popping	None	Existing	None	None	None
Criteria	OK	NG	OK	NG [c]	OK

Notes: (a) ASTM D·2863
  (b) JISA 1321 (Thickness of Test Specimen: 25 mm)
  (c) NG due to large cracks

In Type A, the closed–cell content is high, providing low thermal conductivity. Type A should be able to pass various types of fire tests. JIS–A–1321 (incombustibility test for internal finish material of buildings) is a Japanese fire test method said to be the strictest test in the world. Type A has successfully passed the Model Box Test, one of the items of this severe standard.

The general properties of resol–type foams produced by the block–foaming process are shown in Table 66, while those produced by the spraying process are shown in Table 67 (10, 12).

Table 66: General Properties of Resol–Type Foam Prepared by the Block Foaming Process (12)

Properties	Values	
Density (kg/m³)	60	35
Compressive Strength (kg/cm²)	2.0	1.6
Flexural Strength (kg/cm²)	8.0	5.5
Tensile Strength (kg/cm²)	1.1	1.1
Thermal Conductivity (kcal/mh °C)	0.029	0.025
Thermal Expansion (l/ °C)	3×10^{-5}	3×10^{-5}
Water Absorption (g/100cm²))	2	2
Limiting Oxygen Index	>40	>40
Specific Heat (cal/g °C)	0.48	0.48

Table 67: General Properties of Resol–Type Foam Prepared by the Spraying Process (10)

Properties	Values
Density (kg/m³)	35 − 45
Compressive Strength (kg/cm²)	Less than 1.0
Water Absorption (g/100 cm²)	3 to 4
Adhesive Strength (kg/cm²)	More than 1.0
Thermal Conductivity (kcal/mh °C)	Less than 0.03

The general properties of novolac–type foams are shown in Table 68 (13).

Table 68: General Properties of Novolac–Type Foam (13)

Properties	Values
Density (kg/m³)	40
Compressive Strength (kg/cm²)	1.8
Flexural Strength (kg/cm²)	5.9
Water Absorption (g/100 cm²)	0.5
Thermal Conductivity (kcal/mh °C)	0.024
Specific Heat (cal/g °C)	0.3

Resol–type foams have self–adhesive properties and may be formed into composites with various kinds of facing materials. However, the adhesive strength of these foams is rather inferior when compared with that of polyurethane foam. These foams show comparatively good adhesive strength to such facing materials as plywood, nonwoven cloth, ceramic paper, kraft paper, plaster board, etc. because these facing materials have water–absorptive properties. The adhesive strength between these facing materials and the foam is 0.5 to 0.7 kg/cm². On the other hand, the adhesive strength with steel plate and aluminum plate, is not good, running about 0.1 to 0.3 kg/cm².

Thermal Properties. *Thermal Conductivity.* As stated above, the thermal conductivities of phenolic foams vary remarkable depending on whether they are closed cell or open cell. Generally, the thermal conductivities of foams with 90% or more closed cells are within the range of 0.015 kcal/mh°C; but if they have open cells, the thermal conductivities increase to 0.030 to 0.035 kcal/mh°C. If the foams have 50 to 80% closed cells their thermal conductivities will be an intermediate value between the above two figures. Meantime, the thermal conductivities of foams with 50% or less closed cells will be almost the same as that of open–cell foams.

Thermal Coefficients. The thermal coefficients of phenolic foams are almost the same as those of resol– and novolac–type foams. Examples are as follows: specific heat, 0.38 to 0.42 cal/g°C; flash point, 520° to 540°C; ignition point, 570° to 580°C; softening point, do not soften.

Thermal Resistance. Examples of thermal resistance over a long term are shown in Figure 62. This figure shows the compressive strength of resol–type foams maintained in an air oven for aging.

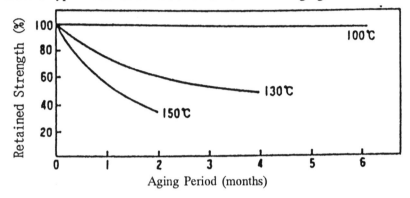

Figure 62: Retention of compressive strength under long–term aging test (12).

Examples of thermal gravimetric analysis (TGA) for phenolic foam are shown in Figure 63. The phenolic foam in this figure is the resol–type foam. In the case of novolac–type foams the values are almost the same (3).

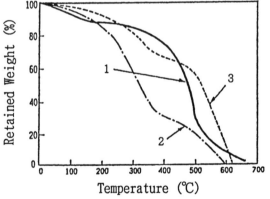

Temperature elevating rate: 10°C/min
Ambient gas: air
Standard material: α–alumina
  1. phenolic foam
  2. polyurethane foam
  3. polyisocyanurate foam

Figure 63. TGA curve (3).

Flame Retardance. The most important reason for phenolic foam being an excellent flame retarder is that the phenolic polymer is easily carbonized and the char part formed as a result is highly stabilized. This mechanism of char–formation is considered that of a multi–aromatic ring with chemically stabilized strong bond formed through a dehydrogenation reaction by heating and oxidation.

When a char layer is formed on the outside surface of the foam the inside virgin foam is protected by the char layer having a high degree of heat insulation. This char–formation mechanism is shown in Figure 64 (14).

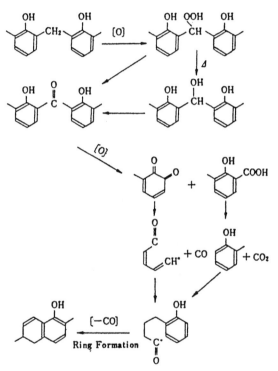

Figure 64. Reaction sequence of char formation in burning of phenolic resins (14)

Examples of phenolic–foam flammability are shown in Table 69 (7). The smoke developed from burning phenolic foam is very small. The result, of course, varies depending solely on the test conditions but an example shows that when phenolic foam is burned the coefficient of light reduction is about 1/10 of that of polyisocyanurate being burned and 1/10 to 1/20 of that of polyurethane foam.

Table 69: Typical Flammability Properties of Phenolic Foam (7)

	Acid catalyst	
	Boric-oxalic	Hydrochloric
Foam density, kg/m³	32	30
ASTM D1692-67T rating	S.E. [a]	S.E. [a]
ASTM E84-60T, Underwriter's Tunnel Test:		
Flame-spread rating	10	
Fuel contribution	23	
Smoke-density rating	0	
Bureau of Mines Flame-penetration Test:	595	
Penetration time, sec	10.3	53 (punked)
Weight loss, %		4.8
ASTM E162, Radiant-panel Test, Flame-spread rating	No ignition	

Note: (a) Self-extinguishing (no ignition).

When a small piece of various kinds of foams are held in hand, and a match or lighter flame is brought near to the small piece, phenolic foam does not burn up and almost no smoke is developed, while other foams burn up with smoke. From this fact it is understood that phenolic foam is excellent in incombustibility and low smoke development.

Drying. Resol–type foams contain 10 to 15% moisture when produced. The source of this moisture is water contained in both resol and catalyst and is the water yielded during the condensation reaction. This moisture is released within 5 to 10 days by leaving the foam in the room until it reaches its equilibrium state. The drying results for different laminates and slab board are shown in Figure 65 (5). These results indicate that it is necessary to leave newly–produced foams in the warehouse for annealing for about one week before shipment.

Chemical Resistance. Both resol– and novolac–type phenolic foams have excellent chemical resistance against inorganic and organic chemicals. However, phenolic foams are unable to resist against highly concentrated inorganic acids and alkalies. Chemical resistance data are shown in Table 70.

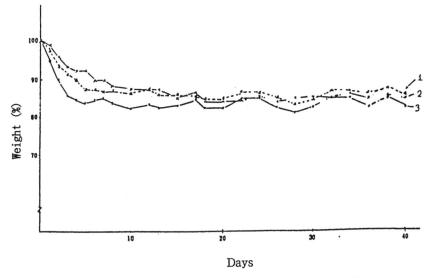

Days

Weight change under drying condition for phenolic foam (5).
   1. Laminate board with craft papers (50 x 900 x 900 mm)
   2. Laminate board with aluminum–foils (50 x 900 x 900 mm)
   3. Slab board without facing (50 x 900 x 900 mm)

Figure 65. Weight change under drying conditions for phenolic foam (5).

Table 70: Chemical Resistance (14–days immersion) (5)

Ingredients (%)	Resol-Type Foam	Novolac-Type Foam
Sulfonic acid (50)	×	×
Sulfonic acid (20)	O	O
Nitric acid (10)	×	×
Nitric acid (5)	O	O
Hydrogen chloride (10)	O	O
Formic acid (10)	O	O
Sodium hydroxide (20)	×	×
Sodium hydroxide (1)	O	O
Ammonia water (14)	O	O
Benzene	O	O
Styrene monomer	O	O
Acetone	O	O
Acetic esters	O	O
Brine	O	O

O: OK
×: NG

Demand and Applications

Demand. As the data for worldwide demand for phenolic foam is unavailable, details of its demand are uncertain. The demand for phenolic foams in Europe during 1978 is given in Table 71 (14). This data shows that the total quantity demanded in Western Europe at that time was 10,000 tons, which did not include the demand from the then communist countries.

Table 71: European Demand for Phenolic Foams in 1978 (14)

Country	Demand (Tons)
Iron Curtain Countries	Over 25,000
France	5,000
West Germany	2,000
United Kingdom	1,000
Italy	1,000
Holland & Belgium	500
Others	500
Total	over 35,000

According to the author's research the demand for phenolic foam in the year 1987 were as follows (5): U.S. and Canada, 28,500 tons; Western European countries (excluding Italy and Spain), 13,500 tons; and Japan, 3,000 tons. Besides the above countries, phenolic foam is also produced in South American countries, Australia, and South Asian countries. It is estimated that worldwide production of phenolic foams exceeded 80,000 tons for that year.

Applications. *Block Foam for Thermal Insulation.*   The foam produced in the large-size block is cut into regular-shape board or pipe covering. The main applications are for thermal-insulating materials for buildings, pipe covering in chemical plants, etc. Examples of pipe-covering use are shown in Figure 66 (5). Block foam is widely used in European countries.

Figure 66. Example of pipe coverings (5).

Sandwich Panels. When making sandwich panels using phenolic foams, it is necessary to increase adhesive bond strength between the foam and the facing material. For this purpose, it is necessary to lower the friability of foam and to increase the adhesive area.

When using metal plates such as steel plate and aluminum plate as facing materials, it is essential to raise the pH of the foam. As a general rule, acid–neutralizing agents are added when producing the foams. When plywood and plaster board are used as facing materials no such problem is encountered.

For making sandwich panels both the resol–type and benzylic ether–type foams can be used. These sandwich panels are used as building materials (partition panels, ceiling panels, wall panels, panels for clean rooms, door panels, panels for prefabricated refrigeration storehouses, etc.).

Continuous Laminated Board. Laminated board with incombustible facing materials (glass fiber, ceramic, alumina–paper), or kraft paper is also widely used as a thermal–insulating building material. For example, laminated board is used for roof, ceiling, wall, and floor applications and for refrigeration storehouses. For producing continuous laminated board, a full–scale double–belt conveyor must be set up. The installation cost of this equipment is very high, and accordingly, there is great demand for this business. Production is promoted in the U.S., Canada, Holland, Belgium, Germany, Italy and Japan. In Korea installation of this type of equipment is planned. Laminated board can be produced with resol–type and benzylic ether–type phenolic foams. Examples of laminated–board products are shown in Figure 67.

Figure 67. Examples of laminate board.

Based on almost the same theory as that of laminated board production, metal–siding board can be manufactured. In metal–siding board steel or aluminum plate is formed to profiled shapes in advance, and phenolic foam and incombustible paper are molded by an integral molding method. This method is widely used in Japan and Korea to produce exterior materials for wooden houses. An example of such material for wooden houses is shown in Figure 68.

Floral Foam. In the U.S. and Western European countries the most popular use of phenol foam for floral applications is a water–soaked foam using a resol–type composition. It is important to maximize the open–cell content in order to improve its water penetration. Floral foam is produced as large block or slab by a batch process. The foam is easily cut and packaged with a minimum of loss due to waste. Floral foam is commonly used for living flowers, but is not used for dried and artificial flowers.

Figure 68. Wooden house employing metal–siding board as exterior material (Courtesy of IG–Industries, Japan).

Spray Foam. Production of foam by a spraying process using a resol–type composition has become possible. Foaming of seamless insulation layer has been carried out by this spraying process, thereby increasing its insulating efficiency. This process is expected to be used as a thermal insulating method for private homes and warehouses.

Novolac Foam. Novolac foam is molded in the manufacturing facilities in a fixed form. The demand for this type of foam has been stabilized. The foam is used as a thermal insulating material for chemical plants, storage tanks, piping for LNG and LPG, oil storage tanks, etc. It is important to note that the acid catalyst does not remain in novolac foam, and for this reason, there is no fear of metal corrosion. Thus novolac foam is an essential thermal–insulating material for chemical plants.

High–Density Foams (Filled Foams). Research to produce high-density foam by adding a large quantity of inorganic filler to novolac- or resol–type compositions is underway. The density of this type foam is 500 to 800 kg/m^3 and the foam has high strength and low combustibility. New uses in the field such as door and picture–frame manufacturing, and materials for furniture, boats, railroad trains, etc., are being promoted.

Miscellaneous.   Phenolic foam has highly effective energy-absorption characteristics, and possible applications are now being studied. Low-density foams are also being studied for use as packaging materials. Meantime, when phenolic foam is carbonized at an elevated temperature under inert atmospheric conditions, cellular carbon can be produced. Cellular carbon may be used as a refractory insulating material.

GRP (glass-fiber reinforced plastic) can be produced by compositing resol-type resin with glass fiber. The GRP has excellent heat resistance and incombustibility, and accordingly, this material is used in the U.S., U.K., France and Japan for use in public traffic facilities, subway trains, tramcars, railway trains, and buildings. According to its narrow definition, the GRP is not strictly a foam, but when curing, a microcellular structure is formed by vaporizing volatile matter such as moisture. Therefore, the GRP can be widely defined as a kind of foam.

A research program to have benzylic ether-type resin react with polyisocyanate for producing polyurethane or urethane-modified polyisocyanurate foam is being promoted. A foam which has intermediate properties between phenolic foam and polyurethane foam can be produced.

Conclusion

Phenolic foam is an excellent material for providing heat resistance and incombustibility, and large volumes of this foam are now being used. Phenolic foams are superior for many applications. However, phenolic foam has the following weak points:

(1)   low productivity
(2)   high friability, low strength
(3)   low yield rate in materials

The cost of raw materials for use in phenolic foams is low, but since this foam has the weak points listed above, it cannot exceed polyurethane foam in cost performance. Resol-type foams and benzylic ether-type foams use chlorofluorocarbon (CFC) as a foamimg agent. It is necessary to develop a new foaming agent which does not destroy the ozone layer in the atmosphere. When these problems are solved phenolic foam applications will make further rapid progress.

REFERENCES FOR PHENOLIC FOAMS

1. Murahashi, S., R. Oda and M. Imoto, *Plastic Handbook,* (Japan) p. 129 (Asakura–Shoten).

2. Knop, A. and W. Scheib (translated by Shoji Seto), *Phenolic Resins,* p. 15–16 (Plastic Age Japan).

3. Iwasaki, K., Chemistry and technology trends in phenolic foams, *Japan Plastics* 38 (2):50–59 (1987).

4. Noda, I., Silicone surfactants for polyurethane, *Kogyo Zairyo* (Industrial Materials Japan) 27 (4):39–42 (1979).

5. Iwasaki, K., Phenolic foams, *Technical Handbook of Foamed Plastics,* p. 169–196, Edited by K. Iwasaki, Joho–Kaihatsu Ltd., Japan (1989).

6. Daiichi Kyogyo Seiyaku Ltd., *Surfactants for Phenolic Foam* (Catalogue).

7. Papa, J. and W.R. Proops, Phenolic foams, *Plastic Foams* Part II, p. 639–673, Marcel Dekker, New York (1973).

8. For instance, Nakamura, T., A. Kuroda, M. Onishi and T. Sasaki (Hodogaya Chemical Industry), *Japan Patent* (Open) *Sho–59–45332* (1984).

9. Iwasaki, K., Unpublished, Japan Patent pending.

10. Bridgestone Ltd., Technical Data.

11. Toho Machinery Co., Technical Data.

12. Mitsui Petrochemical Industries Ltd., Technical Data.

13. Chugoku Kako Ltd., Technical Data.

14. Dvorchak, M.J., Fire Retardant Properties of Phenolic Foam Insulation, *Proceedings of The Fire Retardant Chemical Assn..* p. 17–37 (1985).

15.   Iwasaki Technical Consulting Laboratory Ltd., Technical Data on Phenolic Foams, Unpublished.

16.   Rickle, G.K., and K. R. Denslow, The Effect of Water on Phenolic Foam Cell Structure, *J. Cellular Plast.* 24 (1):70–78 Jan. (1988).

3

THERMOPLASTIC FOAMS

Arthur H. Landrock

INTRODUCTION

This chapter will discuss all types of thermoplastic foams, including rigid, semi–rigid and structural foams.

STRUCTURAL FOAMS (RIGID FOAMS)

Introduction

"Structural foam" is a term originally used for cellular thermoplastic articles with integral solid skins and possessing high strength–to–weight ratios. Currently the term basically covers high–density rigid cellular plastics strong enough for structural applications (1). Thermosetting foams, such as polyurethane and polyisocyanurate, are frequently referred to as structural foams. In general structural foams can be made from virtually any high–molecular–weight thermoplastic organic polymer and will have a cellular core and an integral skin on all sides. The skin is relatively non–porous in relation to the cellular core (2).

Structural-foam construction, when compared to an equivalent amount of conventional foam plastics, results in a 3- to 4- fold increase in rigidity, since rigidity is a function of wall thickness. Chapter 8 on

Methods of Manufacture will cover details of structural foam molding. 1In addition to urethane and other thermosetting foams all thermoplastics can be foamed by the addition of physical or chemical blowing agents. A broad and overlapping division of thermoplastics exists between "commodity" and "engineering" groups of resins used for structural foams. The commodity group consist of the styrenics (polystyrene, styrene–acrylonitrile, etc.), olefins (polypropylene, polyethylene, etc.) and vinyl chlorides (PVCs), while the engineering group includes acetal, ABS, nylon, polycarbonate, polyester and polyetherimide, plus various glass–or carbon–reinforced resins (2).

Structural foams have found markets where it is necessary to conserve energy and natural resources by turning to lighter and less energy–sensitive materials. The automotive and truck industries are being forced to increase mileage relative to fuel consumption, and reduction of vehicle weight by replacing steel with structural foam is the favored way of accomplishing this objective. The Army Materials Technology Laboratory, an organization influencing military materials selection, believes that structural foams can solve cost, performance, and weight–reduction problems in tanks, trucks, aircraft, missiles, weapons, troop support and materials handling. Some of the items under consideration are:

- glass–filled polycarbonate track shoes for tanks
- tops for vehicles
- instrument–package brackets
- nose fairings for missiles

In weapons and troop support the applications are numerous and could include food–service equipment, furniture, rifle and pistol stocks, ammunition boxes, portable shelters, electronics housing, air–drop loading frames, backpacks, and many others (2).

Most structural foam is produced in the form of complete parts by injection molding, rather than bulk raw stock, as in many other foams. Since a complete part is made with a skin on each outer surface, the process is ideally suited for fabrication of parts or components in which light weight and stiffness are required. In all cases with thermoplastic structural foams the resin's original properties of heat and chemical resistance, as well as most electrical properties, remain the same as for the solid resins. The *dielectric constant*, however, is improved over the solid resin, which is the reason foamed polyethylene is used for television cables (3).

With the exception of NORYL® phenylene oxide–based resin most engineering foams are reinforced. Loadings range from 4 to 40% (4). Thermoplastic resins molded in the form of structural foams include the following (5):

- Modified polyphenylene oxide (PPO®) (NORYL®)
- Polycarbonate (LEXAN®) (MERLON®)
- Acrylonitrile–Butadiene–Styrene (ABS)
- Polystyrene (PS)
- Acetal
- Nylon 6,6
- Polyester (Polybutylene terephthalate) (PBT) (VALOX®)
- Polysulfone
- Polyetherimide (ULTEM®)
- Polyethylene (PE) (sometimes considered structural)
- Polypropylene (PP) (sometimes considered structual)

In 1982 the leader in structural plastics used for structural foams parts was *modified polyphenylene oxide* (NORYL®), accounting for almost three–quarters of all material usage. *Polycarbonate* held a strong second materials position, followed by ABS, with polystyrene next (5).

The strength–to–weight ratio of structural foams is reported to be 2 to 5 times greater than that of metal. Another important property is sound dampening (5).

Table 3–1 summarizes the physical properties of typical thermo-plastic structural foams of 1/8" wall thickness with a 20% density reduction over solid plastics (6).

Structural–Foam Types

Phenylene Oxide Alloys (Modified Polyphenylene Oxide): This resin, sold by General Electric Plastics (GE) as NORYL®, is the pioneer in engineering structural foam resins. It has a 205°F (96°C) distortion temperature under load (DTUL) at 66 psi (0.45 MPa) with UL 94 V–O and 5V ratings on flammability, and high mechanical strength. Although its heat–deflection temperature is lower than the other GE engineering structural foams, it exhibits an outstanding combination of heat and impact resistance. Dimensional stability is excellent, with low creep and water absorption. These properties make it very useful for applications such as business–machine housings and their structural bases, weather–resistant electrical enclosures, and lightweight structural components for transportation (7).

Table 3.1: Physical Properties of Thermoplastic Structural Foam (Reprinted from Reference 6, Courtesy of Van Nostrand Reinhold)

(at .250 Wall With 20% Density Reduction)

Property	Unit	Method of Testing	High Density Polyethylene	ABS	Modified Polyphenylene Oxide	Polycarbonate	Thermoplastic Polyester	Polypropylene	High Impact Polystyrene	High Impact Polystyrene w/FR
Specific gravity	lbs./ft.^{3}	ASTM-D-792	.60	.86	.85	.90	1.2	.67	.70	.85
Deflection temperature under load	°F@66 psi °F@264 psi	ASTM-D-792	129.6 93.5	187 172	205 180	280 260	405 340	167 112	189 176	194 187
Coefficient of thermal expansion	in./in./°F $\times 10^{-5}$	ASTM-D-696	12	4.9	3.8	2	4.5	5.2	9	4.5
Tensile strength	psi	ASTM-D-638	1,310	3,900	3,400	6,100	9,910	1,900	1,800	2,300
Tensile modulus	psi	ASTM-D-638		2,500,000	235,000	300,000	1,028,000	79,000	141,160	245,000
Flexural modulus	psi	ASTM-D-790	120,000	2,800,000	261,000	357,000	1,000,000	80,400	200,321	275,000
Compressive strength (10% deformation)	psi	ASTM-D-695	1,840	4,400	5,200	5,200	11,300	2,800	3,447	
Combustibility rating	UL Standard 94°		V-0	V-0	V-0/5V	V-0/5V	V-0	HB	HB	V-0

*This rating is not intended to reflect hazards presented by this or any other material under actual fire conditions.
Material properties given above are typical and vary from supplier to supplier. It is recommended that an end user contact his supplier and/or molder to obtain specific properties for use in a given application.

NORYL® resins are inherently capable of meeting UL94 V-0 standards for flammability without using flame-retardant additives. However, unlike PVC, these alloys are available in a wide range of colors and are not corrosive to processing equipment. These materials offer good performance in creep resistance, dimensional and heat stability, impact, stiffness, and resistance to moisture. These resins, however, will soften or dissolve in many halogenated or aromatic hydrocarbons and, in addition, yield a distinctive odor during processing, which may be irritating to production personnel unless proper ventilation is available (8).

Polycarbonate: This resin, supplied by General Electric (GE) as LEXAN®, Dow Chemical Co. as Calibre®, and by Mobay as MAKRO-LON®, has outstanding impact strength, high heat resistance (deflection temperature of 280°F (138°C) at 66 psi (0.45 MPa), as well as very good flexural characteristics, creep resistance, and processability. Polycarbonate is a good choice for structural components where load-bearing capability at elevated temperatures is a key requirement. It is an excellent alternative to metal for large components in the automotive, appliance, telecommunications, materials-handling, and business-machine industries. Foamable polycarbonate resin combines an unusual blend of rigidity, impact strength, and toughness with UL 94 V-O and 5V flammability ratings (7).

Grades of polycarbonate are available that couple well with glass fibers to improve fatigue resistance and other properties. Polycarbonate offers good colorability and chemical resistance, along with very high rigidity for a resin with such good ductility. It is also one of the most moisture-sensitive thermoplastics, along with nylon. Physical properties are significantly reduced if the resin is not dried well prior to processing. Care is needed in the selection of blowing agents in foam applications since severe reduction in physical properties (up to 30%) has been reported for certain high-temperature chemical blowing agents (8). Solvent resistance of polycarbonates is poor, resulting in stress cracking in cast solid sheet.

Acrylonitrile-Butadiene-Styrene (ABS): ABS is an opaque, amorphous terpolymer of styrene offering impact, heat, and chemical resistance over polystyrenes. It is available in a multitude of grades intended to emphasize key properties. Improved flammability characteristics are possible either by alloying (blending) with PVC or polycarbonate, or by compounding with halogenated additives. In addition, ABS has relatively low mold shrinkage rates, good long-term dimensional stability, and platability. ABS compounds are slightly hygroscopic and should be dried prior to conventional injection molding to avoid splay marks. In structural–foam molding moisture can lead to excessive odor with certain chemical blowing agents. High–flow ABS grades still display relatively stiff flow characteristics and, therefore, like all high–temperature thermoplastics, offer some resistance to foaming. ABS is susceptible to degradation and discoloration upon exposure to ultraviolet (UV) radiation. Modifying the flammability of ABS by means of halogen compounds significantly increases resin cost and decreases color stability, especially in pastels, but to a lesser degree than with polystyrenes (8).

ABS structural foam can be processed by injection molding, through conventional or low–pressure injection machines; by expansion casting in rotational–molding machines or conveyorized–oven systems; or it can be extruded into profiles through conventional extruders. For injection molding and expansion casting the ABS contains a dispersed blowing agent which, when heat is applied, decomposes into an inert gas, forming a cellular structure. With the extrusion grade the ABS pellets are dusted with a blowing agent prior to extrusion. Various ABS grades are also available that meet the approvals needed for business–machine, TV, appliance, and other markets (6).

Expandable ABS is supplied in the form of pellets (0.093 in. cubes) having a specific gravity of 1.04. Depending on the type of expansion process, the finished part may have a specific gravity from as low as 0.3 (20 lb/ft^3 density) up to the density of the solid material (64.9 lb/ft^3). Specific conversion techniques and part size determine the density that can be achieved in a part. Injection–molded parts are normally produced with specific gravities in the range of 0.7 to 0.9 (densities of 46.7 to 60.0 lb/ft^3), and profile extrusion normally produces parts ranging from specific gravities of 0.3 to 0.7 (20.0 to 46.7 lb/ft^3 densities). Structural foam is produced in natural (cream–colored) and in a wide variety of other colors. Because of its excellent balance of properties at reasonable cost ABS structural foam is finding applications in the automotive, furniture, construction, and materials–handling industries (6).

ABS structural foam has excellent buoyancy, a very low (desirable) stiffness–to–weight ratio, good screw and staple pull–out strengths, and creep resistance superior to that of high–impact polystyrene (HIPS) and polyethylene foams. This is particularly important in load–bearing applications, such as pellets, tote boxes, furniture, and parts buried under earth loads (6).

Acetal: This translucent crystalline polymer is one of the stiffest thermoplastics available. It provides excellent hardness and heat resistance, even in the presence of solvents and alkalies. Its low moisture sensitivity and good electrical properties permit direct competition with die–cast metal in a variety of applications. In addition, acetal has extremely high creep resistance and low permeability. Acetal is also available as a *copolymer* (Hoechst Celanese Corp. CELCON®) for improved processability. The *homopolymer* (du Pont DELRIN®) has a very low coefficient of friction and its resistance to abrasion is second only to nylon 6/6. Acetals are frequently blended with glass or TEF-LON® fibers to enhance stiffness and friction properties. Acetal is not particularly weather–resistant, but grades are available with UV stabilizers for improved outdoor performance (8). Acetal, whether homopolymer or copolymer, is not used to any significant degree in forming structural foams.

Thermoplastic Polyester (Polybutylene Terephthalate) (PBT): The outstanding characteristics of this material are its UL 94 V–O and 5V flammability rating, heat–deflection temperature of 420°F (216°C) at 66 psi (0.45 MPa), high flexural strength and modulus, and excellent chemical and solvent resistance. PBT resins are especially suitable for applications requiring a combination of high heat endurance, stiffness, chemical resistance, and moderate creep (7) (9).

PBT resins are translucent, crystalline materials that are conden-sation products of 1,4– butanediol and terephthalic acid. They exhibit a fairly sharp melting point at approximately 440°F (227°C). Specialty grades are available with up to 30% glass reinforcement. Among key properties are low moisture absorption and good chemical resistance. Good frictional coefficients are offered, but are slightly inferior to the acetal. Grades are available with high flow characteristics approaching the olefins, but the heat resistance of these low–melt–viscosity grades decreases rapidly as a function of improved flow. As with other crystalline resins PBT is notch–sensitive. Impact resistance decreases rapidly as a function of corner radii. Thermal decomposition tempera-tures are only slightly higher than the melting point and therefore close attention must be paid to residence time at high temperatures. Mold temperatures below 150°F(66°C) should be avoided to minimize excessive orientation, resulting in relatively long cycles (8).

Polyetherimide: This newly available resin supplied by GE as ULTEM® has been made available in foamable form to provide outstand-ing thermal, mechanical, and electrical properties, together with excep-tional flame resistance. The foamable resin has a heat–deflection temperature of 401°F (205°C) at 66 psi (0.45 MPa). Glass–reinforced grades provide even greater rigidity and dimensional stability while retaining excellent processability (36). The solid resin has excellent flammability characteristics. ULTEM® has a self–ignition temperature

(Setchkin) of 535°C and UL rating of V–O. Flame spread in the radiant–panel test is low (I_s = 2.7) and the LOI is high — 47. NBS smoke–chamber results show a low specific optical density (Ds 4 min = 0.7, Dm = 31, flaming mode), and combustion–product–toxicity studies show results comparable to polystyrene (10).

Polystyrene (PS): Structural polystyrene foams are molded by injection and expansion–casting methods. They have a strong continuous skin and a foam core. Their densities are quite high, ranging from 20 to 40 lb/ft3, in contrast to the 1 to 5 lbs/ft^3 range of the steam–molded and extruded polystyrene foams discussed below (11). Compared to other structural foams, however, the densities are relatively low (see Table 3–1).

Additional Rigid–Foam Types

Discussions on cross–linked vinyl foams, which are rigid, can be found below under Vinyl Foams, and rigid cellular cellulose acetate (CCA) foams and polysulfone foams are discussed in the section under Miscellaneous Foams.

SEMI–RIGID FOAMS

The following are thermoplastic foams that are frequently grouped under structural foams, but do not always have the integral–skin properties of the structural foams described above.

- Polyolefins
    Polyethylene (PE)
    Polypropylene (PP)
    Ionomer (ethylene copolymer)
- Expanded polystyrene (EPS)
- Extruded polystyrene foam
- Vinyls (PVC)
- Cellulose acetate

Polyolefin Foams

Cellular polyolefins contribute many unusual properties to the cellular plastics industry. These foams are tough, flexible and chemical and abrasion resistant. They are known to have superior electrical and thermal insulation properties. Their mechanical properties are intermediate between rigid and highly flexible foams. Densities are 2 lb/ft^3 and higher, approaching that of the solid polymers. The highly expanded polyolefin foams are potentially the least expensive of the cellular plastics. However, they require expensive processing techniques and for this reason their cost per unit volume is higher than that of low–density polystyrene and polyurethane foams. Low–density polyolefin foams are usually considered as those in the density range of 2 to 10 lb/ft^3. They are used for producing extruded planks, rounds, tubes, and special-purpose profiles. Compression–molded items may also be produced from low–density polyolefins. High–density (10 to 40 lb/ft^3) polyolefins were used initially for electrical cable coatings (12).

Low–density polyolefin foams are being widely used in package cushioning. Energy absorption under continued impact provides protection for delicate electronic parts as well as heavy metal assemblies. Polypropylene foam sheeting is used widely to provide protection for polished surfaces. It is particularly easy to fabricate low–density polyolefin foams by cutting and similar methods using band saws, table saws and routers. Slicing–type equipment or electrically heated wires can also be used to produce "sawdust–free" items. Cross–linked compression–molded polyethylene foams are used commonly to produce foam packages for cameras and measuring instruments. These foams offer excellent appearance and good cushioning performance (12).

The production of cellular polyethylene involves only one chemical reaction — the thermal decomposition of a blowing agent at a specific temperature, which action liberates an inert gas. The choice of blowing agent for electrical–service applications is critical because of several unusual requirements. The blowing agent, the gas it liberates, and the residual by–product must not absorb moisture, which would adversely affect the electrical properties of the product. It is also important that the residue left by the blowing agent be nonpolar in order to avoid losses at high frequencies (6).

Because cellular polyethylene is comprised of roughly equal volumes of resin and gas, its properties are different from those of ordinary unfoamed polyethylene. The cellular product has a much lower dielectric constant and therefore lower electrical losses. The composition of polyethylene (dielectric constant 2.3) and an inert gas (dielectric constant 1.0) has a dielectric constant of 1.5. In terms of electrical insulation the lower dielectric constant permits a reduction in space between inner and outer conductors without changing the characteristic impedance. For this reason it is possible to reduce the attenuation by increasing the size of the inner core without increasing the overall diameter. Alternatively, the weight may be reduced by decreasing the size of the inner conductor (6).

Because cellular polyethylene is of the closed–cell type, its permeability to moisture, while several times larger than that of solid polyethylene, is still relatively low. Cellular polyethylene will provide good electrical performance in alternately wet and dry environments. Its moisture resistance is valuable in such applications as antenna lead–in wire for UHF television. This and other high–frequency applications require that the power factor and dielectric constant not be affected by changes in frequency. Variations would result in power losses, e.g., a weakening of the signal brought in by a UHF antenna. Cellular polyethylene satisfies these requirements. The dielectric constant and power factor of the foam are not seriously affected by temperature fluctuations. The power factor of cellular polyethylene is independent of its specific gravity, while dielectric constant increases with increases in specific gravity (6).

Cellular polyethylene is particularly useful in a number of electrical applications, such as in coaxial cables (CATV, military and other) and in twin leads. The foamed material is particularly useful in modern wire–coating equipment because of the ease of handling and the economies it provides in size and weight of insulated conductors (6).

Low–Density Polyethylene Foams: These foams have all the advantages of the base polymer, including excellent resistance to most chemicals, both organic and inorganic. The foams are generally closed–cell. They are classified as "semi–rigid" and "flexible," depending on their densities and shapes. They can be made to feel as soft as low–density flexible urethane foams and as hard as rigid polystyrene foam. There are basically two types of low–density polyethylene foam — extruded and cross–linked. The *extruded foam* is produced in a continuous process by first blending molten polyethylene resin with a foaming or blowing agent (usually a halogenated hydrocarbon gas) under high pressure, conveying this mixture in a temperature–controlled screw extruder though a die opening to a continuous conveyer exposed to atmospheric pressure. When the hot viscous liquid–gas solution is exposed to atmospheric pressure the blowing gas expands to form individual cells. At the same time the mass is cooled to solidify the molten polyethylene, thereby trapping the blowing agent in the interstitial cells. The degree of expansion, cell size, and cell orientation can be controlled by varying the flow rate, heating and cooling temperatures, gas–liquid ratio, and pressure drop through the die opening (6).

Cross-linked low-density polyethylene foam can be produced by batch and continuous processes. The cross-linking is accomplished by chemical or irradiation methods. *Chemical cross-linking* of PE foam is produced in a batch process, and because of production economics is limited to "plank products." The *radiation cross-link process* permits the continuous production of cross-linked PE foam. This process is limited, however, to production of relatively thin cross sections (up to 3/8 inch) or sheet products (6).

The chemical, mechanical, and thermal properties of the extruded and cross-linked low-density polyethylene foams are very similar, although the cell size of the cross-linked foam is generally smaller and more uniform than that of the extruded products. The cross-linked foams also possess a softer "feel" than that of the extruded foams. Polyethylene foam exhibits compression-deflection properties similar to flexible and semi-rigid foams. As would be expected, the compressive strength increases with increase in density. Compressive creep (loss of original thickness with time) is observed more in lower-density polyethylene foam than in high-density foam. Tensile and tear strengths of polyethylene foams are very high compared to other foams. One of the largest uses of polyethylene foam is in the packaging area to protect products from damage during handling, shipping, and storage. Polyethylene foams are excellent energy absorbers. Because of the wide range of dynamic cushioning properties available with different densities of foam, a wide spectrum of packaging requirements can be met with polyethylene foams. These foams provide optimal cushioning characteristics for objects exerting static stresses ranging from approximately 0.4 to over 15 psi (0.003 to over 0.10 MPa). No other flexible or semi-rigid foams commercially available are capable of providing optimal cushioning characteristics in this static-stress range (6).

Polyethylene foams are used extensively in buoyancy applications because of their excellent water-resistant properties. These basically closed-cell foams absorb less than 0.5% by volume of water after being immersed for 24 hours. The low density of the foams also contribute to their buoyancy. The excellent dielectric characteristics of polyethylene are retained when it is expanded into foams. Polyethylene foam is a candidate for many electrical-material uses requiring good properties of dielectric strength, dielectric constant, dissipation factor, and volume resistivity (6).

Polyethylene foam is chemically inert and contains no water-soluble constituents. It is resistant to most chemicals and solvents at room temperature. It is unaffected by contact with fuel oil and heavier hydrocarbons, but exhibits slight swelling when immersed for an extended period of time in gasoline. Acids and alkalies normally do not affect polyethylene foam, but strong oxidizing agents may eventually cause degradation, especially at higher temperatures. At temperatures above 130°F (54°C) it becomes more susceptible to attack by certain solvents (6).

Ultraviolet rays in sunlight cause some degradation of polyethylene over an extended period of time. This degradation is slower in higher–density products. It is first noted as a yellowing of the foam surface. Some degradation of physical properties will be experienced with longer periods of exposure. For applications where long–term performance is required under direct sunlight a protective coating should be used (6).

Ease of fabrication is one of the many advantages of polyethylene foam. It can be skived to precise thickness, cut and shaped to form custom parts, and joined to itself or to other materials without major investment in complex equipment. It can also be vacuum formed. Expanded polyethylene will adhere to itself by the use of heat alone. Hot air, or a plate heated to approximately 350°F (177°C) can be used to simultaneously heat the surfaces of two sections of foam to be joined. Upon softening, the two pieces are quickly joined together under moderate pressure, and an excellent bond formed, with only a short cooling period required. Release of the melted foam is aided by a coating of fluorocarbon resin or silicone dispersion on the heating surface. The foam may also be bonded to itself and to other materials by the use of solvents or commercially available adhesives (6).

High–Density Polyethylene Foams: These foams are defined by the specific gravity (ca. 0.96 g/cc) and melt index of the basic resin. They have been used extensively in the U.S. and Europe, particularly for materials–handling applications, such as boxes, crates, and pallets because, until recently, they were much cheaper than polypropylene (13).

Polypropylene Foams: Polypropylene foams are comparatively recent entries into the structural foam field, supplanting high–density polyethylene foams, but their use is increasing rapidly because of the extreme range of grades and properties available, plus a favorable price advantage, compared with other thermoplastic foams. Glass–reinforced (30% glass) polypropylene foams are commonly used (13).

Low–density *flexible polypropylene* foam film can be extruded in the 0.7 lb/ft^3 (11.2 kg/m^3) range. In spite of its low density the film sheeting consists of a uniform matrix of small closed–cell gas–filled bubbles. This film has outstanding toughness and strength over a wide–

range of temperatures and humidities. Its major characteristics, compared with other packaging films, are its light weight, resistance to tearing, chemical resistance, and moisture–barrier properties. Extrusion parameters are similar to those used for low–density polyethylene (2). A microcellular PP foam of this type (Du Pont MICROFOAM®) has been used as a furniture wrap for use in packaging furniture in interstate commerce. The protective foam is wrapped around the item before insertion into a corrugated carton. Even with movement in the carton the PP wrap will stay with the item it is intended to protect (14).

Polypropylene foam sheeting is specified in Federal Specification PP–C–1797A. There are two types, Type I for general cushioning applications, and Type II for electrostatic protective cushioning applications. These foams are useful from –65°F (–54°C) to 160°F (71°C). The foam sheeting is intended for use as a protective cushioning wrap for low–density items. For high–density items it can be used for protection of surfaces from abrasion. The foam is non–dusting and non–linting. Typical packaging applications are surface protection for optical lenses, equipment with critical surfaces, electrical and electronic equipment, glassware, ceramics, and magnetic–tape rolls. The Preparing Activity for this Federal Specification is the Naval Air Systems Command (15). DuPont, the manufacturer of MICROFOAM® sheeting, has published considerable technical data on this product. DuPont claims that their MICROFOAM® remains flexible and useful over the temperature range from –320°F (–196°C) to 250°F (121°C) (16) (17) (18) (19).

Cross–Linked Foamed Resins: A number of thermoplastic foams cross–linked by radiation are available. The most popular base resin for cross–linked foam is *polyethylene*. These cross–linked foams offer higher stability and mechanical strength, better insulation characteristics, and improved energy–absorption properties. Most of these foams can be thermoformed, embossed, printed, laminated, or punched, using conventional equipment. There are a number of techniques available for producing cross–linked thermoplastic foams. One such technique involves making a mixture that incorporates one or more of the monomers from the group consisting of acrylamide, methacrylamide, acrylic acid, a foaming agent in solid or liquid state at room temperature, a cross–linking agent having more than two polymerizable double bonds in the molecule, and any monomer, if necessary, that is copolymerizable with the monomers mentioned above. The mixture is polymerized to a thermoplastic body containing the foaming agent homogeneously, and the plastic body is then heated to a temperature above the softening point of the plastic in order to make possible, simultaneously, the foaming and the cross–linking reaction between the polymer molecules (6). Of the various polymerization methods, such as polymerization with catalysts or initiators, thermal polymerization, photopolymerization, and ultrasonic polymerization, the radio–induced or the catalytic polymerization, followed by irradiation, give the best results (6) (20).

Ionomer Foams: The term "ionomer" has been used to denote carbonyl–containing polymers or copolymers of olefins associated with monovalent or divalent cations. Although these materials are not cross–linked in the conventional sense, since they are thermoplastic and can be processed in standard thermoplastic processing equipment, they do exhibit characteristics resembling those of partially cross–linked plastics (21).

Ionomer foams are produced by extrusion or foam injection molding of ionomers to produce a tough, closed–cell foam structure. The low melting point (194–205°F) (90–96°C), high melt strength of the resin, and compatibility with nucleating agents or fillers combine to produce systems that are useful in athletic products, footwear, and construction. The melt characteristics provide a tough skin in the low–density foam sheet and a better surface finish in the higher–density injection–molded products. The higher tensile strength and low melt–point characteristics of an ionomer provide strong heat–seal seams for fabricated sections used in packaging applications. Products ranging from 3 to 30 lb/ft^3 are tougher and more solvent resistant than equal–density foams made from polyethylene or polystyrene. The ionomer foamed sheeting may be vacuum–formed, laminated, stitched, glued, and modified for flame–retardant requirements. The resilient nature of the polymer in medium–density sheeting (6 to 9 lb/ft^3) has application in floor and carpet backing, swimming–pool construction, and tennis–court underliner constructions (6).

In foam injection–molded parts the smooth–skinned resilient foam produces a tough energy–absorbing structure that is being used as a wood substitute in athletic products such as lacrosse–stick heads and hockey–stick blades. Protective structures for helmets and automotive parts are under development. All are being produced in densities ranging for 0.35 to 0.7 g/cc (22 to 44 lb/ft^3). In those applications where added stiffness is desirable the use of glass or titanate fiber reinforcement with the foamed structure is very effective (6).

Ionomer foams are competitive with polystyrene, polyethylene, and some urethane structures. Where the combination of light weight, toughness, and/or direct paint adhesion are prerequisites the foamed ionomer systems are more than competitive by reducing the overall cost through lower part weight and less–involved finishing (6). Ionomer foams find application in the construction and transportation industries because of their low–temperature flex strength and good thermal and moisture–barrier properties. These foams, modified with flame–retardant additives, have been used in the aircraft and packaging industries. In a recent application ionomer foam was used as a protective insulation for concrete pours in the construction of the Libby Dam near Libby, Montana. At the dam site the foam sheets were wired to the inside of the steel concrete forms. After the pour the foam remained on the concrete surface for over two weeks, allowing the concrete to cure evenly without cracking, by limiting the heat transmittance from the material (21).

DuPont supplies ionomer under the trade name SURLYN®. In 1983 a new application of the foam was reported. The ionomer foam was being used as pipe–insulation wrap supplied by the Gilman Corporation in Gilman, Connecticut. The Gilman wrap is called TWIST–TITE and it uses SOFTLITE, a low-–density insulating foam made from DuPont's SURLYN® ionomer resin. This soft flexible foam will not absorb moisture or collapse, and contains no irritating fiberglass. It is supplied in four–foot–long spirally wound tubes that can be easily wrapped around pipes of one–half inch to two inches in diameter, as well as tees, valves, hangers, and elbows. No tools or tape are needed. The material can be secured to pipes quickly with convenient plastic ties provided with the pipe wrap. Two varieties are offered: (1) a two–layer foam lamination for indoor pipes, and (2) a foil–laminated wrap suitable for outdoor use (22).

Polystyrene Foams (Low–Density)

This type of foam is available in two forms, *extruded–polystyrene foam* and *expanded polystyrene* for molded foams. Polystyrene foams are light, closed–cell foams with low thermal conductivities and excellent water resistance. They meet the requirements for low–temperature insulation and buoyancy media (6).

Extruded–Polystyrene Foam: This material has been produced for over 40 years and is manufactured as billets and boards. It is made by extruding molten polystyrene containing a blowing agent, under elevated temperature and pressure, into the atmosphere, where the mass expands. The billets and boards can be used directly or can be cut into many different forms. Common tools for fabrication are bandsaws, hot–wire cutters, planers and routers. Boards may have a cut or planed surface, or may have an integrally extruded skin of extremely small cells. Many sizes are available in extruded cellular polystyrene. Billets can be as large as 10 inches x 24 inches x 9 feet long. Boards are available 3/4 inches to 4 inches thick by 24 inches by up to 9 feet long (6).

Applications include low–temperature insulation in freezers, coolers, and other types of refrigerated rooms; truck bodies and railroad cars; refrigerated pipelines; and low–temperature storage tanks for liquified natural gas. Other applications include roof–deck insulation where the foam is placed in the last hot bitumen layer of the roof, which is then covered with gravel or stone to hold it in place. Extruded–polystyrene foam is also used in the insulation of residential housing by using the foam in place of conventional sheathing. The foam is placed over the studs and sill plates, which are sources of thermal shorts in batt–insulation buildings. This type of foam is also used in agriculture to insulate livestock buildings and low–temperature produce–storage buildings (6).

Extruded polystyrene foam boards are also used as core material for structural sandwich panels, especially in the growing recreational–vehicle and motor–home industries. The walls, floors, and roofs of many of these vehicles are constructed of polystyrene foam–core sandwich panels. White and green extruded–polystyrene foams are used in the floral, novelty, hobbycraft, and display fields, either as boards or billets, are fabricated into shapes, such as balls, cones, cylinders, rings, etc. Floating docks, marker buoys, and flotation for small boats are applications for large billets of this foam. Extruded–polystyrene foam sheet in thicknesses of 1/16 to 1/4 inch has found significant usage as a replacement for molded–paper–pulp board in meat and produce trays and egg cartons. The foam sheet is clean, bright in appearance, has excellent cushioning properties, and is nonporous. The foam is extruded as a sheet and is subsequently vacuum–formed into the desired shapes for packaging (6).

Foamed polystyrene sheet and film are manufactured by a tubular film–extrusion process using conventional methods with specially treated expandable–polystyrene pellets, or by injecting a propellant directly into a section of the extruder barrel with standard polystyrene resins and additives. In both techniques the extrudate passes through an annular tubing die and is expanded, either by blowing air inside the tube, or by drawing the tube over an internal sizing mandrel (6).

Expandable Polystyrene (EPS) for Molded Foam: Expandable polystyrene is produced in the form of free–flowing beads, symmetrical shapes, and strands containing an integral blowing agent, such as pentane. When exposed to heat without restraint against expansion these particles "puff" from a bulk density of about 35 lb/ft3 to as low as 0.25 lb/ft^3. In the case of beads intended for molding, this very low density leads to difficulty with collapse during subsequent molding. For this reason the usual limit for pre–expansion is 1.0 lb/ft³. The shapes and strands used as loose–fill cushion packaging are not processed beyond pre–expansion (6).

Molding EPS beads ordinarily comprises two separate steps:

● Pre–expansion of the virgin beads by heat.

● Further expansion and fusion of the pre–expanded beads by heat within the shaping confines of a mold.

In unconfined pre–expansion the translucent beads grow larger and become white in color. Confined and subjected to heat, the pre–expanded beads can produce a smooth–skinned closed–cell foam of controlled density, registering every detail of an intricate mold. To minimize formation of a density gradient and to ensure uniform expansion throughout the molded piece, expandable–polystyrene beads are pre–expanded to the approximate required density by control of time and temperature, since the process of molding does not increase the density (6).

Figure 3–1 is a rough schematic visualization of the steps from the virgin beads to molded foam (23).

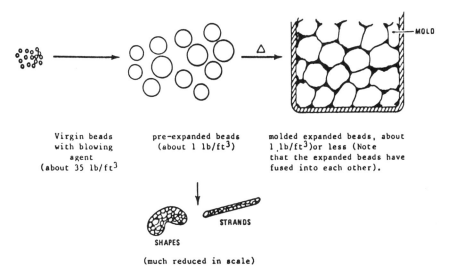

Virgin beads
with blowing
agent
(about 35 lb/ft³

pre–expanded beads
(about 1 lb/ft³)

molded expanded beads, about
1 lb/ft³)or less (Note
that the expanded beads have
fused into each other).

STRANDS

SHAPES

(much reduced in scale)

Figure 3.1. Steps in pre–expansion and final expansion and molding of polystyrene beads (23)

Continuous steam prefoaming is the most widely used method of expanding polystyrene beads. Other methods are *continuous hot–air prefoaming* and *batch prefoaming*. Molding of expandable polystyrene beads requires exposing the pre–expandable or virgin beads to heat in a confined space, as shown in Figure 3–1. A number of conventional heating media are available for the fabricator to use in molding expandable polystyrene. The preferred medium is steam that is directly diffused through the pre–expanded beads in the mold cavity. Other techniques involve conductive heating through the mold wall, with the heat being supplied by steam or some other energy source. The types of molding used are as follows: (6)

- steam–chest molding
- block molding
- foam–cup molding
- steam–probe molding
- continuous–billet molding
- radio–frequency molding

In molding polystyrene foam from beads, when heat is applied to the mold and beads, the beads are expanded enough to fill all the voids between the beads and to heat–seal all the beads together, as shown in Figure 3–1. This process permits the molder to make products with a wide range of characteristics. The density can be easily controlled. It ranges form below 1 lb/ft^3 to above 20 lb/ft^3. Molded products can have a wide range of wall thicknesses. The density of the part is uniform throughout. Most parts are molded with only a very thin skin, but with an additional step in the process it is possible to produce a part with a hard skin. The molded product is stress–free and quickly stabilizes. Molding pressures and temperatures are low (6).

In the low–density range, 0.5 to 1.0 lb/ft^3, expanded polystyrene is used on boats as flotation, in packaging as an energy absorber, in building as insulation, and as a moisture barrier. In the middle–density range, from 1.0 to 4.0 lb/ft^3, the material is used in packaging as a structural support as well as an energy absorber; in the construction field for such application as concrete forms; in the foundry industry as mold patterns; as insulated containers of all sizes and shapes; and in materials–handling pallets. In the high–density range from 5.0 to 20.0 lb/ft^3 the material exhibits almost wood–like properties. Such products as thread spools, tape cores, and furniture parts have been made from these foams (6).

As indicated above, the blowing agent most commonly used on expandable polystyrene is *pentane*, typically about 6.3 – 7.3% by weight. To reduce pentane losses in shipping, expandable beads are ordinarily transported in 5–gallon drums constructed from fiberboard, with an inner liner of aluminum foil. In open containers at 23°C, expandable beads screened through a 10–mesh screen and retained on a 25–mesh sieve decreased their n–pentane content from 7.0 to 6.0% in 6–7 days. Other blowing agents used to some extent are *neopentane* (tetramethylmethane) and *isopentane*, both solid gas–releasing blowing agents (11).

One of the serious limitations of polystyrene foam is its rather low operating temperature of approximately 80°C (176°F).

Vinyl Foams

Foams made from vinyl resins are of two types, *open–cell* and *closed–cell*. The open–cell foams are soft and flexible, while the closed–cell foams are predominantly rigid. Both types are made from *plastisols*, which are suspensions of finely divided resins in a plasticizer. The polymer does not dissolve appreciably in the plasticizer until elevated temperatures are used.

Open–Cell Vinyl Foams: Two methods are used to produce this type of foam, which is the most important type in vinyls: (1) a chemical blowing agent type, and (2) a mechanical frothing process, in which a gas is also used as part of the blowing mechanism. In the preparation of a soft open–cell foam using a chemical blowing agent the plastisol is first chosen for the characteristics desired. The resin and plasticizer ratio in the plastisol should be one in which enough plastisol is present to assure production of a soft, flexible material. To the plastisol is added a paste made of powdered blowing agent dispersed in a plasticizer. The chemical blowing agent (CBA) may be any one of a number of organic materials which, when heated, decompose to liberate a gas, usually nitrogen. One class of materials used for the large majority of vinyl foams is the *azocarbonamides* and other azo compounds that can be used to decompose at temperatures from about 250–425°F (120–220°C). A CBA can be chosen to accommodate practically any required resin temperature. It is extremely important that, at the gas–evolution temperature the viscosity characteristics of the plastisol allow complete and even expansion, and that a uniform cell structure is formed. Factors that influence these properties are the particle size of the blowing agent, the type and quantity of the resin and plasticizer, and the specific type of stabilizer and/or activator used to initiate decomposition of the blowing agent. Soft, very flexible vinyl foams used for garment insulation, upholstery and similar applications are made by this process. Somewhat harder foams used as underlays for rugs and flooring can also be made by this technique, but require different resins and lower plasticizer contents (13). Open–cell chemically blown vinyl foams generally have densities in the range of 10 to 30 lb/ft^3 (23).

The second method of producing open–cell vinyl foams, *mechanical frothing*, is used to produce a large amount of vinyl foams, particularly sheets, such as flooring underlay, wall coverings, and other applications requiring relatively close thickness tolerances.   In this method the plastisol is mixed with a given amount of air in a high–shear, temperature–controlled mixing head. The resulting product, resembling shaving cream, is cast onto a belt or fabric and knifed to a definite thickness. Passage through an oven or heating tunnel then causes fusion of the plastisol. Fusion temperatures for all the plastisol foams range from 257–325°F (125–160°C). The times required may be from 1 to 3 minutes for thin sheets of from 10–20 mils to as high as 7 to 10 minutes for materials as thick as 3/4 inch. Thicknesses over 1 inch present a problem because the insulating properties of the foam prevent the heat from reaching the center of the foam in a reasonable length of time. Dielectric heating can, however, be used very successfully to produce foam as thick as 2 inches in a very short period of time (3).

Closed–Cell Vinyl Foams: These foams are made in a manner somewhat similar to the technique used to produce open–cell foams with chemical blowing agents, except that much higher pressures are used, and the process is accomplished in two steps.   First, the vinyl plastisol containing the blowing agent is placed in a mold in which very little space is left for expansion. The mold is then heated, causing decomposition of the blowing agent and, at the same time, fusion of the foam. This step raises the internal pressure in the mold to anywhere from 200 to as much as 1600 psi (13.8 to 110 MPa). At these high pressures the gas is dissolved in the resin in the form of microscopically small bubbles. The part is then allowed to cool and harden in the mold before being removed (3).

The second step consists of reheating the molded part at approxi mately 200–250°F (93–120°C), at which time the resin softens and the gas expands to form a closed–cell foam. With this technique it is possible to produce foams with densities as low as 2 lb/ft^3, although the usual range is 50 to 10 lb/ft^3. Because of this two–step procedure the process is much slower than the foaming procedure for open–cell materials. For this reason it is used mainly for specialty items such as marine flotation parts, special athletic mats, and similar items. Since the parts made by this process are all made at high pressures these materials are often called "pressure sponge" (3).

Low-density unicellular foam made from this pressure process is available in a range of densities and hardnesses in various sizes ready for fabrication into the final product. Slab stock is used to produce shock-absorbing, flotation, and insulating foams. Typical densities range from 7 lb/ft^3 for shock-absorbing foams, 4 lb/ft^3 for flotation foams, and 6 lb/ft^3 for insulation foams. It is also possible to produce pressure-molded foams in cylindrical and ring-shaped molds (24) (3).

Cross-Linked Vinyl Foams: It is possible to produce a rigid vinyl foam by using a combination of vinyl chloride polymer and monomer, plus maleic anhydride, isocyanate and catalyst, to produce a rigid cross-linked, essentially closed-cell construction of exceptional strength. This type of foam is being laminated with metal and reinforced plastics for use in aircraft construction as a flooring material and in boat construction for hulls. The ingredients are poured into a mold, then heated while confined under pressure. The press is brought to 100°C, and the exothermic reaction results in the maleic anhydride copolymerizing with the vinyl chloride monomer and grafting onto the PVC. The composition, then removed from the mold, is still thermoplastic, since the isocyanate has not yet reacted. The plastic is then exposed to hot water or steam, thereby causing the isocyanate to liberate CO_2, which acts to expand the plastic mass. After expansion is completed the water then reacts with the grafted maleic anhydride, and the resultant maleic acid reacts with the isocyanate and cross-links it (25).

Miscellaneous Foams

Cellular Cellulose Acetate (CCA): Cellular acetate was one of the first rigid foams finding commercial use in aircraft manufacture in World War II and in specialty construction uses. This foam was produced for many years by the Strux Corporation, but Strux stopped making it in 1975. Currently it is supplied on a limited basis by Deltex Associates. (The foam supplied by Deltex is called CCA (Strux) and can be obtained from DELTEX Associates, Inc., P.O. Box 71, Freehold, N.J., (201) 442-2396 or 462-8106). The production of the foam involves extrusion of flaked cellulose acetate, employing a mixture of acetone and ethyl alcohol as blowing agent and comminuted mineral of 100 to 325 mesh as nucleating agent. The solvent mixture also acts as a solvent for the cellulose acetate above 70°C and as a plasticizer during the hot extrusion. The extrusion process leads to the formation of unicellular foam by the sudden release of the superheated, pressurized molten polymer mass as it emerges from the extrusion die. The extruded foam has a denser outer skin, which may be removed by cutting off the outer edges (21).

Cellulose acetate foam is extruded in the form of boards and rods. The foam cannot be molded, and so end–use shapes must be made by machining. Densities are in the range of 6 to 8 lb/ft^3 (96 to 118 kg/m^3). The foam has excellent strength, low water absorption, and resistance to attack by vermin and fungus. The service–temperature range is broad (– 57° to 177°C), and the foam will tolerate short exposures to 194°C. It has good electrical properties, making it useful for electronics and x–ray equipment (21). Hydrogen bonding is responsible for CCA's insolubility in water and its inertness to most solvents and resin systems. It can also be steam–formed to achieve desired shapes (26).

The most widely used applications reported in 1973 included rib structures in the fabrication of lightweight reinforced–plastic parts and as a core material in sandwich construction, bonded to metal, wood, or glass.   Other applications are as reinforcements for aircraft–control surfaces, radome housings, filler blocks under fuel cells, tank floats for indicating devices, and ribs, posts, and framing in houses and shelters. Due to its buoyancy characteristics cellulose acetate foam has been used in lifeboats, buoys, and other flotation devices (21).

Polysulfone Foams: Polysulfone resins were introduced by Union Carbide Corporation in 1965 to produce stable and self–extinguishing thermoplastics. The polysulfones contain the sulfur atom in its highest excitation state, and hence are resistant to oxidation.   The structure is as follows:

Up to 1973 very little information had been published on these foams. France has produced polysulfone foams from expandable beads which resemble polystyrene beads in their processing. In a typical formulation 1– butene and sulfur dioxide are copolymerized in such a way that the copolymer contains excess unpolymerized 1–butene, which volatilizes to cause foaming. Water, ethyl hydroxyethyl cellulose, 1– butene, sulfur dioxide, and a solution of isoprophyl peroxydicarbonate in ethyl maleate are stirred at 250 rpm for 6 hours at 48°C and heated at 75°C for 1 hour to give pearls 0.5 to 2 mm in diameter, which are then washed and dried at 25°C. These pearls, which contain 9.5% unpolymerized 1–butene, are foamed at 100°C to give cellular polysulfone of about 3.1 lb/ft^3 (49.6 kg/m^3) density (21).

Molded articles from cellular polysulfone, coated with polystyrene, epoxy resin, or vinyl resin, are reported to have improved hardness and surface rigidity. They can also be used as insulating or packaging materials, or, when modified with urea–formaldehyde resin, as flowerpots that are permeable to air and water. These foams have excellent self-extinguishing and low smoke–generating properties (21).

REFERENCES

1. Whittington, L.R., *Whittington's Dictionary of Plastics,* 2nd Edition, Sponsored by the Society of Plastics Engineers, Technomic, Westport, Connecticut (1978).

2. *Foams, edition 2, Desk–Top Data Bank*®, edited by M.J. Howard, The International Plastics Selector, Inc., San Diego, California (1980).

3. Schwartz, S.S. and Goodman, S.H., *Plastic Materials and Processes*, Van Nostrand Reinhold, New York (1982).

4. Wood, A.S., "Structural Foams are Tougher Now; Reinforcement Opens Big New Options," *Modern Plastics, 52* (12): 38–40 (December 1975).

5. Wehrenberg II, R.M., "A Bright Outlook for Structural Foams," *Materials Engineering, 95* (6): 34–39 (June 1982).

6. *Plastics Engineering Handbook of the Society of the Plastics Industry, Inc., 4th Edition*, edited by J. Frados, Van Nostrand Reinhold, New York (1976). (*Author's Note:* The 5th Edition, edited by M.L. Berins, was published in 1991).

7. General Electric Company, Specialty Plastics Division, Structural Foams Resins Section, *The Handbook of Engineering Structural Foam*, 67 pp., Publ. SFR–3B (June 1981) (This handbook is now out of print).

8.   Marchetti, R., "Selecting Resins for Thermoplastic Structural Foam," *Plastic Design and Processing, 19* (4): 20–27 (April 1979).

9.   General Electric Company, Plastic Operations, Engineering Structural Foam Products Section, *Engineering Structural Foam – Today's Technology for Tomorrow's Designs*, 9 pp., Publ. SFR–29 (1983).

10.   Landrock, A.H., *Handbook of Plastics Flammability and Combustion Toxicology,* Noyes Publications, Park Ridge, New Jersey, (1983).

11.   Ingram, A.R. and Fogel, J., Chapter 10, "Polystyrene and Related Thermoplastic Foams," in *Plastic Foams, Part II*, edited by K.C. Frisch and J.H. Saunders, Marcel Dekker, New York (1972).

12.   Sundquist, D.J. Chapter 5, "Polyolefin Foams," in *Plastic Foams, Part I*, edited by K.C. Frisch and J.H. Saunders, Marcel Dekker, New York (1972).

13.   Norgan, N.R., "Thermoplastic Structural Foams – A Review," *Cellular Polymers, 1* (2): 161–177 (1982).

14.   Anonymous, "Foam Cushioning for Shipping Furniture," edited by Joel Frados, *Plastics Focus, 6* (15) (June 10, 1974).

15.   Federal Specification PP–C–1797A, "Cushioning Material, Resilient, Low Density, Unicellular, Polypropylene Foam," 1 September 1982. This specification supersedes the earlier Military Specification MIL–C–81823(AS) (1971).

16.   Du Pont Company, "MICROFOAM® Sheeting," *Technical Bulletin* A–80786, March 1972. Revised July 1972, 1 page.

17.   L.A. De'Orazio, "Polypropylene–Foam Sheet," *Modern Packaging, 44* (3): 62–65 (March 1971).

18.   Closek, B.M. and Patterson, D., Jr., "Impact Test for Cushioning," *Modern Packaging, 44* (10): 66–67, 70–71 (October 1971).

19.   Du Pont Company, "MICROFOAM® Sheeting, etc.," Publication A–73432, Revised November 1971, 2 pages.

20.   Readdy, A.F., "Applications of Ionizing Radiations in Plastics and Polymer Technology," *PLASTEC Report 41* (March 1971). Available from NTIS and DTIC as AD 725 940.

21.   Frisch, K.C., Chapter 15, "Miscellaneous Foams," in *Plastic Foams, Part II,* edited by K.C. Frisch and J.H. Saunders, Marcel Dekker, New York (1973).

22.   Anonymous, "News in Brief. Energy–Saving Pipe Insulation," *Du Pont Magazine, 77* (1): 28 (January/February 1983).

23.   Landrock, A.H., "Handbook of Plastic Foams," *PLASTEC Report R52* (February 1985). Available from NTIS as AD A156758.

24.   Skochdopole, R.E. and Rubens, L.C., "Physical Property Modifications of Low–Density Polyethylene Foams, *Journal of Cellular Plastics, 1*: 91–96 (January 1965).

25.   Werner, A.C., Chapter 6. "Polyvinyl Chloride Foams," in *Plastic Foams, Part I,* edited by K.C. Frisch and J.H. Saunders, Marcel Dekker, New York (1972).

26.   Deltex Associates, Inc., P.O. Box 71, Freehold, N.J., *Technical Bulletin No. C–1079* on CCA (STRUX). Telephone No. is (201) 442–2396 or 462–8106.

4
ELASTOMERIC FOAMS

Arthur H. Landrock

INTRODUCTION

This chapter will briefly discuss elastomeric foams (other than urethane foams). The reader will note that the emphasis of this handbook is on plastic foams (cellular plastics), but it was felt by the author that some discussion of the more well-known elastomeric foam types would be helpful.

General

Rubber products with a cellular structure have been used widely for many years. The earliest developments of these products predated World War I. The two forms of *natural rubber—raw rubber*, and *latex*, form the basis for different product types, one being blown dry rubber, and the other foamed and dried latex. Blown sponge and latex foam are distinctly different materials, although the end–products may appear similar and have some overlapping applications.

Sponge Rubber

In the trade, sponge rubber refers to both open– and closed–cell materials produced by sheeting, molding, and extrusion from compounded gum rubber, generally using a blowing agent. This usage is not technically correct, because in ASTM D 1056 *sponge rubber* is defined as "cellular rubber consisting predominantly of open cells made from a solid rubber compound," and *expanded rubber* is reserved for closed–cell compounds (sometimes called unicellular rubber). Sponge rubbers are made by incorporating into the compound an inflating agent, such as sodium bicarbonate, which gives off a gas that expands the mass during the vulcanization process. Sponge rubbers are manufactured in sheet, strip, molded, or special shapes. Expanded rubbers are made by incorporating gas–forming ingredients in the rubber compound, or by subjecting the compound to high–pressure gas, such as nitrogen. Expanded rubbers are manufactured in sheet, strip, molded, and special shapes by molding or extrusion. Styrene–butadiene rubber (SBR), and natural, polyisoprene, and neoprene rubbers are the materials used in forming sponge or expanded rubber (1) (2).

Cellular Rubber

Cellular rubber is a generic term and includes latex foam produced chiefly by aeration of compounded latex, which is of necessity open–cell because water must be removed rapidly during processing (1). ASTM D 1055 defines *flexible cellular rubber* as "a cellular organic polymeric material that will not rupture within 60 seconds when a specimen 200 by 25 by 25 mm is bent around a 25–mm diameter mandrel at a uniform rate to produce 1 lap in 5 seconds in the form of a helix at a temperature between 18° and 29°C." The structure of latex foam rubbers consists of a network of open or interconnecting cells. Latex foam rubbers are made from rubber lattices or liquid rubbers. They are manufactured in sheet, strip, molded, or specific shapes (3).

Comparison of Cellular–Rubber Products

The numerous flexible and semi–rigid forms of foam rubber and sponge provide a broad spectrum of performance characteristics adaptable to consumer needs. These materials compete with cushioning or padding materials such as cloth, fiberfill polyester, hair, jute, etc., as well as urethane, polyvinyl chloride, and other plastic foams. From a performance point of view it is not always clear how to measure comfort, durability, and appearance, while parameters such as cost, weight, compression resistance, recovery, rebound, and modulus are more readily evaluated. "Feel" is therefore a factor in assessing these cellular products. *Low–density latex foam* has maintained a high quality of performance for comfort cushioning. Good indentation–load response gives it a feel of deep comfort. This feeling comes from the fine–cell structure, the resilient nature of the rubber, and the open cells for heat and moisture transfer. Latex foam is adapted to coating applications, while *open–cell sponge*, generally made at higher densities, is not as readily processed for coatings. Nevertheless, latex foam and sponge tend to compete in a few product areas, such as automotive–seating applications. Intrinsically, the two materials can perform essentially the same functions by proper selection of formulation variables. For this reason other factors, such as fabrication economics and design features, determine the material of choice (1).

At comparable densities sponge rubber will generally have better flex, tear, and abrasion qualities than latex foam, even at higher filler loadings, due to mastication of ingredients compared with coalescence of latex particles. Sponge rubber will usually have superior water resistance because emulsifiers are not required for processing. Latex foam has the advantage of light color and less objectionable odor. It usually recovers faster from deflection than sponge, since the rubber is not oil–modified. Closed–cell expanded rubber is different from the open–cell product. Water is sealed out in weather stripping and flotation uses. Closed cells are usually advantageous in gasketing and insulation applications (1).

TYPES OF ELASTOMERIC FOAMS

Neoprene

Neoprene, or polychloroprene, is a synthetic rubber discovered by the Du Pont Company in 1931. It is an organic polymer composed of carbon, hydrogen, and chlorine in the ratio of 55:5:40. Its relatively high chlorine content was responsible for the early recognized resistance of the polymer to burning. Practical use of this property was not developed until procedures for making foam structures from neoprene latex were developed in the 1940's. The U.S. Navy adapted the material to make neoprene foam mattresses that reduced the fire hazards in the crews' quarters of naval vessels. For many years neoprene has been the only material to meet Navy specifications for this application.

Polychloroprene begins to release hydrogen chloride (HCl) on heating to 430°F (221°C). The evolution of other gases, including carbon monoxide (CO) and carbon dioxide (CO_2), and visible smoke continues as the temperature increases to 700°F (371°C). By the time this temperature is reached substantially only carbon residue (C) remains in the form of a rigid char. Because of the high chlorine content of the polymer neoprene has a relatively low heat of combustion (4.7 k cal/g) and a high oxygen–index value (Limiting Oxygen Index or LOI = 36). The burning characteristics of polychloroprene are altered and generally improved by the presence of fillers and chemical additives. Neoprene foam compositions generally contain antimony oxide as a flame retardant and hydrated aluminum oxide to release water vapor and thereby delay ignition. Other additives, such as char promoters and smoke retardants, may also be used (4).

In the early 1980's a neoprene–type foam (VONAR) made by Du Pont was developed specifically for use as a comfort cushioning material for critical applications where low flammability was required, such as in transport aircraft seat cushions. This material is designed for use over flame–retardant polyurethane foam and is believed to provide 50 seconds more of evacuation time in the event of a post–crash fire. VONAR has a density of 10.66 lb/ft^3, compared to 7–8 lb/ft^3 for standard neoprene foam, and 1.2 – 2 lb/ft^3 for polyurethane foam (5).

Silicone Foams

Three types of silicone foams are discussed briefly below — silicone elastomeric foams, silicone rubber sponge, and room–temperature–foaming silicone rubbers.

Silicone Elastomeric Foams: These lightweight rubbery foams are made by mixing two components. The mixing requires only 30 seconds, and the currently available materials must be poured immediately because expansion begins promptly upon blending. Negligible pressure is generated and the articles formed can be removed from the molds within five minutes, at which time the foam will have developed about 80% of its ultimate strength. Maximum strength is developed after 24 hours. When cast against glass cloth or asbestos paper the foam adheres strongly. Finished pieces can be easily bonded to each other, or to metal, by appropriate silicone adhesives (6).

Silicone Rubber Sponge: This material is prepared by blending catalysts, fillers, and blowing agents into a high–molecular–weight, linear siloxane polymer ($\equiv$SiOSi$\equiv$). Cure is accomplished by heating in the presence of small amounts of organic peroxides, such as dibenzoyl peroxide. The high–molecular–weight polymer is intensively mixed with high–surface–area silica fillers, plasticizers, and colorants to form the rubber base. A chemical blowing agent suitable for silicone rubber sponge must not affect the vulcanization or the physical properties, and it must be readily dispersable in the unvulcanized rubber base. It should decompose to form a suitable gas at or near the rubber vulcanization temperature. There are two methods used for sponging and vulcanizing the silicone rubber, *cold forming* and *hot pressing*. Silicone rubber sponge can be used wherever thermally stable, flexible, cushioning–type thermal insulation is needed. It is currently used for pads in high–pressure plastic or metal–forming equipment. A soft, fine–cell silicone elastomer sponge is currently being used in localized scleral indentation treatment of retinal detachment of the eye. The product is packaged sterile, measures 8 cm. in length, and is available in either cylindrical or oval shapes. The sponge is soft, resilient, inert, and non–toxic and, therefore, exhibits minimal cellular or vascular response. It will not adhere to the ocular tissue and does not harden or break down on long–term implantation (7).

Room–Temperature–Foaming Silicone Rubbers: Liquid silicone rubber prepolymers that foam and cure at room temperature are available. These products are foamed by the liberation of hydrogen from the reaction:

$$\equiv\!SiH \ + \ HOSi\!\equiv \ \xrightarrow[\text{Stannous Octate}]{H_2O} \ 2 \equiv\!SiOH \ + \ H_2$$

Cross–linking proceeds through silanol condensation to form siloxane cross–link and water. The prepolymers generally used in this type of foam are silanol–end–blocked copolymers or mixtures of homopolymers prepared from the hydrolysis of difunctional chlorosilane monomers. Additives normally used in silicone rubbers, such as oxidation inhibitors, fillers, plasticizers and pigments, may be compounded into the prepolym–er system. These foams are supplied in two–component systems. The rubber base should be well stirred just before the catalyst is added to ensure uniform color and cell structure in the finished foam. Normally about 5 to 8 parts of catalyst is mixed with 100 parts of base, using vigorous stirring. Foaming starts as soon as the catalyst is added, and the foam mixture should be poured into place within 60 seconds from the time of catalyst addition. The catalysts contain stannous octoate and may cause skin irritation. This type of catalyst gradually loses its reactivity when exposed to air, so containers should be kept tightly closed when not in use. The foam becomes tack free in 2 to 5 minutes and is firm enough to handle within 5 to 10 minutes. Maximum strength is reached after a 24–hour cure at room temperature (7).

These foams range in density from 9–20 lb/ft^3 and contain about 90% open cells.   Like other silicone rubber materials, RTV foams maintain their resiliency over a temperature range of −75° to 200°C (−103° to 392°F), have excellent electrical properties, and resist weathering and aging.   Medical–grade RTV foams are physiologically inert.   These materials are used whenever a resilient, low–density silicone rubber foam is required, such as for: (1) mechanical shock, vibration damping, or cushioning;   (2) thermal insulation;   (3) foam–rubber parts;   and   (4) lightweight electrical insulation.   Silicone rubber RTV foams are used in several medical applications which require nontoxic, low–exotherm materials that foam and cure at room temperature.   Useable foam rubber pads for limb prostheses were prepared by the injection of the catalyzed foam into the socket between the stump and the prosthetic device.   The resilient silicone foam enables pressures to be evenly distributed over the entire distal end of the stump, resulting in an intimate, comfortable fit of the prostheses.   Silicone rubber RTV foams have also been evaluated as diagnostic enemas.   The catalyzed material was injected into the colon, foamed in place, and the foam allowed to expel without assistance.   The silicone foam accurately recorded the shape of the colon, and surface details, such as polyps, constrictions, and lesions were reproduced on the foam surface (7).

REFERENCES

1.   Zimmerman, R.L. and Bailey, H.R., Chapter 4, "Sponge Rubber and Latex Foam," in *Plastic Foams, Part 1,* eds. Frisch, K.C. and Saunders, J.H., Marcel Dekker, New York (1972).

2.   ASTM D 1056–85, "Standard Specification for Flexible Cellular Materials—Sponge or Expanded Rubber," in Vol. 09. 02, *Annual Book of ASTM Standards.*

3.   ASTM D 1055–90, "Standard Specification for Flexible Cellular Materials — Latex Foam," in Vol. 09, 02, *Annual Book of ASTM Standards.*

4.   Morford, R.H., "The Flammability of Neoprene Cushioning Foam," *Journal of Fire and Flammability, 8*; 279–298 (July 1977).

5.   Hardy, E.E. and J.H. Saunders, Chapter 14, "New High Temper-
        ature–Resistant Plastic Foams," in *Plastic Foams, Part II,* eds.,
        Frisch, K.C. and Saunders, J.H., Marcel Dekker, New York
        (1973).

6.   *Plastics Engineering Handbook of the Society of the Plastics
        Industry, Inc.*, 4th Edition, ed. Frados, J., Van Nostrand Reinhold,
        New York (1976) (*Authors Note*: The 5th Edition, edited by M.L.
        Berins, was published in 1991).

7.   Vincent, H.L., Chapter 7, "Silicone Foams" in *Plastic Foams,
        Part I,* eds. Frisch, K.C. and Saunders, J.H., Marcel Dekker, New
        York (1972).

5

MISCELLANEOUS AND SPECIALTY FOAMS: *(Epoxy Foams, Polyester Foams, Silicone Foams, Urea–Formaldehyde Foams, Polybenzimidazole, Foams, Polyimide Foams, Polyphosphazene Foams, and Syntactic Foams)*

Arthur H. Landrock

This chapter will discuss thermosetting foams not covered in Chapter 2. It will also present additional material of interest on syntactic foams.

EPOXY FOAMS

Epoxy foams are made by free–foaming processes and may be cured with or without the use of heat. A wide selection of formulations, resulting in closed– or open–cell structures, is available. Prefoamed boards and sheets can be fabricated and can be worked similarly to wood. Epoxy foams are tough and strong; have excellent resistance to chemicals, solvents and water; have excellent dielectric constants ranging from 2 to 7 at 50–100 cycles per second; and are dimensionally stable (1). Their minimal shrinkage can be attributed to the fact that no by–products are formed during cure (2).

Epoxy foams are rigid. Some are predominantly of closed–cell structure and some of open–cell structure. Current efforts may lead to the development of special shock–resistant and semi–rigid, or even flexible types, and of a type applicable to surfaces by spraying. Epoxy resins are frequently used as binders for microscopic, hollow gas–filled beads of phenolic, urea, and polystyrene resins (see Syntactic Foams). In addition to prefoamed sheets, boards and planks, epoxy foams are available in pack–in–place and foam–in–place systems (2).

Epoxy foams are made in two types. One is a *powder foamant* that contains powdered epoxy resin, such as Shell Epon 1031; powdered diaminodiphenylsulfone, the hardener; and a powder chemical blowing agent (CBA). The mixture is heated and liquified, and the CBA causes foaming; next the curing agent causes hardening. The other type of epoxy foam is a *liquid foamant* that uses liquid resins, curing agents, blowing agents, and surfactants. The curing agents usually develop a highly exothermic reaction so that little, or in some cases no, external heat is required. The foams range from 2 lb/ft^3 (32kg/m^3) to as high as 38 lb/ft^3 (608 kg/m^3). Like the urethanes, the epoxy foams may be handled a short time after cooling, perhaps as little as 15 minutes. To achieve full strength, however, the materials should be postcured for several hours, as recommended by the manufacturer of the resin. The low–density foams (2–7 lb/ft^3) (32–112 kg/m^3) are usually serviceable up to 82°C (180°F), while the higher–density foams (>7 lb/ft^3) (112 kg/m^3) have been used up to 200°C (392°F). The excellent electrical character- istics and adhesive properties of epoxy foams have led to their use in many electrical and electronic applications as potting and encapsulating materials (3).

POLYESTER FOAMS

Until recently little information has been published on foams made from unsaturated polyesters. In early 1983 Ahnemiller published a paper describing the production of foamed unsaturated thermosetting polyesters to provide cost and weight reductions, increased stiffness, and improved insulating characteristics. A foaming system using a sulfone– hydrazide chemical blowing agent (CBA) produces a uniformly closed– cell foam structure with a good stiffness–to–weight ratio, excellent density reduction, and significant labor savings. These foams have been prepared for use as core materials to provide the reinforced plastics (RP) industry with a new high–modulus material. Densities of 0.43 to 0.63 g/cc (430 to 630 kg/m^3) were obtained. Foams of 0.5 g/cc (500 kg/m^3) were laminated to two 1–1/2 oz glass mats to provide laminates of high flexural modulus (avg. 713,000 psi) (4917 MPa) and high flexural strength (16,500 psi) (4).

SILICONE FOAMS

Three types of silicone foams are available—premixed powders, room–temperature–curing resins, and elastomeric foams. The latter are discussed in Chapter 4 on Elastomeric Foams. The older type, *premixed silicone foaming powders*, is expanded by means of a blowing agent that decomposes into nitrogen gas and an alkaline by–product at about 300°C (572°F). The silicone resins used are solventless polysiloxanes with a melting point of 120–140°F (40–60°C). In the presence of an appropriate catalyst they become thermosetting through the condensation of hydroxyl groups. To make a foam, the powder containing the resin, blowing agent, and fillers is simply heated above 320°F (160°C). After the resin liquifies, the blowing agent decomposes. Nitrogen gas formed expands the resin, while the amines given off act as catalysts for the condensation of the resin. Expansion and gelation are in this way synchronized so that the resin gels at maximum expansion. There are three types of foaming powder, as follows:

Type A – can be foamed to densities 10–14 lb/ft^3 (160–224 kg/m^3)

Type B – can be foamed to densities 12–16 lb/ft^3 (192–256 kg/m^3)

Type C – can be foamed to densities 14–18 lb/ft^3 (224–288 kg/m^3)

Types A and B can be foamed in place. Type C can be foamed satisfactorily only as a block or sheet, but is stronger than the others, and retains more compressive strength at high temperatures. Foamed structures produced from Type A are the most resistant to thermal shock. Samples have been cycled repeatedly between room temperature and 600°F (316°C) for 72 hours with only slight dimensional change. The total loss of weight is less than 10% under these conditions. Similar to Type A in most respects, Type B retains a considerable amount of its compressive strength at elevated temperatures, especially if postcured for 48 hours at 480°F (249°C). In many applications this foam will cure further and become stronger with use. Molds for these silicone resins may be made of metal, wood, glass, etc. They do not require preheating (2).

The second and newest type of silicone foams, *room–tempera–ture–curing resins*, is based upon chemical reaction between two silicone components in the presence of a catalyst. The reaction is slightly exothermic, but temperatures reached seldom exceed 150°F (60°C), even in very large pours. Hydrogen gas is liberated as the expanding agent, but the quantity is small and has not presented any explosive hazards. The reaction is complete in 15 minutes, and maximum strength is developed in 24 hours. These materials are supplied in the form of two liquid components which are blended in a high–speed mixer for 30 seconds and then poured. Expansion of 7 to 10 times the initial volume is complete within 15 minutes, but the foam remains soft and tender for about 2 hours. After 10 hours the foam is hard enough to be cut and handled. Its density is usually between 3.5 and 4.5 lb/ft^3 (56 and 72 kg/m^3), depending on the geometry of the cavity. Finished foams of this type can be used continuously at 600–650°F (316–343°C). They have low thermal conductivities (0.28 BTU/ hr/°F/ft^2/in), good electrical–insulation properties, and low water absorption. The heat of reaction and the expansion pressure are so low that molds of cardboard or heavy paper can be used (2) (5).

The excellent high–temperature resistance of silicone foams has led to their use as molded components for aircraft and as insulating materials for electrical/electronic equipment (1).

UREA–FORMALDEHYDE (UF) FOAMS

Urea–formaldehyde foam is made by a condensation reaction in which the urea–formaldehyde resin is mixed with air, an aqueous detergent, and an acid catalyst. The foam is usually generated by mixing the air in with the resin and the catalyst in an application machine. Depending on the strength and amount of catalyst, the foam, which resembles a froth on emerging from the applicator gun, will set up, with no exothermic heat, in 10–120 seconds. Full hardness and complete drying, however, will take from several days to a month or more, depending on temperature, humidity, and permeability of the surrounding walls. Prior to setting up, the foam can be made to flow into crevices or around pipes, or it can be hand–trowelled to the desired shape or surface condition (3).

The finished foam is a material with an approximate density of 0.7 lb/ft^3 (11.2 kg/m^3), with very low compressive strength. The dried foam has a nominal K–factor of 0.2 BTU/hr/ft^2/°F/in. at 21°C (70°F) and

0.18 BTU/hr/ft^2/°F/in. at 2°C (35°F). When used at thicknesses of 1-1/4 to 2 inches (3.2 to 5.1 cm) or more it is reported to be a good sound insulator (sound transmission 40 decibels). One of its most interesting properties is its relative noncombustibility, since a temperature in excess of 650°C (1200F°) is required to cause the material to decompose. Upon removal of the source of ignition the material immediately stops burning or decomposition (3) (5).

Urea-formaldehyde apparently, at least in some formulations, can produce highly toxic combustion products compared to other cellular materials. The University of Utah found it only slightly less toxic than phenolic foam, although with nonflaming combustion (830°C) it was found to be the fastest material to incapacitate test animals. Cyanide was found to be the causative agent of death. CO, CO_2, and ammonia are also produced in combustion, all of which are toxic. University of Pittsburgh studies showed UF foam to be only slightly less toxic than PTFE (solid, not foam) (6).

Urea-formaldehyde foams are usually brittle structures with low compressive strength (under 50 psi or 0.34 MPa). The term "frangible" may be applied to them. They are open-cell, sponge-like foams that can absorb large quantities of water. These foams also exhibit thermal and acoustical insulating properties common to low-density foams. For example, their thermal conductivities range within the values quoted for polystyrene foam (0.24 – 0.33). This is the result of their low density and small cell size (5).

Foamed-in-place UF insulating foams must be injected into a cavity under pressure to guarantee a complete fill with minimum voids and cracks. Because of shrinkage during cure and drying UF foams lose insulation efficiency if not formulated properly. Since these foams are water-absorbing they lose their insulating efficiency at high humidities (5).

Urea-formaldehyde foams have been used to raise sunken vessels, contain fires in coal mines, and package lightweight objects. They are also being used for wound dressings since they aid in healing (1). Sterilization of plasticized UF foams make them useful as surgical sponges and dressings where absorbency and ventilation are required. The small amount of residual formaldehyde provides bactericidal properties and promotes healing. Urea-formaldehyde foams have been modified and used as replacements for wood pulp in absorbent bandages (5).

Applications other than thermal insulation and medical uses include: packaging insert cushions, cellular concrete, low-density cores for structural panels, buoyant filler materials (when encapsulated with barrier coatings or films), agricultural applications, intumescent coatings, and expanded adhesives. An application that has consumed a large volume of urea–formaldehyde foam in the U.S. is its use in cut–flower displays. A large amount of water is absorbed by the foam. Since the foam retains water for long periods of time without dripping or drainage the benefit to fresh–cut flower displays is obvious. For some reason, perhaps as a result of soluble nutrients in the presence of small amounts of formaldehyde, cut flowers are reported to remain fresher with urea–formaldehyde foam than is otherwise possible. European developers have concentrated on insulating foams and recently on agricultural applications with BASF's "Hygromull." In the latter case soil bacteria decompose the readily wettable foams to yield nutrient nitrogen. A combination of the two characteristics (bacterial action and affinity for water absorption) has resulted in UF foams being used to retain water in arid soils and to supply plant foods. The UF foams made by BASF and other glycol modified UF foams have been developed for these applications (5).

In recent years there has been considerable attention given to the problem of liberation of formaldehyde vapors from UF foam insulation applied within buildings. A large number of individuals have complained of becoming sick upon continued exposure to these systems. As a result its use for thermal insulation was banned in several states and the Consumer Products Safety Commission (CPSC) ruled against its use nationally. The current status of restrictions on the use of UF foams for building thermal insulation is not known by this writer. There are ways to minimize the likelihood of free formaldehyde being given off from UF foam insulations, and undoubtedly much of the problem has been caused by unqualified applicators who do not use proper controls (7).

Polyurea foams discussed in Chapter 2 above can be used for retrofitting of cavity walls.

POLYBENZIMIDAZOLE (PBI) FOAMS

Polybenzimidazoles are condensation products of diphenyl esters and aromatic diamines. They have good adhesive properties and high thermal and oxidative stability. They are used for matrix resins, adhesives, coatings and foams. Their fire resistance is excellent. The limiting oxygen index (LOI) of plastic forms of PBI is 41. Foaming occurs in these resins as a result of the liberation of phenol and water. They exhibit excellent resistance to ignition and form large amounts of char under appropriate conditions (6).

PBIs have been shown to have outstanding thermal resistance, and the polymerization reaction is such that foaming can be achieved, along with an increase in molecular weight. Cross–linking can also be introduced. For convenience in foaming a prepolymer approach is generally used. The reaction of diphenyl isophthalate with 3,3′–diaminobenzidene is an example. Such a prepolymer is a solid that can be handled as a powder. Heating under appropriate conditions will cause foaming, due to further liberation of phenol and water as blowing agents. In addition, the powder can be combined with reinforcing fibers, such as graphite, and with microballoons for further density control. Cross–linking can be obtained thermally, or by introducing higher functionality in the reactants (8).

PBI foams of 12–80 lb/ft^3 (192–1280 kg/m^3) density were prepared in 1968 by blending the prepolymer powder with carbon fiber and silica microballoons in a mold, which was then heated in a press at 120°C and 15 psi (0.10 MPa) for 30 minutes. Foaming occurred, but curing was not complete. An additional heating of about 0.5°C/min. to 315°C (599°F) and holding at that temperature for 2 hours gave a moderate degree of cure. Postcuring was accomplished by raising the temperature at 0.5 to 1°C/min. to 450°C (842°F) in an inert atmosphere and holding at that temperature for 2 hours. The foam was then cooled to 260°C (500°F) or less in an inert atmosphere. Density was presumably controlled largely by the volume of microballoons used (8).

PBI foams have the best thermal stability of the three foams—polyisocyanate, PBI and PI—but generally are less resistant to oxidation at elevated temperatures than are the PI (polyimide) foams. The PBI foams are the most expensive of the three types. So far their fabrication technology appears to be limited to the molding of small– to moderate– sized items (8).

POLYIMIDE FOAMS

These foams were discussed in Chapter 2 primarily from the point of view of their polymer chemistry. The discussion following will emphasize fabrication, properties, and applications.

Because of their excellent thermal and flame–resistance polyimides have been used for films, coatings, adhesives, molding compounds, and most recently, as *rigid foams*. These polymers have been foamed by two general methods, the first being the more conventional reaction of aromatic dianhydrides with aromatic diamines. The second method utilizes the reaction between dianydrides and polyisocyanates. Foams prepared from proprietary liquid dianhydride–diamine components give densities in the range of 6 to 20 lb/ft³ (96 to 320 kg/m³). These foams contain both open and closed cells and do not have as good thermal stability as foams prepared from solid polyamic acid prepolymers. The solid prepolymer is converted to foam by powdering, placing in a mold, and heating to foam and cure. Densities of 23 lb/ft³ (368 kg/m³) to 71.3 lb/ft³ (1141 kg/m³) were prepared in this way. The dielectric constants of these foams at RT were about 2–3, and tan δ was about 0.002 to 0.02 for the 34.5 – 69.3–lb/ft³ (552–1109 kg/m³) density range and 0.1– to 1000–kHz frequency range. Water–vapor permeability was low and the foams did not ignite when tested by the procedure in ASTM D 1692–59T (*Author's note*: This method has been discontinued) (8).

In the case of using aromatic polyisocyanate in making polyimide foams, crosslinking is increased by adding a highly functional polyol to react with part of the isocyanate so that the foam contains some urethane structure. A typical foam prepared in this way would have a density of 3.97 lb/ft³ (63.5 kg/m³) and a K–factor of 0.26 Btu/(h) (ft²) (°F/in). Foams of this type have been prepared in a range of densities of 2.5 to 18.5 lb/ft³ (40 to 296 kg/m³) with compressive strengths of 25 to 1340 psi (172 to 9239 kPa) (8).

Polyimide foams are formed by the liberation of either water or CO_2. They have outstanding thermal stability, low flammability, and high char formation. Structural stability ranges from –320° to 700°F (–196° to 371°C). They are nonigniting in air and produce little or no smoke in a propane–air flame. NASA recently undertook a program designed to develop ambient–curing, fire–retardant, nontoxic polyimide foams. In 1982 the Solar Division of International Harvester[1] reported on the development of an open–cell polyimide foam designed to withstand high temperatures and flame. The foam is resistant to open flame and is claimed to produce no toxic or detectable smoke emission. It is flexible and resilient at temperatures from –300° to 500°F (–184° to 260°C). The foam is also easier to cut, fit, or install for applications such as thermal insulation, acoustical damping, cushioning materials, and weight–sensitive structures. It is made in densities from 0.6 to 1 lb/ft³ (10 to 16 kg/m³). It is sold under the trade name SOLIMIDE and is used in applications such as insulation aboard the Space Shuttle Columbia and as cushion-

---

[1]   Later (1983) Imi–Tech Corp., Elk Grove Village, IL

ing/insulating material for electronics in naval vessels. The foams are produced in microwave ovens (6).

A 0.5 lb/ft³ (8 kg/m³) polyimide flexible foam developed by SOLAR Turbines International was evaluated for cushioning and flammability characteristics by the Air Force Packaging Evaluation Agency at Wright–Patterson AFB in Ohio in 1980. The results showed that the dynamic cushioning characteristics were equivalent to 0.5 lb/ft³ (8 kg/m³) polyurethane foam at 72°F (22°C), superior at −40°F (−40°C), and the polyimide foam was found to have improved reusability. When exposed to a flame of a Bunsen burner the foam was considered non–burning, with little smoke emission or odor. This project was carried out as part of a joint program with the Navy Logistics Engineering Group.

POLYPHOSPHAZENE FOAMS

These are inorganic phosphorus polymers—that is, polymers with no carbon in the backbone chain. The *phosphonitrilic polymers*, more exactly named *polyphosphazenes*, have the general formula:

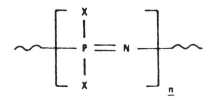

The cyclic oligomers, where n̲ is 3 or 4, have been most intensively investigated. When X is a strongly electronegative group such as Cl, F, or SCN, the cyclic oligomers may be converted to high polymers by heating at 200–300°C. The resulting materials have rubberlike properties and are either insoluble or difficultly soluble. The structure of this "inorganic rubber" is not yet known with certainty. Above 300°C a depolymerization takes place with formation of an equilibrium mixture of oligomers and polymers. Because of its hydrolytic instability the dichlorophosphazene polymer is not suitable for use as a plastic material. Phosphazene high polymers fall into two general categories: linear–type polymers and cyclo–matrix polymers. It is important to distinguish between these two classes because the physical and chemical properties of the two types are often strikingly different. Linear–type polyphospha-zenes are elastomeric, rubbery, or thermoplastic materials. Cross–linked cyclo–matrix polymers are usually rigid, high–melting, thermally stable materials.   Their structure consists of an interconnected lattice of phosphazene rings held together by bifunctional ligands. A few polymers with intermediate structure, i.e., cyclolinear polymers, are also known (9).

Polyphosphazene compounds have been used in a variety of flame–retardant applications.   The phosphazenes have great utility as flame retardants because of (1) the high percentage of phosphorus present, (2) the simultaneous presence of large amounts of nitrogen, and (3) the possibility of incorporation of halogens at the same time (10).

Poly(aryloxyphosphorus) elastomers have two different pendant groups attached to the polymer backbone in a non–regular fashion, as shown in Figure 5–1.

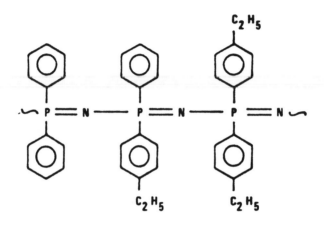

Figure 5.1: Non–regular substitution on phosphazene backbone.

Foams of poly(arylphosphazene) were prepared in a 1974 study by using a mixed–peroxide curing system and the blowing agent azodicarbonamide.   Precure was accomplished by confinement of the sample in the mold under pressure for up to 8 minutes at 230–260°F (110–127°C).   Final expansion and cure were accomplished in a forced–air oven for 5 to 60 minutes at 285–350F° (140–177°C) where a two–fold linear expansion was obtained. The resultant closed–cell foam had a density of 7 lb/ft^3 (112 kg/m^3) and was quite stable to thermal aging at 300°F (149°C) for several hundred hours while maintaining flexibility and density with little change in compression resistance.   Commercial foams for the same insulation application embrittled within 24 hours at 300°F. The limiting oxygen index (LOI) was 27 to 28 for unfilled and uncured homopolymers and copolymers. Halogenated compounds had LOIs of 38 to 65, with bromine more effective than chlorine.   Rigid and flexible foams are possible. The foams extinguish flame immediately upon removal of the flame source, and depending on the length of burn time, the foams largely retain their structure and develop a char with some durability. Poly(arylphosphazene) foams also have low smoke generation and low flame–spread–index values (10).

A 1980 Naval Research Laboratory report provides data on the toxic combustion products evolved by burning polyphosphazene foams, with and without fire retardants. Toxicants produced were CO, CO_2, 2–chlorobutane, 1–chlorobutane, benzene, toluene, and trichloroethylene (for foams not coated with fire–retardant paints) (11).

In 1981 Lieu *et al* at the University of Pittsburgh reported on studies comparing the flammability–toxicity of polyphosphazene and polyurethane foams, using the newly proposed Potential Hazard Index (PHI). In these studies polyphosphazene ranked less hazardous than polyurethane foams. Toxic gases evolved by polyphosphazene included CO, HCN, and NO_2, with relative proportions dependant on which formulation was under study (12).

SYNTACTIC FOAMS

Syntactic foams have been discussed in Chapter 2 on Thermosetting Foams. The discussion following is intended to cover points not made in Chapter 2.

Syntactic foams are made using a resin matrix to which has been added hollow spheres of various materials. The resultant product is a foamlike material made without the use of a blowing agent. The most common matrix resins are epoxies and polyesters, although urethanes, PVC plastisols, and phenolic resins have also been used. Indeed, any polymer that can be made liquid, either before final polymerization or by heat, can be used as the binding resin. In syntactic foams the resin matrix is the continuous phase and the hollow spheres the discontinuous phase.

The foam elements in syntactic foams may be glass *microspheres*, with diameters ranging from 20 to 200 microns, or other spheres with diameters ranging from 0.05 mm to 2 cm, which are classed as *macrospheres*. The materials used to form the spherical particles are glass, phenolic resins, silica, and naturally occurring materials such as perlite and coal dust. The most commonly used materials are glass and phenolic spheres. The syntactic foams are made by simply mixing the micro– or macrospheres into the catalyzed resin until the desired consistency is obtained. In most cases the materials are mixed to a patty–like state, or, if a casting material is desired, to a state in which the material can just be cast. The usual ratio of filler to resin is approximately 60% filler by volume (3).

Densities of syntactic foams range from 10 to 40 lb/ft^3 (160 to 640 kg/m^3). Syntactic foams are being used as core materials in sandwich structures in construction of aircraft and hulls and decks of boats, and are being investigated for use in roofing structures (2).

Syntactic silicone foams have also been formed. They cure at room temperature or at slightly elevated temperatures and become flexible insulators and ablators. They can be cut, trowelled, molded, or air–injected into place, and cured pieces can be bonded to substrates by using RTV silicone adhesives (13).

Glass microspheres are preferred to phenolic microspheres for deep–submergence applications. The glass microspheres can be coated to improve particle–to–resin adhesion and to enhance dispersibility. Glass bubbles also have the advantage of having considerably higher hydrostatic strength than other hollow spheres (5).

Currently syntactic foams used for deep–submergence applications range around 34 lb/ft^3 (544 kg/m^3), with a capability of extended service to 20,000–foot (6,562–meters) ocean depths where the hydrostatic force approaches 4.5 tons/in^2. Design criteria call for 30 lbs buoyancy per cubic foot at 20,000–foot depths. Syntactic foams of 44 lb/ft^3 (704 kg/m^3) have been used successfully at ocean depths as low as 36,000 feet (11,811 meters). Other typical applications of syntactic foams include providing buoyancy for deep–ocean–current metering devices, instrument packages for electronic gear, anti–submarine warfare units, cable buoys, offshore drilling operations, and any other deep–submergence uses requiring high–reliability buoyancy material. Also, some acoustical properties of syntactic foams are similar to those of sea water. Thus, the foam has found utility as "acoustic windows" (14).

In 1977 the National Materials Advisory Board (NMAB) of the National Academy of Sciences held a workshop on Acoustic Attenuation Materials Systems. The sponsoring agency was the Office of Naval Research, and the proceedings were published by the NMAB Committee on Structural Applications of Syntactic Foam (15).

REFERENCES

1.   Anonymous, *Foams—Edition 2*, Desk–Top Data Bank®, The International Plastics Selector, Inc., San Diego, CA, (1980).

2.   *Plastics Engineering Handbook of the Society of the Plastics Industry, Inc.*, 4th Edition, edited by J. Frados, Van Nostrand Reinhold, New York (1976). (*Author's note*: The 5th Edition, edited by M.L. Berins, was published in 1991).

3.   Schwartz, S.S. and Goodman, S.H., *Plastic Materials and Processes*, Van Nostrand Reinhold, New York (1982).

4.   Ahnemiller, J., "Process and Performance Optimization of Chemically Foamed Polyesters," *Plastics Engineering, 39* (2): 22–23 (February 1983).

5.   Benning, C.J., *Plastic Foams: The Physics and Chemistry of Product Performance and Process Technology*, Vol. 1: Chemistry and Physics of Foam Formation, Wiley–Interscience, New York (1969).

6.   Landrock, A.H., *Handbook of Plastics Flammability and Combustion Toxicology,* Noyes Publications, Park Ridge, New Jersey (1983).

7.   Staff written, "The Screws Are Off Urea Formaldehyde Foam," Industry and Market News, *Modern Plastics, 60* (6): 136, 138 (June 1983).

8.   Hardy, E.E. and Saunders, J.H. Chapter 14, *New High Temperature Resistant Plastic Foams*, Part II, edited by K.C. Frisch and J.H. Saunders, Marcel Dekker, New York (1973).

9.   Sander, M. and Allcock, H.R., "Phosphorus–Containing Polymers," articles in *Encyclopedia of Polymer Science and Technology,*, Vol. 10, pp. 123–144, edited by N.B. Bikales, Wiley–Interscience, (1969).

10. Sicka, R.W., Thompson, J.E., Reynard, K.A. and Rose, S.H., "Poly(aryloxyphosphazene) Foam Insulation Materials," pp. 244–249 in Proceedings, *6th National SAMPE Technical Conference, Vol. 6, Materials on the Move,* held at Dayton, OH, Oct. 8–10, (1974). (Work supported by the Naval Ship Systems Command (NAVSHIPS 03421) and The Firestone Tire and Rubber Co.).

11.   Eaton, H.G., Williams, F.W. and Tatem, P.E., "Combustion Products Evaluation from Hull Insulation Materials Coated with Fire Retardant Paints," Naval Research Laboratory, *Interim Report, NRL–8403*, Sept. 8, (1980). AD AO89973 (PLASTEC 38572).

12.   Lieu, P.J., Magill, J.M. and Alarie, Y.C., "Flammability–Toxicity Ratings of Some Polyphosphazene and Polyurethane Foams," *Journal of Combustion Toxicology, 8*:   282–259 (November 1981).

13.   Vincent, M.L., Chapter 7, "Silicone Foams," in *Plastic Foams*, Part I, edited by K.C. Frisch and J.H. Saunders, Marcel Dekker, New York (1972).

14.   Johnson, R.W., "Here's What's Happening in High Performance Syntactic Foams," *Plastics Design and Processing, 13* (4): 14–17 (April 1973).

15.   National Materials Advisory Board, Committee on Structural Applications of Syntactic Foam, Proceedings of the Workshop on Acoustic Attenuation Materials Systems, Report NMAB–339, (1978) (Distribution unlimited) (PLASTEC 29483).

6

SOLVENT CEMENTING AND ADHESIVE BONDING OF FOAMS

Arthur H. Landrock

INTRODUCTION

The information given in this brief chapter covers plastic foams, both thermoplastic and thermosetting, and elastomeric foams, although very little discussion is given in the latter subject.

SOLVENT CEMENTING

Solvent cementing is a process in which thermoplastics are softened by the application of a suitable solvent or mixture of solvents and then pressed together to effect a bond. The resin itself, after evaporation of the solvent or solvents, acts as the adhesive. Many thermoplastic resins are more amenable to solvent cementing than to conventional adhesive bonding (1). In many cases mixtures of solvents give better results than the individual solvents. With solid plastics, if evaporation rates are too fast due to excessive volatility of the solvent, crazing or blushing often occurs. This is not a problem with cellular materials, however. Often small amounts (1–7%) of the substrate to be bonded are dissolved into the solvent to aid in gap filling and to accelerate curing. Solvents for cementing may be brushed, sprayed, or the foam surfaces may be dipped in the solvent. The foam parts should be held firmly in place during cure.

Most thermoplastic foams can be solvent cemented. However, some solvent cements will collapse thermoplastic foams. The best way to determine if such a problem exists is to try it. In cases where the foam collapses due to softening of the foam cell walls it is desirable to use water–based adhesives based on SBR or polyvinyl acetate, or 100%–solids adhesives. In general, the relatively amorphous thermoplastics, such as the cellulosics, polycarbonate, and polystyrene are easier to solvent cement than the crystalline materials, but there are exceptions.

When two dissimilar plastic foams are to be joined, which is rarely done, adhesive bonding is generally preferable because of solvent and polymer incompatibility problems. Solvents used to cement plastics should be chosen with approximately the same solubility parameter (δ) as the plastic to be bonded. The solubility parameter is the square root of the cohesive energy density (CED) of the liquid solvent or polymer. CEDs of organic chemicals are primarily derived from the heat of vaporization and molecular volume of the molecules, and are expressed as calories per cubic centimeter (cal/cm^3). Literature sources provide data on δ's of a number of plastics and resins (2) (3) (4).

Thermoplastic Foam Substrates

Cellular Cellulose Acetate: A number of solvents can be used by themselves in cementing cellulose acetate. The following formulations involve mixtures of several solvents, including bodying resins (dope–type cements) (5).

- cellulose acetate, 18% by wt.
  acetone, 55% by wt.
  methyl CELLOSOLVE, 20% by wt.
  methyl CELLOSOLVE acetate, 7% by wt.

- cellulose acetate, 15% by wt.
  methyl CELLOSOLVE, 25% by wt.
  methyl ethyl ketone, 60% by wt.

- cellulose acetate, 10% by wt.
  acetone, 60% by wt.
  methyl CELLOSOLVE, 30% by wt.
  (This is a good "dope" for *quick bonding*.)

Acrylonitrile – Butadiene – Styrene (ABS): ABS is conventionally bonded with a dope cement containing 15–25% ABS resin in a blend of ketone solvents, e.g., acetone, methyl ethyl ketone, and methyl isobutyl ketone, or tetrahydrofuran or methylene chloride (5).

Acetal Homopolymer (DELRIN®): In general, solvent cements are relatively ineffective with DELRIN®, a highly crystalline polymer. Methylene chloride, methylene chloride mixed with ethylene chloride, and DELRIN® in methylene chloride have been used, however (6).

Acetal Copolymer (CELCON®): As with DELRIN®, this is a highly crystalline polymer with excellent solvent resistance. It is somewhat more amenable to solvent cementing than the homopolymer, however. The solvent cement recommended by the manufacturer, Hoechst Celanese, is *hexafluoroacetone sesquihydrate*, available from Allied Signal, Inc. This solvent is a severe eye and skin irritant, however, and should be handled with care (7).

Polyvinyl Chloride (PVC): A formulation that has worked satisfactorily in both rigid and flexible PVC is as follows (5):

	Parts by Weight
PVC resin (medium mol. wt.)	100
tetrahydrofuran	100
methyl ethyl ketone	200
organic tin stabilizer	1.5
dioctyl phthalate (plasticizer)	20
methyl isobutyl ketone	25

Care must be taken in handling this formulation because of the slightly toxic nature of the tetrahydrofuran. The resin solids content of this formulation is over 22%. So it is a heavy–bodied cement.

Polycarbonate: Solvent cementing is the most common method of bonding polycarbonate. Bonding can be carried out with specific solvents, mixtures of solvents, and mixtures of polycarbonate and solvents. Methylene chloride, when used by itself, has an extremely fast evaporation rate and is recommended for fast assembly of polycarbonate parts. A solution of 1–5% polycarbonate resin in methylene chloride has a decreased evaporation rate. Parts bonded with methylene chloride are usable at elevated temperatures after approximately 48 hours, depending on the bonding area. Ethylene dichloride is also used (5).

Polystyrene: Polystyrene may be bonded to itself by solvent cementing, conventional adhesive bonding, thermal welding, spin welding, ultrasonic welding, or electromagnetic bonding.   Solvent cementing is the most effective method, however. Solvents recommended are (8) (9) (10):

methylene chloride	ethylene dichloride
ethyl acetate	trichloroethylene
methyl ethyl ketone	

All these are fast–drying (20 seconds or less) solvents, and in solid plastics would present a crazing problem. However, in foamed plastics this is not a problem. *Aromatic* hydrocarbon solvents, such as toluene and xylene, should not be used since they would cause a collapse of the cellular material.

Polysulfone: Polysulfone can be solvent cemented with chlorinated hydrocarbons. A solution of 5% resin in methylene chloride can be used to bond polysulfone to itself. High pressures (500 psi) are required for *solid plastics* (5).

Modified Polyphenylene Oxide (NORYL®): Solvent cementing is by far the simplest and most economical method of joining this material to itself. The resin can be readily softened and dissolved by some aromatic and chlorinated hydrocarbons. The latter are preferred because of their better solubility and stronger bond formation. Solvent–cemented parts of NORYL® are less sensitive to thermal shock than adhesive–bonded joints. The bonds are unaffected during extended aging or prolonged exposure to steam. The solvent should be applied in a thin uniform layer and the parts should be rapidly positioned and clamped. This material requires very little solvent for softening the surface to be bonded. Excess soaking is undesirable. The best results are obtained by applying the solvent to only one mating surface. In solid materials a four–minute holding time in the clamp at 400 psi is recommended (5).

The following solvents and solvent combinations are recommended:

> methylene chloride
> chloroform/carbon tetrachloride (95/5)
> chloroform + 2% NORYL®
> ethylene dichloride
> xylene/methyl isobutyl ketone (25/75)
> trichloroethylene with 1–7% NORYL® for bodying
> methylene chloride/trichloroethylene (85/15)

Polybutylene Terephthalate (PBT): Solvents recommended by General Electric for their VALOX® thermoplastic polyester are (11):

> hexafluoroisopropanol
> hexafluoroacetone sesquihydrate

The solvent is brushed on the mating surface and dried under pressure. These solvents are toxic and should be applied only in areas of positive ventilation.

Polyetherimide (ULTEM®):   Methylene chloride, with or without a 1–5% solution of ULTEM®, is recommended.   Moderate pressures of 100–600 psi for 1–5 minutes are required (12).

ADHESIVE BONDING

Adhesive bonding of plastic and rubber foams is used where it is not possible to use solvents or thermal means of dissolving or melting both adherend surfaces. Examples of such bonding are: polystyrene to metal, polycarbonate to phenolic, polyethylene to itself (no solvent can be used to solvent cement polyethylene because it is highly insoluble). Before attempting to adhesive bond any material attention should be paid to the problem of adherend surface preparation. If the surfaces are not properly prepared, usually by solvent cleaning, chemical oxidation, and/or roughening, the bonds will not be durable.   There are a number of sources of information in this area. The details of individual procedures are too lengthy and complex to be presented here.

Thermoplastic Foam Substrates

Some solvent cements and solvent–containing adhesives, such as pressure–sensitive adhesives (PSAs), will collapse thermoplastic foams by dissolving the cell walls. In such cases water–based adhesives based on SBR or polyvinyl acetate, or 100%–solids adhesives are often used (1).

Acetal Copolymer (CELCON®): Two types of non–solvent adhesives are used, structural and non–structural.   Most structural adhesives are based on thermoset resins and require the use of a catalyst and/or heat to cure. This type of adhesive is normally used in applications which require maximum bond strength and minimum creep of the adhesive joint under sustained loading. Many structural adhesives can be used continuously at temperatures up to 350°F, which is higher than the recommended continuous–use temperature of the copolymer. Structural adhesives recommended are:

> epoxy (to 160F°)
> polyester with isocyanate curing agent (to 200°F)
> polyvinyl butyrate, modified with a thermosetting
>   phenolic (to 250°F)
> cyanoacrylate (to 181°F)

Non–structural adhesives used on CELCON® are usually one–component, room–temperature–curing systems based either on thermoplastic resins or elastomeric materials dispersed in solvents. They are normally used in applications which will not be exposed to temperatures over 180°F. Neoprene rubber adhesives are examples. (7).

Acetal Homopolymer (DELRIN®): Adhesives recommended for this material include (6):

> polyester with isocyanate curing agent
> rubber–based adhesives
> phenolics
> epoxies
> modified epoxies
> vinyls
> resorcinol
> vinyl–phenolic
> ethylene vinyl acetate
> cyanoacrylate
> polyurethane

Acrylonitrile–Butadiene–Styrene   (ABS):   Conventional adhesives recommended include epoxies, urethanes, thermosetting acrylics, elastomers, vinyls, nitrile–phenolics, and cyanoacrylates (8) (13).

Cellular Cellulose Acetate: Conventional adhesives recommended include polyurethanes, synthetic resins, thermoplastics, resorcinol–formaldehyde, nitrile–phenolic, and rubber–based materials (8).

Polyvinyl Chloride (PVC): With PVC plasticizer migration to the adhesive bond line can cause difficulties, especially in the softer, highly plasticized materials. Adhesives must be tested for their ability to resist the plasticizer. *Nitrile–rubber adhesives* are resistant to plasticizers. *Polyurethanes* and *neoprenes* are also used. Even rigid PVC contains up to 5% plasticizer. Most vinyl materials are fairly easy to bond with elastomeric adhesives. The highest bond strengths for rigid or semi–rigid PVCs are obtained with two–component, RT–curing *epoxies*. Other adhesives recommended for rigid PVCs are (8):

polyurethanes	nitrile–rubber phenolic
modified acrylics	polyisobutylene rubber
silicone elastomers	nitrile rubber
anaerobic structural adhesives	neoprene rubber
polyester–polyisocyanates	epoxy polyamide
polymethylmethacrylate	polyvinyl acetate

Polycarbonate: Conventional adhesives recommended include epoxies, modified epoxies, polyurethanes, acrylics, RTV silicones, cyanoacrylates, one–part elastomers, some epoxy–polyamides, and hot melts (13).

Modified Polyphenylene Oxide (NORYL®): Conventional adhesives recommended include epoxies, polysulfide–epoxies, silicone, synthetic rubber, acrylics, cyanoacrylates, and hot melts (14).

Polystyrene: Although polystyrene is usually bonded by solvent cementing, it can be bonded with vinyl acetate/vinyl chloride solution adhesives, acrylics, polyurethanes, unsaturated polyesters, epoxies, urea–formaldehyde, rubber–base adhesives, polyamide (Versamid–base), poly–methylmethacrylate, and cyanoacrylates. The adhesives should be medium–to–heavy viscosity and room–temperature and contact–pressure curing. An excellent source is a Monsanto Company technical informa–tion bulletin which recommends particular commercial adhesives for bonding polystyrene to a number of different surfaces. Adhesives are recommended in the fast–, medium–, and slow–setting ranges (10).

Polyethylene and Polypropylene: Acceptable bonds have been obtained between treated polyolefin surfaces with polar adhesives, such as epoxies, or solvent cements containing synthetic rubber or phenolic resin. The solvent adhesives are applied to both surfaces and the solvents allowed to evaporate before the parts are joined. Recommended epoxies are the anhydride–cured and amine–cured types. Also suitable is a two–component, polyamide–modified epoxy compound. Other adhesives that provide adequate bond strength to treated polyolefins include styrene–unsaturated polyester and solvent–type nitrile–phenolic (15).

Ionomer: Adhesives recommended for duPont's SURLYN® ionomer are epoxies and polyurethanes.

Nylons (Polyamides): There are a number of nylon types, but the most important and most widely used is nylon 6/6. The best adhesives for bonding nylon to nylon are generally solvents. Various commercial adhesives, especially those based on phenol–formaldehyde (phenolics) and epoxy resins, are sometimes used for bonding nylon to nylon, although they are usually considered inferior to the solvent type because they result in a brittle joint. Adhesives recommended include nylon–phenolic, phenolic–nitriles, nitriles, neoprene, modified epoxy, cyanoacrylate, modified phenolic, resorcinol–formaldehyde, and polyurethane. Bonds in the range of 250–1000 psi (1.7–6.9 MPa) have been obtained with solid nylons (8) (13).

Polyetherimide: A wide variety of commercially available adhesives can be used in bonding polyetherimide to itself or to dissimilar materials. Among these are polyurethane [(cure at RT to 302°F (150°C)], RTV silicones, hot melts (polyamide types) curing at 401°C (205°C) and epoxies (non–amine type, two–part) (12).

Polybutylene Terephthalate (PBT):   Commercial adhesives recommended include modified epoxies, cyanoacrylates, acrylics, polyurethanes, silicone, and polyesters.

Polysulfone: A number of adhesives have been found useful for joining polysufone to itself or to other materials.   These include 3M Company's EC 880 solvent–base adhesive, EC 2216 room–temperature–curing epoxy two–part paste, Bloomingdale Division, American Cyana-mid Company BR–92 modified epoxy with DICY curing agent, or curing agent "Z" (both spreadable pastes), vinyl–phenolics, epoxy–nylons, epoxies, polyimide, rubber–based adhesives, styrene polyesters, resorcin-ol–formaldehyde, polyurethanes, and cyanoacrylates. The EC 880, EC 2216, and the two BR–92 adhesives are recommended by the polysulfone manufacturer, Union Carbide (16) (17).

Thermosetting Foam Substrates: Most thermosetting plastics are not particularly difficult to bond. Obviously, solvent cementing is not suitable for bonding thermosets to themselves, since they are not soluble. In some cases solvent solutions can be used to join thermoplastics to thermosets. In general, adhesive bonding is the only practical method of joining a thermoset to itself or to a non–plastic material. Epoxies or modified epoxies are the most widely used adhesives for thermosets (1).

Polyurethanes: Rigid urethane foam can be adhered to most materials through the use of adhesives. In this manner the foam can be bonded to glass, metals, gypsum board, plastics, paper, wood and brick. Hot melts (not over 250°F melting temperature) may be used, along with solvent contact adhesives which can be flash–dried to permit rapid bonding of the foam. Pressure–sensitive rubber emulsion adhesives may also be used, but they have the disadvantage of being considerably slower–drying than the rubber solvent contact adhesives. Solvents in solvent contact adhesives must be removed by evaporation before the adhesive joint is completed because they cannot evaporate through the foam. For high–temperature applications thermosetting adhesives, such as polyesters or epoxies, must be used. Other thermosetting adhesives used include phenolics, resorcinol, resorcinol–phenolic, internal–setting asphalt, mortar modified with additive, and mineral cements. Flexible polyurethane foams can be bonded with butyl, nitrile and polyurethane adhesives (18).

Epoxies: For maximum adhesion to epoxies, epoxy adhesives with primers are used. In general, the curing temperature should be as high as possible, allowing for the distortion of temperature. In this way maximum strength and heat resistance are obtained. Fast–bonding nitrile–phenolics with curing resins can give excellent bonds if cured under pressure at temperatures of 300°F. Bonds of lower strength can be obtained with most rubber–based adhesives. Other adhesives used are: modified acrylics, polyesters, resorcinol–formaldehyde, phenol–formaldehyde, polyvinyl formal–phenolic, polyisobutylene rubber, polyurethane rubber, neoprene rubber, melamine–formaldehyde, silicone and cyanoacrylates (1).

Polyester: Adhesives used include neoprene or nitrile–phenolic, epoxy, epoxy–polyamide, phenolic–epoxies, polyesters, modified acrylics, cyanoacrylates, polyurethanes, butyl rubber, polyisobutylene, neoprene, and polymethylmethacrylate (1).

Phenolic: Adhesives recommended are neoprene and urethane elastomer, epoxy and modified epoxy, phenolic–polyvinyl butyral, nitrile–phenolic, polyester, cyanoacrylates, polyurethanes, resorcinols, modified acrylics, polyvinyl acetate, and urea–formaldehyde (1).

Silicone: Silicone resins are generally bonded with silicone adhesives, either silicone rubber or silicone resin (1).

Urea–Formaldehyde: Adhesives used are epoxies, nitrile–phenolics, phenol–formaldehyde, resorcinol–formaldehyde, furan, polyesters, butadiene–nitrile rubber, neoprene, cyanoacrylate, and phenolic–polyvinyl butyral (1).

Syntactic Foams: Adhesives should be selected based on the resin matrix, which is usually *epoxy* or *phenolic* (18).

REFERENCES

1.   Landrock, A.H., *Adhesives Technology Handbook*, Noyes

Publications, Park Ridge, New Jersey (1985).

2.   Skeist, I., "Choosing Adhesives for Plastics," *Modern Plastics, 33* (9): 121–130, 136 (May 1956).

3.   Miron, J. and Skeist, I., Chapter 41, "Bonding Plastics," *Handbook of Adhesives,* 2nd Edition, ed. I. Skeist, Van Nostrand Reinhold, New York (1977).

4.   Barton, A.F., CRC *Handbook of Solubility Parameters and Other Cohesion Parameters*, CRC Press, Boca Raton, Florida (1983).

5.   Landrock, A.H., Effects of Varying Processing Parameters in the Fabrication of Adhesive Bonded Joints, Part XVIII, "Adhesive Bonding and Related Joining Methods for Structural Plastics—Literature Survey," *Picatinny Arsenal Technical Report 4424* (November 1972).

6.   E.I. DuPont de Nemours & Co., Plastics Dept., Wilmington, Delaware, *DELRIN® Acetal Resins Design Handbook*, A–67041 (1967).

7.   Celanese Plastics Co., Chatham, NJ, *The CELCON® Acetal Copolymer Design Manual*, (undated).

8.   Petrie, E.M., Chapter 10, "Plastics and Elastomers as Adhesives," *Handbook of Plastics and Elastomers*, edited by C.A. Harper, McGraw–Hill, New York (1975).

9.   Chapter 27, "Joining and Assembling Plastics," *Plastics Engineering Handbook of the Society of the Plastics Industry, Inc.* 4th Edition, ed. by J. Frados, Van Nostrand Reinhold, New York (1976).

10.   Monsanto Plastics and Resins Co., St. Louis, MO, *Fabrication Techniques for LUSTREX® Styrene Plastics*, Technical Bulletin 6422, 8–940–0277–6 (undated).

11.   General Electric Company, Plastic Operations, Pittsfield, Massachusetts, *VALOX® Resin Design Guide* (undated).

12.   General Electric Company, Pittsfield, Massachusetts. *The Comprehensive Guide to Material Properties, Design, Processing and Secondary Operations, ULTEM® Polyetherimide Resin,* ULT 20 (undated).

13.   Cagle, C.V., Chapter 19, "Bonding Plastic Materials," *Handbook of Adhesive Bonding,* ed. by C.V. Cagle, McGraw Hill, New York (1973).

14.   General Electric Company, Plastics Dept., Selkirk, New York, *NORYL® Resin Design Guide*, CDX–83.

15.   Anonymous, "3 Prime Factors in Adhesive Bonding of Plastics," *Plastics Design and Processing*, 8(6): 10–22 (June 1968).

16.   Union Carbide Plastics, Polysulfone News No. 10, *Adhesives for Polysulfone,* BA–107–17, (undated).

17.   Union Carbide Plastics, *UDEL® Design Engineering Data Handbook*, Section 6 – Fabrication, p. 32 Adhesive Bonding (undated).

18.   Landrock, A.H., *Handbook of Plastic Foams, PLASTEC Report R52*, (February 1985).   Available from National Technical Information Service (NTIS) as AD A156 758.

7

ADDITIVES, FILLERS AND REINFORCEMENTS

Arthur H. Landrock

INTRODUCTION

Additives are substances added to other substances. In the plastics industry the term is most often employed for materials added in relatively small amounts to basic resins or compounds to alter their properties (1). According to this definition, blowing agents (foaming agents) would not be additives, since there would be no plastic foam without them (except in the case of syntactic foams). Blowing agents are, however, covered in this chapter. They are of particular interest currently because of their serious effects on the environment.

The discussion in this chapter does not cover all types of additives used in plastic foams. The following terms not discussed in any detail are defined as follows:

Accelerator (promoter) – a substance used in small proportion to increase the reaction rate of a chemical system (reactants plus other additives) (2).

Activator – a substance used in small proportion to increase the effectiveness of an accelerator (2). A material that speeds up a reaction in unison with a catalyst; an activator often starts the action of a blowing agent; used almost synonymously with *accelerator* or *initiator* (3).

Antioxidant – a substance used to retard deterioration caused by oxidation (2).

Processing aids – additives such as viscosity depressants, mold–release agents, emulsifiers, lubricants, and anti–blocking agents (3).

The topics to be covered in some detail in this chapter are: antistats, blowing agents, catalysts, fire retardants, mold–release agents, nucleating agents, reinforcements, stabilizers, and surfactants. These topics are presently in alphabetical order as a matter of convenience. The reader should be aware that there are a number of additives used in plastic foams that serve dual functions. These will be noted in the following text.

ANTISTATS (ANTISTATIC AGENTS)

Antistats are chemicals which impart a slight to moderate degree of electrical conductivity to normally insulative plastics compounds, thereby preventing the build–up of electrostatic charges on finished items. Antistats may be incorporated in the materials before molding. These materials function either by being inherently conductive, or by absorbing moisture from the atmosphere. Examples of antistatic additives include the following (1) (4):

- long–chain aliphatic amines and amides
- phosphate esters
- quaternary ammonium salts
- polyethylene glycols
- polyethylene glycol esters
- ethoxylated long–chain aliphatic amines
- glycerine
- polyols

Plastics compounders are generally more interested in using *internal antistats* rather than external applications. There are two types of *internal antistats* — conductive fillers (carbon black, carbon fibers, metals) compounded into the resins to form a conductive path, and the other type which is a material that, with limited compatibility in the resin matrix, migrates to the surface. There its hydrophilic group attracts ambient moisture, providing a path for dissipating the electrostatic charge. In a few instances *cationic surfactants* function as antistats by providing ions at the surface, rather than by exercising hygroscopicity. Some antistats may also function as lubricants, reducing the surface friction that builds up the electrostatic charge (5).

A material used as antistat for urethane foams is reported to also reduce corrosion risk, and *neoalkoxy titanates* and *zirconates* have been found to be effective antistats for polyolefins, polystyrene, and polyesters (6).

In September 1991 Statikil, Inc., Akron, OH announced the reformulation of its Statikil antistatic agents, which no longer contain the ozone-depleting 1,1,1-trichloroethane (6).

BLOWING AGENTS (FOAMING AGENTS)

General

Blowing agents are the particular agent which cause plastics to foam. There are two types in common use:

1. gases introduced into the molten or liquid plastic material.

2. chemicals incorporated in the plastic which, at a given temperature, decompose to liberate gas.

In either case, the gas, if evenly dispersed, expands to form the cells in the plastic. There are a number of different ways to bring about the formation of cells, depending on the gas being used, the chemical blowing agent (CBA) the type of plastic resin, and/or the particular process being used (7).

One of the desirable attributes of foam blowing agents is a low K-factor, referring to the thermal-insulating properties of the plastic foam. The K-factor indicates the thermal conductivity of the foam in BTUs per hour, per square foot, per inch of thickness, under a thermal difference of 1°F. In general, plastic foams have K-factors ranging from 0.15 to 0.35 (0.02 to 0.05 W/m·K) at room temperature. As the temperature increases the K-factor increases. When test temperatures are not stated it is assumed the K-factor refers to room-temperature conditions. Foams with lower K-factors have superior thermal-insulating properties. Another method of classifying foams is in accordance with the R-factor, widely used in the refrigerator industry today. R indicates the resistance of the material to the transmission of heat. R is the reciprocal of K (R=1/K). Thus, the higher the R-factor the better the insulating properties of the plastic foam (7).

General Production Methods for Blowing Foams

Plastic foams are generally made using any of seven different blowing methods (7):

- Incorporating a chemical blowing agent (CBA) into the polymer to form a gas by decomposition at an elevated temperature. These CBAs are usually in the form of fine powders that can be evenly dispersed in either a liquid resin, or mixed with molding pellets. The blowing gas evolved is usually *nitrogen* liberated from organic materials called azo compounds. A typical CBA is *azodicarbonamide*, also called azobisformamide (ABFA). CBAs are available which decompose at temperatures from 100°C (230°F) to as high as 280°C (537°F). A CBA is available to match any polymer melting point or processing temperature desired.

- Injecting a gas, usually nitrogen, into a molten or partially cured resin. The gas may be injected into the resin, either in the barrel of an extruder or injection press, or into a large mass in an autoclave. In either case, when the pressure is decreased, the gas expands and forms the cellular structure.

- A bifunctional material, such as an isocyanate, may be combined with a polyester or other liquid polymer. During the polymerization reaction to form a solid polymer the isocyanate also reacts to liberate a gas which forms the cells. This is the basis of some *polyur-ethane foam* techniques.

- Volatilization of a low-boiling liquid, either by the heat liberated by an exothermic reaction, or by externally applied heat. Commonly used liquids are chlorofluoro-carbons (CFCs). This is the most widely used technique in the production of *rigid polyurethane foams*. However, due to the ozone depletion problem in the stratosphere, they must be phased out and industry is presently searching for alternative blowing agents.

- Whipping air into a colloidal–resin suspension and then gelling the porous mass. This is how *foamed latex rubber* is made.

- Incorporating a nonchemical, gas–liberating agent into the resin mix. When heated, the mix then releases a gas. This material might be a gas adsorbed onto the surface of finely divided carbons.

- Expansion of small beads of a thermoplastic resin by heating an internally controlled blowing agent, such as pentane. This technique is used to expand polystyrene beads used in making plastic cups, packaging, and mannequin heads.

Chemical Blowing Agents (CBAs)

These agents are solid compounds (usually powders), but occasionally liquids, that decompose at processing temperatures to evolve the gas that forms the cellular structure. The most important selection criterion is the *decomposition–temperature range*, which must be matched to the processing temperature of the polymer being used. The decomposition reaction of the CBA must take place when the polymer is at the proper melt viscosity or degree of cure. Activators that can lower the blowing agent's decomposition temperature are available, thus affording greater flexibility to the formulator. It is also necessary to consider the amount of gas being liberated and the type of gas (and how it can affect the end product).

CBAs can be used in almost any thermoplastic and can be either inorganic or organic. The most common CBA is *sodium bicarbonate*, but its use is limited in plastics because its decomposition cannot be controlled as can the organic CBAs. The following are the most popular organic CBAs for plastics usage (8):

- ABFA, azodicarbonamide, or 1,1–azobisformamide. Widely used for foaming HDPE, PP, HIPS, PVC, EVA, acetal, acrylic, and PPO–based plastics. Decomposes at 400°–415°F (204°–213°C). Non–plateout grades are available to eliminate formation of cyanuric acid, which can attack molds.

- OBSH, p,p'–oxybis(benzenesulfonyl hydrazide). Relatively low–temperature decomposition at 315°–320°F

(157°–160°C).   Commonly used in LDPE, EVA and PVC.

- TSSC, p–toluene sulfonyl semicarbazide. Intermediate high–temperature decomposition at 442°–456°F (228°–236°C). Used with HDPE, PP, ABS, HIPS, rigid PVC, nylon, and modified PPO.

- THT, trihydrazine triazine. Can be used at high processing temperatures, (527°F or 275°C). High exothermic decomposition results in fine, uniform cell structure and good surface appearance, like ABFA. Ammonia–generating, which may present problems.

- 5–PT, 5–phenyltetrazole. Efficient, decomposes at 460°–480°F (238°–249°C). Decomposition gases are almost all nitrogen. Used with ABS, nylon, PC, thermoplastic polyester, and other high–temperature resistant plastics.

- A high–temperature cyclic peroxyketal peroxide cross-linking agent for polyethylene has been found to function as a blowing agent as well. This is another example of dual–function additives. Activated by thiodipropionate antioxidants, it evolves CO_2 and should be useful in making crosslinked polyethylene foams.

Physical Blowing Agents

This group changes from one form to another during processing (from liquid to gas, for example) (8):

> *Compressed gases*—Most common gases used are nitrogen, air and carbon dioxide. These gases are dissolved under pressure in the resin and produce foam upon release of the pressure. The use of nitrogen in injection–molded foam products is typical. The nitrogen is injected under high pressure. When the pressure is relieved the gas becomes less soluble in the polymer and forms cells.

> *Volatile liquids*—These foam the resin as they change from a liquid state to a gaseous state at the high temperature of processing. The most important materials in this category are fluorinated aliphatic hydrocarbons (chloro-

fluorocarbons or chlorofluoromethanes). These blowing agents have been used extensively in both rigid and flexible polyurethane foams. They can also be used in polystyrene, PVC and phenolic foams.

Flexible polyurethane foams are blown with water, methylene chloride, and chlorofluorocarbons (CFCs). Carbon dioxide from the water/isocyanate reaction functions as the blowing agent. The methylene chloride and CFCs assist in the blowing and contribute properties such as added *softness* and *resilience*. The CFCs also contribute to the *insulation* properties of rigid urethane foams.

Chlorofluorocarbon Liquids (CFCs)

The major advantage of these agents is that they become gaseous at well–defined temperatures and controlled rates, providing product quality and contributing to some improved performance characteristics. However, the effect of CFCs on the environment is under debate. These liquids, odorless and innocuous as they are, are linked to the ozone hole in the stratosphere. The industry is searching for feasible, environmentally and economically acceptable alternatives. Production levels of CFCs have been frozen and gradual phase–out is underway (8).

The earliest polyurethane foams were water (CO_2) blown. In the late 1950s CFC–11 was discovered to be an excellent blowing agent for polyurethane foams, especially low–density foams. The development of the Ozone Depletion Theory in the late 1970s and its further refinement in the 1980s linked CFCs to a reduction of ozone in the upper atmosphere. As a result of the concern of such ozone reduction causing an increase in ultraviolet (UV) radiation at ground level the world community produced the "Montreal Protocol on Substances that Deplete the Ozone Layer" in late 1987 (9).

Up to the present time, many communities and nations are accelerating the phase–out of CFCs by shortening the original timetable of the Montreal Protocol and taxing the use of CFCs. Currently the use of CFCs is limited to 1986–usage levels. It is hoped that two of the major candidates to replace CFC–11, HCFC–141b and HCFC–123, will be fully commercialized by 1993 (9)

At the Polyurethanes World Congress in Nice, France in 1991 it was reported that the Montreal Protocol was approved by 93 nations in June 1990, and that suppliers have been scrambling to meet its mandate of complete phase–out of CFCs by the year 2000. It was brought out at this Congress that HCFCs appear to be the most promising replacement for CFCs in rigid polyurethane foams. The most promising HCFCs were thought to be HCFC–141b (dichlorofluoroethane), HCFC–123 (dichloro-trifluoroethane), and HCFC–134a (tetrafluoroethane). However, these compounds are only stopgaps because of their chlorine content. For this reason a number of different agents are being tested as alternatives to chlorine–based blowing agents (10).

It was brought out at the Nice congress that there is a problem of compatibility of blowing agents with refrigerator–liner materials, commonly ABS and high–impact polystyrene (HIPS). Certain blowing agents cannot be used without causing stress cracking of the liners. So far HCFC–141b and HCFC–123 are the best of the CFCs from the point of view of refrigerator–liner compatibility. Many foam suppliers feel that carbon dioxide (CO_2) is the alternative blowing agent that will ultimately be most widely used in rigid polyurethane foams. However, some attendees felt that CO_2 is not suitable for refrigerator liners because of its detrimental effect on the foam's K–factor. But HCFC–123 and HCFC–141b also have a negative effect on K–factor (10).

It was reported by the *New York Times* in October 1991 (11), that the ozone layer in the Antarctic stratosphere was measured as 110 Dobson units, compared with the normal value of about 500 Dobson units. Dobson units measure the atmosphere's ability to absorb and block certain wavelengths of light coming from the Sun, notably ultraviolet (UV) radiation. The low value of 110 Dobson units was the lowest ever recorded in 13 years of data collection by the TOMS instrument. Seasonal ozone holes are signs of a worldwide depletion of stratospheric ozone. Public health experts fear that the increasing intensity of UV radiation that now penetrates the atmosphere may greatly increase the incidence of skin cancer and cataracts, and could significantly diminish the output of global crops and the marine food chain (11).

Evidence has been rapidly accumulating since the late 1980s that the main cause of *stratospheric ozone depletion* has been the presence of chlorofluorocarbon (CFC) chemicals released into the air by human activity. These substances are widely used as refrigerants, solvents and foaming agents in plastics insulation. Because they are highly resistant to chemical attack, CFCs remain in the earth's atmosphere for many years, eventually drifting up into the stratosphere where they are broken down by UV radiation. The chlorine and oxygen compounds formed by this chemical breakdown then destroy the natural stratospheric ozone (11).

Table 7.1: CFC and HCFC Blowing Agents for Plastic Foams in Use in 1991 (12)

	CFC-11	CFC-12	CFC-141b	HCFC-123	HCFC-134a
			--------Blowing Agent-------->		
Chemical Formula	CCl_3F	CCl_2F_2	CCl_2FCH_3	$CHCl_2CF_3$	CF_3CH_2F
Chemical Name	Trichloro-fluoromethane	Dichlorodi-fluoromethane	Dichlorofluoro-ethane	Dichlorotri-fluoroethane	Tetrafluoro-ethane
Molecular Weight	137.4	120.9	116.95	152.91	102.03
Main Uses in Plastic Foams	(Main CFC blowing agent used to date) Rigid foams	Rigid foams by frothing process because of low B.P.. (021.6°F)	Rigid and flexible foams	Rigid Foams	All-purpose foam (rigid and flexible)

Although the world's major users and producers of CFCs have agreed to phase out their use by the end of the century, some scientists and conservationists agree that ozone depletion has reached a crisis and that a more urgent global ban on these chemicals is essential (11).

Table 7–1 will provide some useful information on CFC and HCFC blowing agents that have been used in the past and on those blowing agents that are suggested to replace them (12).

Carbon Dioxide (CO_2): Until 1958 when halocarbons were first used as blowing agents for urethane foams carbon dioxide (CO_2) was the blowing agent used. The CO_2 was liberated by the isocyanate–water reaction shown below (13).

$$2R\text{---}NCO + H_2O \rightarrow R\text{---}NH\overset{\overset{\textstyle O}{\|}}{\text{---}C}\text{---}NH\text{---}R + CO_2\uparrow$$

isocyanate   water   disubstituted urea

The CO_2-blown rigid urethane foams had the following disadvantages over the CFC–11 blowing agent (14):

- K–factor of about 0.25 compared to 0.11 for CFC–11, requiring about twice as much insulation as CFC–11.

- The induction period before foaming is smaller with CO_2 because of the latent heat of vaporization of the CFC–11.

- The gelation rate of the expanding foam is decreased, thereby preventing thermal pressure cracks and charring of the foam in large applications.

- The compressive strength of the CFC–11–blown foam is increased by about 30 percent over the CO_2-blown foam.

- The moisture vapor transmission (MVT) of the CFC–11 blown foam is reduced (3.5 perms vs. 5.5 perms for the CO_2-blown foam).

- The CFC–11 blown foam has better adhesion to metal.

- The edge of CFC–11 blown foam is not friable.

- The CFC–11–blown foam has a higher proportion of closed cells (about 90% vs. 85% for CO_2–blown foam).

- The cost of the foam is reduced.

In 1991 Vandichel and Appleyard (15) described a new promising approach for the production of "soft" flexible slabstock urethane foam blown exclusively by CO_2 generated by the water–isocyanate reaction. These workers found that by the addition to the formulation of certain *hydrophilic materials* a substantial hardness reduction is obtainable, thereby permitting a considerable reduction, or even total elimination, of CFC–11 from some "conventional" foam formulations. The hydrophilic additive is called CARAPOR™ 2001. An example is a foam produced with an ILD value of 80N at a density of 21.5 kg/m³ (1.34 lb/ft³) (15).

Flexible Foams: CO_2 obtained *in situ* by the reaction of water with isocyanate has been the chief blowing agent for all commercially produced flexible urethane foams. The amount of water and tolylene diisocyanate (TDI) used determines foam density, providing most of the gas formed is used to expand the urethane polymer. Because water participates in the polymerization reactions leading to the expanded cellular urethane polymer, it has a very pronounced influence on the properties of foams. For better control of the foaming process most foam manufacturers employ distilled or deionized water (16).

In addition to water, auxiliary blowing agents may be included in the foam formulation to further reduce the foam density (16) (17). These agents can be used in addition to, or as part replacement for the water in developing special foam properties. An example is the use of methylene chloride or CFC–11 in either polyether– or polyester–based systems for softening the resulting foam. A number of other volatile solvents are known to have been used also.

See also the discussion of the work of Vandichel and Appleyard above (15).

The amount of water used in flexible urethane foam formulations, together with the corresponding amount of TDI, largely determines the foam density. As the amount of water increases, with a corresponding increase in TDI, the density decreases. If water content is increased without increasing the TDI, foams may be obtained with coarse cells and harsh textures. Lower tensile and tear strengths and compression moduli result, while the compression set tends to increase. Another important effect of too much water is poor aging characteristics. Too little water, on the other hand, will result not only in higher densities than desired, but also in slower curing and may cause shrinkage in the foam (17).

Rigid foams: Tables 7–2 and 7–3 provide interesting information on blowing agents used in rigid urethane foams. Table 7–2 (13) shows the advantage of CO_2 over air and the advantages of the CFC blowing agents over both air and CO_2. Note that the CFCs have about half the thermal conductivities of CO_2. It can also be seen that the thermal conductivities of the CFCs do not increase in the same proportion as air or CO_2 as the temperature rises (13). The effect of aging on the K–factors of rigid urethane foams blown with different blowing agents is shown in Table 7–3 (18). The high density (high molecular weight) of the fluorocarbon gas (CCl_3F or CFC–11 as it is now called) causes it to be a poor conductor of heat. Fortunately the permeability of the fluorocarbon through the cell walls of common polyurethane foams is extremely slow so that the fluorocarbon gas and its excellent insulating properties are retained almost indefinitely (19).

Another factor of critical importance in foam processing is the *viscosity* of the reactants. Most polyether polyols have high viscosities, and it is difficult to carry out high–speed mixing with these components with low–viscosity polyisocyanates. When halocarbon blowing agents are added to the polyether polyol component the viscosity is reduced to that of a thin liquid, thereby facilitating pumping, mixing and metering. The halocarbons also have a high degree of hydrolytic stability and hydropho-bicity (19).

Most rigid polyurethane foams are produced in the 2 lb/ft^3 (32 kg/m^3) range. CO_2–blown foams cannot be made with reliably low densities. The lowest practical limit is about 4 lb/ft^3 (64 kg/m^3). Halo-carbon–blown foams also provide better physical properties than CO_2–blown foams. The greater uniformity of the halocarbon–blown foams is, in part, responsible for their superior physical properties. In addition, the polyisocyanate residue from reaction with water is deleterious in several respects. Foaming conditions are less critical with halocarbons because of the absorption of the heat of reaction by the halocarbons (13).

CFC–12 halocarbon (CCl_2F_2) is especially useful in the frothing process (see Chapter 8). Since its boiling point at 1 atm. (–21.62°F) (29.6°C) is very low it immediately vaporizes when the foam ingredients are discharged from the mixing head. This vaporization produces a foam of low density to overcome the pressures exerted by the liquid ingredients which must expand 30–fold to reach densities of about 2 lb/ft^3 (32 kg/m^3). CFC–11 blowing agent is also included in froth formulations to obtain the final density (13).

Table 7.2: Comparison of Thermal Conductivities
of Blowing Agents Compared with Air (13)

Blowing Agent	Temperature		
	32°F	68°F	86°F
Air	0.168*	0.180*	---
CO_2	0.101	0.117	---
CCl_3F (CFC-11)	0.054	0.057	0.058*
$CHCl_2F_2$ (CFC-12)	0.068	0.064	0.067
CCl_3-CCl_2F_2 (CFC-113)	0.046	0.051	0.054

* BTU/°F ft^2 hr in. (To obtain SI units of W/m·K multiply by 0.144)

Table 7.3: Thermal Conductivities of Rigid Polyurethane Foams Containing Different Blowing Agents (18)

Blowing Agent	Mol. Wt.	Wt % to give 2 lb/ft$^{(3)}$	Initial K-factor$^{(4)}$	Final K-factor$^{(4)}$
FREON-11 (CFC-11) (CCl$_3$F)	137	16	0.11	0.14
Methylene Chloride$^{(1)}$	85	12	0.14-0.15	0.24
n-Pentane	72	9	0.18	0.20
CO$_2$	44	1.7$^{(2)}$	0.20-0.23	0.24
Air	29	---	0.24	0.24

(1) Properties extrapolated from blends with FREON-11 (foams with 100% methylene chloride cannot be made)

(2) As water

(3) Estimated

(4) BTU·in/hr°F ft^2 (to obtain SI units of W/m·K multiply by 0.144)

Fluorocarbon blowing agents act as moderating agents and do not produce additional crosslinking in rigid foams. These blowing agents are inert and are retained in the polymeric structure of the foam. They are also non–flammable and have a very low order of toxicity (20).

Structural Foams (Thermoplastic):   The design of some structural–foam molding equipment permits the direct addition of nitrogen gas blowing agent to the polymer melt. The gas is forced into a solution and remains there until the external pressure on the melt is relieved as it enters the mold cavity. When the pressure is removed from the melt the gas immediately expands, driving the melt to the extremities of the mold. Although nitrogen is low in cost and clean in terms of chemical residue, many commercial operators are phasing it out because of poor cell structure, as well as the fact that most modern equipment is not modified to accept direct nitrogen gas addition (12).

On the other hand, the use of chemical blowing agents (CBAs) has increased. These agents generate the gas necessary for structural foam by either decomposition or chemical reaction in the melt. Factors to consider when selecting a blowing agent for a particular thermoplastic structural foam are (21):

- chemical compatibility
- cost
- convenience
- safety
- the exothermic or endothermic nature of the reaction
- the nature and quantity of gas released
- cell structure

The time that must elapse before a molded part is painted is called the "degas time," and the type of gas generated by the CBA plays an important role in determining the degas times. If a part is painted before it is completely degassed the paint will blister. Blisters result from the blowing–agent gas, which is under greater than atmospheric pressure in the core of the part, and migrates through the foam structure to reach equilibrium with the surrounding atmosphere. Nitrogen is very sluggish in its permeability characteristics. All other factors being equal (level of internal gas pressure, part thickness, and type of paint systems), both carbon dioxide- and hydrogen–generating blowing agent can more quickly reach equilibrium with the surrounding atmosphere than can nitrogen–generating blowing agents (21).

The endothermic or exothermic nature of a blowing agent affects processing economics. One of the factors that frequently controls the cycle time of structural–foam parts, and therefore the number of parts that can be produced in an hour, is the *post blow*. Post blow results when a part is removed from a mold before the core is sufficiently cooled, causing swelling in areas with the greatest thickness. This swelling is caused by the presence of pressurized gas in the core of the part (21).

If a blowing agent is *exothermic* it increases the temperature in the core of the part, which, in turn, softens the thermoplastic material and increases the gas pressure from the blowing agent. *Endothermic* blowing agents, on the other hand, draw heat from the core during gas generation, helping to solidify the cell wall and decrease internal gas pressure. The combination of lower core pressure and higher–modulus structure allows the parts to be removed from the mold after a shorter time period. The economic advantage is obvious (21).

With respect to *chemical compatibility*, it is essential to select a blowing agent and polymer combination that will not produce undesirable side reactions affecting properties or appearance (21).

Depending on the conditions of foaming and mainly on the amount and type of blowing agent used, structural foams can range from almost solid resin to a material with a density approximately 75 percent of the solid resin. As a rule, thin–walled parts require a higher percentage of blowing agent to obtain the same density as a thicker part. This is because of the cooling effect of the metal mold walls on the resin which inhibits, to some extent, bubble formation in material contacting the metal surfaces (7).

CATALYSTS

General

Catalysts are substances which cause or accelerate a chemical reaction when added to the reactants in minor amounts, without being permanently affected by the reaction. A negative catalyst (inhibitor, retarder) decreases the rate of reaction (1).

Accelerators (promoters) are substances which hasten a reaction, usually by acting in conjunction with a catalyst or a curing agent. Accelerators are sometimes used in the polymerization of thermoplastics, but are used most widely in curing systems for thermosets and natural and synthetic rubbers. Accelerators are sometimes called *cocatalysts* (1).

Catalyst types used with urethane foams are as follows (22):

Tertiary amines

N–ethylmorpholine	(least active)
N–methylmorpholine	increasing
N–methylpiperazine	activity
Trimethylene diazine	(most active)

The prime function of tertiary amines is to catalyze the reaction of isocyanate with water to form CO_2 gas for blowing the polymer.

Metal salts

Stannous octoate (esp. for flexible foams)
dibutyltin dilaurate (DBTDL) (esp. for rigid foams)
tin mercaptides, such as dioctyltin mercaptides

Hybrids

amine/tin   (cocatalysts)

Isocyanurate foams require such catalysts as potassium compounds—for example, potassium octoates, and amines such as 2, 4, 6–tris–(N, N–dimethylaminomethyl) phenol (22).

Reaction–injection–molded (RIM) urethane foams are using various tin/amine catalysts, with some special variations developed expressly for this processing technique. Amine/amine and tin/tin combinations are also under development (22).

Because of environmental concerns about using chlorofluorocarbon (CFC) blowing agents, methylene chloride is being used as a blowing agent. A number of catalysts have been developed that are particularly compatible with methylene chloride. Several of these new catalysts work on the *delayed–action principle* to avoid splitting of the foams (22).

The reaction between an isocyanate and the hydroxyl group in a polyol will take place without a catalyst, but at too slow a rate to be practical. Without a catalyst a foam may expand, but it may not cure adequately to give good physical properties. The urethane reaction can be catalyzed by basic materials such as the *tertiary amines* (20).

Rigid Urethane Foams

In rigid urethane foaming systems using the CO_2 formed by the water–isocyanate reaction a balance of the relative rates between the urea and urethane reactions is necessary. If the urethane reaction is not fast enough the gas will not be trapped and no foam will be formed. On the other hand, if the urethane reaction is too fast the polymer will set up before the gas is formed and a high–density foam will result. This reaction is much less controllable by catalytic action than the urethane reaction (20).

Tertiary amines alone can be used as catalysts, but for some applications, such as spraying, more speed is desirable. Metal salts, particularly tin salts, accelerate the foaming reactions, and can be used alone or in combination with the tertiary amine–type catalysts. Tin catalysts of importance for *rigid urethane foams* are stannous octoate and dibutyltin dilaurate. Stannous octoate will hydrolyze rapidly in the presence of a basic catalyst with loss of activity. Masterbatches containing stannous octoate and moisture are stable for only a few hours at room temperature. Resin masterbatches containing dibutyltin dilaurate may stay stable for months. For this reason this catalyst is preferred for foaming systems packaged for use at other locations or plants where the resin masterbatch is not used immediately (20).

Delayed–action catalysts have been made successfully. Buffered amine catalysts, where the activity of the amine has been reduced by the presence of an acid, have also been used. Acidic materials can be used to retard the urethane reaction. Hydrogen chloride and benzoyl chloride have been used in combination with amine–type catalysts to control reaction rates. A small percentage of acid can increase foaming time from 2.2 to 6 minutes (20).

Temperature can also be used to control the urethane foaming reactions. Some delayed–action rigid urethane foaming systems have been made by premixing all of the foaming ingredients at temperatures down to $-300°F$ $(-184°C)$. When these systems are heated, foaming of the mass takes place (20).

The vapors of tertiary amine catalysts are irritating and contact with the skin can cause dermatitis. The catalyst can produce severe irritation by contact with the skin. Their vapors are also irritating. Care must be taken to ensure that these materials in solid, liquid or vapor form do not come in contact with the human body (20).

For CO_2–blown foams tertiary amines are adequate. For solvent–blown foams, however, a more reactive catalyst is necessary because of the cooling effect of the solvent (such as CFC–11). A synergistic action exists between the catalysts and tertiary amines (23)(24).

Rigid urethane foam systems, because of their greater degree of crosslinking, build gel strength so rapidly that tertiary amines are adequate for one-shot or prepolymer systems using polyethers or polyesters. The structure of the tertiary amine has a considerable influence on the catalytic effect and also on its usefulness for foam production. The catalytic strength generally increases as the basicity of the amine increases and as steric shielding of the amino nitrogen decreases. The tertiary amine catalysts provide satisfactory foaming with either the one-shot polyester or the polyether prepolymer systems, both of which are relatively high in initial viscosity (17).

The density and moldability of urethane foam is greatly influenced by the choice of the catalyst system and concentration, due to the effect of catalysts on the foaming rate (25)(26).

An increase in catalyst concentration generally produces an increase in tensile, shear and compressive strength, as well as a reduction in cure time. There is an upper limit to catalyst concentration, however, above which the urethane foam tends to fissure and crack. An increase in catalyst concentration usually produces a decrease in K-factor when fluorohydrocarbon blowing agents are used. This is because a low K-factor (better thermal insulation) results from a good retention of fluorohydrocarbon by the foam, which implies a negligible permeation of fluorohydrocarbon through the foam cell walls. An increase in catalyst concentration would be expected to produce a more tightly crosslinked polymer structure and thereby reduce the diffusion constant and possibly the solubility of the fluorohydrocarbon (27).

Flexible Urethane Foams

According to Woods (28), the rapid growth of high-resilience (HR), cold-cure, microcellular (and rigid) foam manufacture, all of which require higher levels of catalysis than conventional slabstock flexible foams, has caused a rapid growth of the catalyst market and an increased rate of new-product development. Most polyether slabstock foam manufacturers use *stannous octoate* as the main polymerization catalyst, together with a *tertiary amine* to control the rate of blowing and foam rise. Medium-density foams, i.e., from ca 20-30 kg/m³ (1.25-1.88 lb/ft³), depending on the type of machine and process, made with predominantly water blowing, are commonly made using the low-cost N,N-dimethylaminoethanol. Soft high-density foams and those made with a high level of auxiliary blowing agents, CFC-11 or methylene chloride, will require the addition of more active tertiary amine catalysts. A proportion of DABCO, a bis-dialkylaminoalkylether or an aliphatic tertiary polyamine is used, often at a low level (28).

The residual odor of *tertiary amines* is objectionable in many applications, such as bedding, upholstery, and bottle wadding. This odor may be minimized by using volatile catalysts such as N–methyl morpholine, the structure of which is shown as follows:

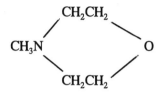

Woods suggests a list of 36 tertiary amine catalysts (28).

FIRE RETARDANTS (FLAME RETARDANTS)

General

Fire retardants or flame retardants, as used here, are materials that reduce the tendency of plastics to burn. They are usually incorporated as additives during compounding, but sometimes are applied to surfaces of finished articles (1).

Fire–retardant chemicals available commercially for plastics can be divided into two general classes, *unreactive additives* and *reactive monomers* or crosslinking agents. The unreactive additives are generally added to the polymer during processing, but do not react chemically with the other constituents of the composition. The reactive types, on the other hand, are generally reacted with the polymer structure at some processing stage. The ideal fire–retardant additive should be inexpensive, colorless, easily incorporated into the polymer composition, compatible, stable to heat and light, efficient in its fire–retardant properties, nonmigrating, and have no adverse effect on the physical properties of the polymer. It should also be non–toxic. Unfortunately, most presently available additives seldom meet all these requirements (29).

Additive Fire Retardants

These systems are generally composed of both organic and inorganic materials acting synergistically to provide an optimum balance of flame retardance, physical properties, and cost. Additive fire retardants are generally incorporated by compounding and are useful in a variety of polymer systems. These materials are generally used for *thermoplastic resins*, although there are exceptions. With few exceptions additive retardants are used to fire retard flexible polyurethane foams (30).

Halogenated polymers, both brominated and chlorinated, have been developed to yield better polymer compatibility, improve physical properties, and long–term–aging characteristics in many thermoplastic resins, particularly the high–performance engineering thermoplastics, such as nylon, polybutylene terephthalate (PBT) and polyethylene terephthalate (PET). These materials still use antimony oxide as a synergist to achieve the desired flame resistance (31).

Antimony oxide by itself is essentially useless as a fire–retardant additive. However, in combination with other materials, it is by far the most widely used antimony–containing flame retardant additive. It is generally used with bromine– or chlorine–containing compounds (32).

Chlorine compounds used as fire retardants include the following (32)(33):

- polychloroprene
- polyvinyl chloride (PVC)
- chlorinated polyethylene
- chlorosulfonated polyethylene
- tris (chloroethyl) phosphate
- tris (dichloropropyl) phosphate
- methyl pentachlorostearate
- cycloaliphatic chlorine–containing flame retardants
- chlorendic anhydride

Bromine compounds are also used as fire retardants. These compounds are about twice as effective as chlorine compounds on a weight basis, so that significantly lower concentrations are needed. However, bromine compounds are higher in cost than chlorinated compounds and are generally less stable under exposure to heat and light (29). Those compounds containing aromatic bromine are significantly more stable to heat and hydrolysis than the aliphatic type. Examples are decabromodiphenyl   oxide   (DBDPO),   tetrabromobisphenol   and tetrabromobisphenol A. A pentabromodiphenyl oxide blend is available for *urethane foams* and polyesters (34). Aliphatic bromine–type additives are used as flame retardants in plastic foams (polyurethane and polysty-rene (33).

Halogenated *organophosphorus compounds* are used widely in flame–retarding polyurethane foams (31). Flame retardance by phospho–

rus compounds is believed to be due to thermal degradation to nonvolatile acids that promote char formation in the plastic, especially those plastics that contain hydroxyl or oxygen groups (35).

Alumina trihydrate (ATH), also known as hydrated alumina, is unique among low–cost fillers and extenders due to its high proportion (34.6%) of chemically bound water. This water of hydration is stable and unreactive at the processing temperatures of many plastics. When heated to approximately 220°C (428°F) or higher by exposure to a flame front, the hydrate decomposes into its constituents with the absorption of a considerable amount of heat (36). Alumina trihydrate is an extremely important fire retardant for use in plastics. Its usage is close to 50% of all the additive fire retardants used for plastics (37). Applications include carpet backing for SBR latex and polyurethane foam. In most applications loadings are high (over 40 parts per hundred of basic resin) (32). ATH is a dry light powder that functions by absorbing heat, by evolving steam to dilute the combustible gases being generated, and by providing a nonflammable char barrier between the heat source and the material. It also acts as a smoke suppressant (33).

Reactive Fire Retardants

These systems contain functional groups allowing them to be incorporated directly into the polymer structure through chemical reactions. The main advantage of this type of fire retardant is the permanence of the fire retardancy imparted. In most cases chemically reacting the fire retardant into the polymer essentially eliminates long–term migration of the fire retardant. Reactive fire retardants are primarily used in unsaturated polyesters, epoxy resins and *polyurethane foams*. Two of the most popular reactive retardants are tetrabromobisphenol A and dibromomonononeopentyl and tetrabromophthalic glycol. Others include chlorendic acid and anhydride, tetrabromophthalic and tetrachlorophthalic anhydride, and diallyl chlorendate (29).

Reactive polyols which contain halogen groups, phosphorus, or both, are offered by a number of suppliers for flame–netardant *urethane–foam* applications. These materials can be used alone, or with other flame retardants as synergists. Although reactive flame retardants may appear to be more costly initially, in the long run they may be found to be less expensive than the additive types (31).

Uses of Fire Retardants in Specific Foam Types

Rigid Polyurethane Foams.   In general, fire retardance is imparted to polyurethane foams by the chemical incorporation of halogen and/or phosphorus compounds into the material.   Chemical modification of the polyol with phosphorus or a phosphorus–chlorine composition is currently employed in the commercial preparation of fire–retardant foams. Halogen alone is not satisfactory because the high viscosity of the polyols containing sufficient halogen (typically 25–30% by wt. of chlorine) for effective fire retardance significantly impairs the processing properties (38).

The use of phosphorus in fire–retardant polyurethane foams leads to a high char formation, combined with easy processing, because of the relatively low density of most phosphorus compounds. This combination of desirable properties has made phosphorus compounds, with or without halogen, the most widely used fire retardants for polyurethanes. Reactive phosphorus compounds such as FYROL 6®, are used extensively. FYROL 6®, supplied by Stauffer Chemical Co., is widely used for polyurethane foams (38).

Buszard and Dellar in the UK compared three polyurethane foams using the BS4735 horizontal burn test and the DIN 4102–B2 vertical burn test covering a wide range of polyol types.   The flame retardants used were as follows (39):

- trismonochloropropyl phosphate (TMCPP), containing 9.5 % phosphorus and 32.5 % chlorine by weight.

- dimethyl methylphosphonate (DMMP). Although relatively expensive, it has a phosphorus content of 25% and was found to be the most effective of the additive flame retardants studied.

- REOFLAM® 306 (Ciba–Geigy). This is a proprietary flame retardant containing 15.2% phosphorus and no halogen. It combines a good level of flame retardance with good processing and anti–scorch properties, and has excellent compatibility with aromatic polyester and halogenated polyols and halocarbon blowing agents.

Nonreactive additive fire retardants that act as fillers or plasticizers may also be used for polyurethane foams. The most commonly used example is tris (2,3–dibromopropyl) phosphate. Nonreactive additives have not been used extensively because of their fugitive nature and their tendency to migrate from the cellular plastic under many conditions of extended use (38). Alumina trihydrate is also used as a fire retardant for polyurethane foams, especially in conjunction with the synergist Sb_2O_3 (36).

The reader interested in fire retarding rigid urethane foams for building construction should refer to the Society of the Plastics Industry (SPI) in Washington, DC for specific recommendations and warnings as to the use of polyurethane foam in such applications. It must be noted that spray–applied polyurethane foam *must be protected by a nonburning barrier* approved by the appropriate building codes. Without such barriers polyurethane foam will spread fire rapidly once ignited, even if it is fire retarded. With a few exceptions, all model codes require that foam plastic insulation be covered by a thermal barrier equal in fire resistance to 1/2–inch gypsum board, or be used only in sprinklered buildings.

Flexible Polyurethane Foams. Methods for retarding flexible polyurethane foams are essentially the same as for rigid foams. Flexible urethane foams, however, burn readily even when fire retarded (38). Creyf and Fishbein have discussed advances in flexible polyurethane foam technology from a fire–resistance viewpoint. These UK workers reported that the most recent development in flexible urethane foam fire retardants remain the so–called *aluminum hydrate* (plus other synergists) *impregnated foams*. These foams have been found to be really resistant to large ignition sources, e.g., 180 grams (0.40 pounds) of burning newspaper. The aluminum hydrate is added either by post treatment, or in a one–shot operation (40).

Polystyrene Foams. The blue–colored fire–retarded polystyrene foam used for construction applications is reported to contain acetylene tetrabromide and copper phthalocyanine (38). Polystyrene foam is highly flammable, with a limiting oxygen index (LOI) of 19.5, and the same flame–retardant treatments are applied as for solid plastics. Free–radical initiators are especially useful since they promote depolymerization and hence melting and dripping. Most flame retardants for foamed polystyrene have this effect — the foam melts away faster than the flame front can follow it. The most frequently used fire retardants in general are brominated cycloalkanes or brominated aromatics with aliphatic side chains. These retardants are used without antimony oxide. With foamed polystyrene, as opposed to the solid form, abundant use is made of inorganic fillers such as calcium carbonate, silicates, and glass fiber, which may, however, have adverse effects on properties such as density or toughness. On the other hand, phosphorus compounds are much less widely added to foams then to solid plastics. One of the most common individual additives is tris (2,3–dibromopropyl) phosphate (38a).

Polyolefin Foams. Antimony oxide/chlorowax combinations or other aliphatic chlorine sources are generally used to fire retard polyolefin foams. Because of the higher processing temperatures required (up to 280°C or 536°F), when using azo blowing agents, the antimony oxide/chlorine system is inadequate, and in these cases phosphorus–containing fire–retardant systems are preferred (38).

Polyvinyl Chloride (PVC) Foams. *Rigid PVC foam* is inherently fire retardant because of the high chlorine content (56.7%). *Flexible PVC foams* present increased fire hazards because of the plasticizers they contain. Flammable plasticizers used include alkyl phthalates, as dioctyl phthalate. Non–burning types include alkyl aryl phosphates (phosphate esters). The latter types should improve resistance to ignition and reduce flame spread when compared to the usual phthalate plasticizers (38) (41).

Phenolic Foams. Cured phenolic resins have good thermal stability and high tendency to char in an intensive fire. Even after the removal of the ignition source the foams often smolder and char until they are almost completely consumed. This phenomenon, called *punking*, is claimed to be overcome by the addition of boric acid/oxalic acid and ferric/aluminum chloride as the foaming catalysts. The addition of antimony compounds is also reported to decrease punking (38).

The Naval Ammunition Depot (NAD) at Crane, Indiana has developed improved fire–retardant phenolic foams containing blends of boric/oxalic acids as catalysts, as described above. The resultant foams were found to be extremely efficient fire barriers due to their high heat absorptivity, the amount of carbon and/or coke produced during pyrolysis, and the adhesion of the char to the burned materials. Other advantages of the foam during flaming and nonflaming pyrolysis are its low smoke emissions and lack of toxic fumes other than carbon monoxide. It takes one hour to reach 230°F (110°C) when a 13 lb/ft^3 (208 kg/m^3) phenolic foam specimen 2.9 inches (7.4 cm) is exposed to a fully developed fire (41).

Urea–formaldehyde Foams. While urea–formaldehyde (UF) foams can be rated as difficult to burn, blending of UF with another polymer can decrease the resistance of the foam to burning. Fire retardants, including phosphorus and boron compounds, have been added to decrease the flammability of UF foams (42). According to Frisch (42) phosphonates, furfuryl alcohol and ethylene glycol have been used as fire retardants.

MOLD-RELEASE AGENTS (PARTING AGENTS)

General

A mold-release agent is a lubricant, often wax, silicon or fluorocarbon fluid, used to coat a mold cavity to prevent the molded part from sticking to it, and thus to facilitate its removal from the mold. Mold-release agents are often packaged in aerosol cans for convenience in application (1).

Mold-release agents create a barrier between the resin and the metal mold, preventing adhesion to the mold. The same additives that promote mold release are sometimes used to reduce viscosity and ease flow, thereby improving surface detail and ensuring more complete filling of small complex cavities. Mold-release agents may also have an antistatic function (43).

External Mold Releases

Legislation restricting the use of chlorofluorocarbons (CFCs) and volatile hydrocarbons in mold releases is accelerating the trend toward water-base and chlorine-free emulsions. An unusual new development is a PTFE release film designed to replace conventional spray-on and wipe-on methods. This new technology provided by Chemfab/Chemical Fabrics Corp, Merrimack, NH provides chemical inertness and the non-stick properties of PTFE polymers with intrinsic heat resistance up to 400°F (204°C). The film is applied by hand into the mold. Once the part is formed the mold-release film can be easily peeled off for disposal. Costs range from the high teens to $20/lb. This system offers many advantages—no residue left on the mold or part, no sprays to contend with, and no ventilation problems. It is also cost effective and offers reduced variability and improved consistency (43).

Mold-release additives are available in several forms: powder, flake, liquid, or paste. High-bulk-density powders and pills have been introduced, and non-dusting grades are available. They may be added at the reactor stage of processing, or later by the compounder using melt-compounding extruders or mills. Powders may also be dusted on pellets (43).

Paraffins, Hydrocarbon Waxes. These low-melting (65°-75°C) (149°-167°F) waxes function as external lubricants. They range from short to long carbon chains, with some branching. Straight-chain products are harder; branched-chain materials are softer. With branching these high-molecular-weight paraffins lose some crystallinity characteristics.   Because of the remaining microcrystals they are referred to as microcrystalline waxes or microwaxes (43).

Polyethylene Waxes. These are low–molecular–weight (2,000– 10,000) long–chain hydrocarbons, slightly branched, with melting points from 100° to 130°C (212°–266°F). High molecular weight makes them more effective than paraffins at high processing temperatures.   These waxes find application in a wide range of thermosets and thermoplastics because of their compatibility and good flow properties (43).

Water–Base Mold Releases.   Water–base release agents are being used more widely. Solvent–base agents evaporate too quickly in warm areas.   One example is a hydro–alcoholic solution of saponified fatty acids claimed to produce a dry–film coating. Axel Plastics Research Laboratories in Woodside, NY has a new product, Mold Wiz H40–3U, for flexible polyurethane foam. This material is heated to evaporate the water and is then applied with high–pressure spray equipment (44).

Semi–Permanent Mold Releases.   Considerable research and development is currently underway with this type of release agent. Percy Harms Corp. in Wheeling, IL has a Slide Dura Kote for thermosets and thermoplastics.   Dura Kote is a solvent dispersion of air–drying release resin claimed not to transfer from mold to molded part, nor to discolor. By baking the release agent onto the mold a harder, more durable film coating is produced. Dexter Frekote Mold Release Products, Seabrook, NH has introduced a CFC–free line of semi–permanent products called Frekote NC.   The solvent carrier has been modified by replacing trichlorotrifluoroethane with aliphatic hydrocarbons.   This product is claimed to have improved wetting, film formation, and drying. Stoner, Inc., Quarryville, PA has a new release agent, K464, for urethane foams (44).

Chapter 8 on Methods of Manufacture discusses the use of mold–release agents in molding polyurethane foam parts.

NUCLEATING AGENTS (NUCLEATORS)

Nucleating agents are chemical substances which, when incorpo– rated in polymers, form nuclei for the growth of crystals. In foams, nucleating agents act by forming many small bubbles, rather than fewer larger bubbles (45). Colloidal silicas and micro–expanded silicas are commonly used nucleating agents, as are dry air and nitrogen (28).

The action of nucleating agents in forming small bubbles in foam formation may be likened to the use of "boiling chips" added to distillations to insure a steady, even flow of small gas bubbles during distillation. Materials such as "dust," some finely dispersed silicone oils, silica, and other finely divided solids may serve as nucleating agents in foam formation. Gases also serving as nucleating agents for foams include carbon dioxide, air, nitrogen and butane, Inert nucleating agents such as silicon dioxide (SiO_2), diatomaceous earth, calcium silicate, carbon black, boron nitride, clay, and titanium dioxide (TiO_2) are usually only moderately effective, or even ineffective as nucleating agents in foam formation. Many reactive materials which decompose or rearrange exothermally have been found to be highly effective nucleating agents in producing foams of densities greater than 20 lb/ft^3 (320 kg/m^3). These materials are less effective in producing foams of lower densities. These nucleators have one common feature: by their exothermic reactions they provide many localized hot spots, and are therefore termed "dynamic nucleators." An excellent example is azodicarbonamide, which evolves mainly nitrogen during its decomposition, and is probably more exothermal in decomposition than any other chemical blowing agent. This brief discussion is from an excellent, although not recent, presentation by Saunders and Hansen (46).

Effective nucleating agents for extruding polystyrene foam sheet include the following (47):

- mixture of sodium bicarbonate and citric acid
- nitrogen gas
- aluminum silicate
- water from hydrated salts
- amino acids
- ammonium citrate
- acid phosphate baking powders
- potassium metasilicate
- mixtures of nitrogen-releasing blowing agents and calcium oxide
- mixtures of magnesium silicate and silica
- fluorocarbon polymers
- carbon dioxide

Kumar and Weller (48) have very recently produced microcellular polycarbonates using carbon dioxide as the nucleating gas.

REINFORCEMENTS

Urethane Foams

The development of reinforced foam materials has been underway since about 1961, although most of the progress in this area has been made since 1976. In early applications continuous–filament glass mats were incorporated into molds, particularly with polyurethane. Since 1965 the *reaction injection molding (RIM)* process has been used, especially n the automotive industry. In this process the reacting materials, usually polyurethanes, are transferred into a mold where they usually react to form crosslinked polymers. Molding pressures are very low, of the order of 150 psi (1 MPa) compared to injection molding (ca 3000 psi or 21 MPa), or compression molding (ca 100–1000 psi or 0.69–6.9 MPa), and tooling costs are, for this reason, very low. In reinforced reaction injection molding (RRIM) the reinforcements may range from particulate fillers to fibrous materials such as glass, carbon, or KEVLAR® aramid fibers (49).

Thermoplastic Structural Foams

Thermoplastic structural foams (TSFs) have been commercially available for a number of years and are widely used in the machine–housing and automotive markets. These foams are discussed in Chapter 3 on Thermoplastic Foams. That chapter mentions the use of glass reinforcements to some degree. Incorporation of reinforcements into structural foams follows the practice generally used for thermoplastics, namely the use of a masterbatch of reinforced polymer usually prepared by a compounding extruder or a crosshead extender. The reinforcement used in these foams is usually 3– to 6–mm chopped–glass fibers treated with an appropriate coupling agent suitable for the particular polymer matrix being reinforced. In general, the properties of reinforced thermoplastic structural foams reflect the attributes required of RRIM. The advantages are as follows (49):

- Thermal–expansion coefficients are reduced
- Deflection temperature under load (DTUL) is increased
- Creep is reduced
- Modulus (stiffness) is increased
- Impact properties are improved

Applications for reinforced structural foams are growing, but currently they are used mostly for very large moldings, or for molding operations at high temperatures. One example is a washing–machine tub. Reinforced thermoplastic structural foam was used in this application to provide stiffness with good high–temperature–creep resistance (49).

In reinforced thermoplastic foams, where glass fiber contents usually range from 10–30% by weight for the same voids content, the inclusion of high–density glass reinforcement (2.5–2.6 sp. gr.) somewhat increases the overall foam densities in comparison with unfilled structural foams. The stiffness and strength (flexural and tensile) of structural foams reinforced with even a comparatively low percentage of glass fiber can equal, or even exceed, these properties of the solid unreinforced parent polymers, such as nylon or polypropylene, despite the fact that the foam may contain up to about 30% voids. The heat deflection under load (DTUL) and the extensibility of reinforced structural foam are also closely comparable to those of the reinforced parent polymer. For this reason, reinforced structural foams can be greatly superior to the solid, unreinforced polymers in these respects (50).

In structural foams the reinforcing effect of the fibrous filler comes about essentially as it does in unfoamed moldings. The reinforce- ment improves, to a great extent, the strength and stiffness, as well as other properties of the solid material (the cell walls) of the foam. The effects are further enhanced by the tendency of the reinforcing fibers to become oriented in the course of cell formation and solidification of the cell–wall material. Finally, superimposed on these local effects is the overall strengthening effect of the reinforced structural skin in which the reinforcing fibers are likely to be oriented. In the mold the glass fibers tend to align themselves within and along the cell walls when the cells are being formed. This local anisotropic effect increases the strength and stiffness of the cell walls (50).

In a properly foamed molding the foam cells are roughly spherical and the cell walls have no particular directional alignment. They are essentially randomly oriented. The local orientation and anisotropic reinforcement effect of the fibers around individual cell walls are integrated into, and become additional factors in, a general enhanced reinforcing effect that is isotropic in the molding considered as a whole (50).

STABILIZERS

Stabilizers are agents used in compounding some plastics to assist in maintaining the physical and chemical properties of the compounded materials at satisfactory values throughout the processing and service life of the material and the parts made from the material. Some additives that function as stabilizers include the following (1)(51):

- emulsion stabilizers
- viscosity stabilizers
- antioxidants
- UV stabilizers
- light stabilizers
- antiozonants
- biocides
- fungicides
- heat stabilizers
- surfactants

The term "stabilizer" is obviously a general term covering a wide range of additives intended to maintain certain properties at desired levels.

SURFACTANTS

General

The word "surfactant" is a widely used contraction of "surface-active agent," a compound that alters the surface tension of a liquid in which it is dissolved (1). Surfactants impart stability to polymers during the foaming process. They help control cell structure by regulating the size, and to a large degree, the uniformity of the cells. In urethane foams the choice of surfactant is governed by factors such as polyol type and method of foam preparation (13).

Flexible Foam Surfactants

Many different types of surfactants have been employed for flexible foams, but usually *nonionic and anionic surfactants* and *silicones* are used. The choice of surfactant depends upon the method of foam preparations and the desired end use (e.g., fine, regular–celled cushions vs. large, irregular sponges). The most widely used surfactants for polyether–based urethane foams are *silicones*, such as the polyalkylsilox-ane–polyoxyalkylene copolymers. Conventional polydimethylsiloxanes of relatively low viscosity, 10 to 100 cps at 25°C (77°F), have been used for *polyether* urethane systems prepared and foamed as prepolymers. The silicones perform differently in *polyester* urethane foam systems, causing an unstable foaming situation. Silicone oils can be used in very small amounts to enlarge the cell size of polyester foams. Stabilizers used for polyesters are ionic, such as sulfonated castor oil and other natural oils, amine esters of fatty acids, and long–chain fatty acid partial esters of hexatol anhydrides. In polyester systems non–ionic surfactants are used to help modify the viscosity of the polymer during foaming and to provide control of reaction rates (16).

In the prepolymer method for one–shot polyether flexible urethane foams the primary role of the silicone surfactant is to lower surface tension and to provide film (cell–wall) resilience. Resilient films prevent the collapse of the foam during foam rise and continue to stabilize it until the foam is self–supporting. A secondary, but nevertheless important role of the silicone surfactant is cell–size control. The silicones can be added to the formulation in any of the 2 to 6 streams usually fed to the mixing head in the one–shot process. Usually, however, the silicone is metered separately, in combination with the polyol, or added as a wa-ter/amine/silicone mixture. It can also be added in the fluorocarbon blowing agent (52).

Rigid Foam Surfactants

Polyester– and polyether–based rigid urethane foams generally require a surfactant, whether expanded with CO_2 from the water-isocyanate reaction, or with an inert blowing agent such as fluorocarbon. Without surfactant the foam may collapse or have a coarse cell structure. Castor–oil–based systems generally do not require surfactants, but better results will be obtained if they are used (20).

Surfactants used in rigid foams range from ionic to non–ionic types to the silicones. Anionic surfactants (those which contain active hydroxyl groups) should not be added to an isocyanate–containing foam component (20).

The most widely used surfactants are copolymers based on dimethyl polysiloxane and polysiloxanes. Some of these silicones are prepared with ethylene and propylene oxides. Some silicones contain Si-

O–C linkages and are hydrolytically unstable; others do not contain a silicone–carbon bond and are stable. In general, higher–viscosity silicone copolymers are more efficient and will provide foams with finer cell structures. Surfactants are used at the 0.5% to 1% level in rigid urethane foams. With too little silicone foam, cell structure is large. Too much silicone does not affect the foam properties, but is wasteful. There is no known health hazard with the use of silicones (20).

Surfactants other than silicones used in one–shot rigid urethane foams include Spans (long–chain fatty acid partial esters of hexatol anhydrides), Tweens (polyoxyalkylene derivatives of hexatol anhydride partial long–chain fatty acid esters) and Emulphors. These act in a manner similar to the polydimethylsiloxanes. The choice of surfactant to give optimum results may vary with the foam system used, e.g., as the polyol or isocyanate is changed (52). The use of these non–silicone surfactants usually imparts a higher proportion of open cells. In addition to a high content of closed cells the use of silicone copolymers in rigid polyether foams also leads to higher–strength properties than when other surfactants are used (18).

REFERENCES

1.   Whittington, L.R., *Whittington's Dictionary of Plastics*, 2nd Edition, Sponsored by The Society of Plastics Engineers, Technomic Publishing Co., Westport, Connecticut (1978).

2.   Hilyard, N.C. and Young, J., Chapter 1, "Introduction," in *Mechanics of Cellular Plastics*, N.C. Hilyard, Editor, Macmillan, New York (1982).

3.   *Plastics Engineering Handbook of the Society of The Plastics Industry, Inc.*, 4th Edition, ed. by J. Frados, Van Nostrand Reinhold, New York (1976). (Author's Note: The 5th edition, edited by H.L. Berins, was published in 1991).

4.   Finck, H.W., Chapter 12, "Antistatic Agents," *PLASTICS ADDITIVES*, 2nd Editon, R. Gächter and H. Müller, Hanser Publishers, Munich, Vienna, New York (Distr. by Macmillan Publishing Co., New York) (1987).

5.   Staff Written, "Additives and Modifiers—Antistatic Agents,"
      Plastics Compounding 1991/92 Redbook, 14(4):24–26
      (July/August 1991).

6.   Staff Written, "Additives 1991," *Plastics Engineering*, 47 (9):32
      (September 1991).

7.   Schwartz, S.S. and Goodman, S.H., Chapter 15, "Foam Process-
      es," in *Plastic Materials and Processes*, S.S. Schwartz and S.H.
      Goodman, editors, Van Nostrand Reinhold, New York (1982).

8.   Staff Written, "Additives and Modifiers — Blowing Agents,"
      Plastics Compounding 1991/92 Redbook, 14 (4):26–28 (Ju-
      ly/August 1991).

9.   Krueger, D.C., Christman, D.L., Hass, J.D. and Reichel, C.J.
      (BASF Corp.), "Low Density Rigid Foams without the Use of
      CFCs," *Journal of Cellular Plastics*, 27(3):252–264 (May/June
      1991).

10.   Monks, R. (Assoc. Editor), "Environmental Issues Top Agenda at
      Polyurethanes Conference," *Plastics Technology*, 37(10):57–62+
      (September 1991).

11.   Browne, M.W., "Worst Ozone Hole Stirs Health Fear," *New York
      Times*, (10 October, 1991).

12.   Allied Signal Inc., Technical Bulletin 1–90 GENETRON®
      (1990).

13.   Ferrigno, T.H., *Rigid Plastics Foams*, 2nd Edition, Reinhold, New
      York (1967).

14.   E.I. duPont de Nemours & Co., Inc., FREON Products Div.,
      Wilmington, DE, "FREON Blowing Agents for Polyurethane
      Foams," FREON Tech Bulletin BA–1 (September 1959).

15.   Vandichel, J.-C.N.E. and Appleyard, P., "Reduction of CFC–11
      Usage in Flexible Polyurethane Foams through Modifications to
      Polymer Morphology," *Journal of Cellular Plastics*, 27(3):279–
      294 (May/June 1991).

16.   Gemeinhardt, P.G., Section X–Handbook of Foamed Plastics, Ed., Bender, R.J., Lake Publishing Corp., Libertyville, Illinois (1965).

17.   Saunders, J.H. and Frisch, K.C., Polyurethanes — Chemistry and Technology, Party I, Chemistry, Interscience Publishers, New York (1962).

18.   Frisch, K.C., "Relationship of Chemical Structure and Properties of Rigid Urethane Foams," *Journal of Cellular Plastics*, 1(2):325–330 (April 1965).

19.   Bauer, A.W., "Relationships between Blowing Agents in and Physical Properties of Rigid Polyurethane Foams," Wayne State University (Detroit) – Polymer Conference Series – *Cellular Plastics*, Lecture Notes, (May 24–28, 1965).

20.   Stengard, R.A., "Chemistry and Prepolymer Techniques," Part 1 in Section IX – Rigid Urethane Foam, *Handbook of Foamed Plastics*, Ed., Bender, R.J., Lake Publishing Corp., Libertyville, Illinois (1985).

21.   Brewer, G.W., "Properties of Thermoplastic Structural Foams," pp., 508–513 in *Engineering Materials Handbook, Vol. 2 – Engineered Plastics*, ASM International, Materials Park, Ohio (1988).

22.   Staff Written, "Additives and Modifiers – Urethane Foam Catalysts, *Plastics Compounding 1991/92 Redbook*, 14(4):56–58 (July/August 1991).

23.   Frisch, K.C., "The Chemistry of Rigid Urethane Foams," Wayne State University (Detroit), Polymer Conference Series – *Cellular Plastics*, Lecture Notes (May 24–25, 1965).

24.   Wolfe, H.W., Jr., "Tin Catalyst Activity in Urethane Foam," E.I. duPont de Nemours & Co., Inc. Tech. Bulletin, (February 24, 1966).

25.   Dickert, E.A., Himmler, W.A., Hipchen, D.E., Kaplan, M., Silverwood, M.A., and Zettler, R., "Molding of One–Shot Rigid

Urethane Foam," *Chemical Engineering Progress*, 59(9):33–38 (September 1963).

26. Ritzinger, G.B., "Catalyst: Its Effect on Foaming Characteristics and Physical Properties of Rigid Foams," du Pont Co. Brochure A–26302 (June 3, 1960).

27. Simonds, H.R., *Source Book of the New Plastics*, Vol 2, Reinhold, New York (1961).

28. Woods, G., *Flexible Polyurethane Foams—Chemistry and Technology*, Applied Science Publishers, London and New Jersey (1982).

29. Hendersinn, R., "Fire Retardancy (Survey)," in *Encyclopedia of Polymer Science and Technology*, Bikales, N.B., Ed., Supplement, Vol. 1, pp. 270–339, Wiley–Interscience, New York (1977).

30. Staff Written, "A Designer's Guide to Flammability," *Plastics Design Forum*, 1(2):16–21 (May/June 1976).

31. Avento, J.M., "Flame Retardants," *Modern Plastics Encyclopedia 1981–1982*, 58(10A):187–190 (October 1981).

32. Levek, R.P., "Flame Retardant Additives for Plastics," pp. 259–275, in *Additives for Plastics*, State of the Art, Vol. 1, ed, Seymour, R.B., Academic Press, New York (1978).

33. Frados, J., *Plastics Compounding 1981/82 Redbook*, pp. 39–40.

34. von Hassell, A., "Flame Retardants: Signs of a Thaw in R&D," *Plastics Technology*, 26(8):71–74 (July 1980).

35. Staff Written, "Additives and Modifiers – Flame Retardants," *Plastics Compounding 1991/92 Redbook*, 14(4):36–38 (July/August 1991).

36. Bonsignore, P.V., "Alumina Trihydrate as a Flame Retardant for Polyurethane Foams," *Journal of Cellular Plastics*, 17(4):220–225 (July/August 1981).

37.   Staff Written, "Flame Retardants Offer Something Extra," *Modern Plastics*, 57(2):58–59 (February 1980).

38.   Report of the Committee on Fire Safety Aspects of Polymeric Materials, National Materials Advisory Board, National Research Council, National Academy of Sciences, *Fire Safety Aspects of Polymeric Materials, Vol. 1 — Materials State of the Art*, Publication NMAB 318-1, Technomic Publishing Co., Westport, CT (1977).

38a.   Cullis, C.F., "Combustion of Polyolefins", *Oxidation and Combustion Reviews*, 5:83–133 (1971) (As cited by Reference 38 above).

39.   Buszard, D.L. and Dellar, A.J., "The Performance of Flame Retardants in Rigid Polyurethane Foam Formulations," in *Fire and Cellular Polymers*, Eds., Buist, J.M., Grayson, J.J., and Wooley, W.D., Elsevier Applied Science Publishers, New York (1986).

40.   Creyf, H. and Fishbein, J., "Advance of Flexible Polyurethane Foam Technology," in *Fire and Cellular Polymers*, Eds., Buist, J.M., Grayson, S.J. and Wooley, W.D., Elsevier Applied Science Publishers, New York (1986).

41.   Naval Ammunition Depot, Crane, Indiana 47522, "Fire Retardant Foam," undated brochure (contact M. Philip Smith or Mr. Al Smith).

42.   Frisch, K.C., "Urea–Formaldehyde Foams," Chapter 12 in *Plastic Foams, Part II*, K.C. Frisch and J.H. Saunders, Editors, Marcel Dekker, Inc., New York (1973).

43.   Staff Written, "Additives and Modifiers — Mold Release Agents," *Plastics Compounding 1991/92 Redbook*, 14(4).46, 48 (July/August 1991).

44.   Staff Written, "Special Report – Chemicals and Additives – External Mold Release," *Modern Plastics*, 69(9):94–96 (September 1991).

45.   Foams — Desk–Top Data Banks®, Edition 2, M.J. Howard, Editor, International Plastics Selector, Inc., San Diego, California, a subsidiary of Cordura Publications (1980).

46.   Saunders, J.H. and Hansen, R.H., Chapter 2 — "The Mechanism of Foam Formation," in *Plastic Foams, Part I*, Eds., Frisch, K.C. and Saunders, J.M., Marcel Dekker, New York (1972).

47.   Ingram, A.R. and Fogel, J., Chapter 10 — "Polystyrene and Related Thermoplastic Foams," in *Plastic Foams, Part 1*, K.C. Frisch and J.H. Saunders, Editors, Marcel Dekker, New York (1972).

48.   Kumar, V. and Weller, J.E., "Microcellular Polycarbonate — Part 1: Experiments on Bubble Nucleation and Growth," in *Proceedings, ANTEC 91 — Annual Technical Conference (49th), Society of Plastics Engineers (SPE)*, Technical Papers, Vol. XXXVII, pp. 1401–1405 (May 1991).

49.   Methuen, J.M. and Dawson, J.R., Chapter 8—"Reinforced Foams," in *Mechanics of Cellular Plastics*, Ed., Hilyard, N.C., Macmillan, New York (1981).

50.   Titow, W.V. and Lanham, B.J., Chapter 10—"Reinforced Foams," in *Reinforced Thermoplastics*, Applied Science Publishers, London, U.K. (1975).

51.   *Additives for Plastics—Desk–Top Data Bank®*, Edition 1, Rex Gosnell, Technical Consultant, International Plastics Selector, Inc., San Diego, California, a subsidiary of Cordura Publications (1980).

52.   Saunders, J.H. and Frisch, K.C., *Polyurethanes—Chemistry and Technology, Part II*, Technology, Interscience Publishers, New York (1964).

8

METHODS OF MANUFACTURE

Arthur H. Landrock

INTRODUCTION

This chapter will discuss the following manufacturing techniques: molding (reaction injection molding, liquid injection molding, slabstock molding), spraying, frothing, laminating, structural–foam molding, syntactic foam preparation, and foam–in–place techniques.

There is no sharp line of demarcation between the manufacturing methods described in this chapter. For example, some writers consider "foam–in–place" or "pour–in–place" varieties of molding. The technique of "frothing" may be used in "pour–in–place" methods. Slab stock prepared by continuous slab production is used in lamination (1).

MOLDING

Foam molding operations are those in which a liquid mixture of foam components is poured into a cavity to form a cellular object or shape. The molded item is later removed after setting or curing. In the case of expandable polystyrene beads the pre–expanded or virgin beads are poured into a mold and heated to form the desired object. In this case, liquids are not used, although the free–flowing beads might be considered as a "fluid."

A number of flexible foam components are molded, rather than fabricated from slabstock. For intricately shaped articles, molding results in savings up to 15%, compared to slabstock. Rigid foam can also be molded, although such applications are fewer in number. In a typical low–pressure molding operation a mold is preheated and coated with a mold–release agent. A preheated amount of liquid mix is poured into the mold, the mold is closed and the ingredients foamed to the mold configuration. After curing and cooling, the part is stripped from the mold. the scrap rate is seldom more than 5–10% (2). Mold temperatures are usually in the range of 150°–160°F (66°–71°C). The best results are usually obtained when the mold–release agent is spray–applied to the clean, warm mold just before each pour. Care must be taken, however, to remove all solvents from the mold–release compound before the foam components are poured into the mold cavity (3).

Curing of the molded polyurethane foam part is usually carried out in two stages, a precure of 15–20 minutes of about 270°F (132°C), permitting removal of the part from the mold, and a final cure of 60 minutes at the same temperature (3). Mold release of polyurethane foams can be difficult since uncured polyurethane has well–known adhesive properties. Two basic types of mold–release agent are used. The first requires the hot molding to be stripped from the mold. Mixtures of paraffin and microcrystalline waxes are used for this technique, in which hot wax releases the part from the mold. The molds must be heated and coated with wax before each filling. There is a tendency, however, for the paraffin wax to be slowly oxidized by the repeated heating. For this reason a release agent containing an antioxidant should be used. The breakdown products formed have no release properties, and it is important to use a thin layer of wax each time a molding procedure is carried out (4).

The second type of release agent, such as polyethylene waxes, is used if the mold must be stripped away when cold. In this case the foam comes away from the release agent. Mold–release problems can be reduced by adding small amounts of dibutyl tin dilaurate, which promotes curing at the surface of the mold and thus improves mold release (4).

Slabstock molding (discussed below) requires very little curing. A bank of infrared heaters suspended above the conveyor is often sufficient to facilitate curing. In conventional molding, however, the exotherm generated is not sufficient to cure the foam, and external heat must be applied. Microwave curing permits a reduction in curing time from 20 minutes to 4 minutes. Plastic molds are used with a gel coat of epoxy resin containing iron powder. The plastic molds must be cooled to an even temperature before refilling, however, and this step may introduce problems. There is evidence to suggest that foam cured with microwaves has properties slightly superior to foam cured by conventional heating, especially in compression set. Dielectric heating has also been developed for use in curing (4).

Reaction Injection Molding (RIM)

This technique was developed in Germany in the 1960's for molding urethanes. It is used currently for rigid urethane foam, flexible microcellular foam, and rigid microcellular foam. The process embodies high–pressure–impingement mixing of the liquid components before they are injected into the mold. RIM has advantages over the standard low–pressure mechanical–mixing systems in that larger parts are possible, mold cycles are shorter, there is no need for solvent–cleaning cycles, surface finishes are improved, and rapid injection into the mold is possible (5).

The RIM process involves a high–pressure delivery of two or more liquid urethane components to a very small mixing chamber where they are continuously mixed and injected into a closed mold at rates approaching 650 lb/min. The liquid components are heated to maintain low viscosities. The heart of the system is the mixing chamber, where the liquid components must be thoroughly mixed without imparting turbulence. Continuous delivery of the components to the mold is accomplished by high–volume, high–pressure recirculating pumps from liquid–storage tanks. Automatic controls are used to maintain precise flow and temperature of the resins (5).

Since mold pressures in RIM processes are usually below 100 psi (0.69 MPa), mold–clamp–pressure requirements can accordingly be low. Most RIM equipment is in the 25–100 ton range. Depending upon production quantity and quality requirements, molds for RIM may be made from aluminum–filled epoxy for low production, or from cast or machined aluminum or chrome–plated steel for longer runs. Molds must be heated to 120-160°F (49-71°C). Applications include automobile bumpers, radio and TV cabinets, furniture and business–machine housings (5).

Liquid Injection Molding (LIM)

LIM is a variation of the RIM process. The major difference is in the manner in which the liquid components are mixed. In the LIM process the entire shot is mixed in a chamber *before* injection into the mold, rather than being continuously mixed and injected, as in the RIM process. Accordingly, LIM is used to mold smaller parts, under 5 pounds (2.27 kg), which are below the capacity of RIM equipment currently available. LIM allows higher–viscosity materials, such as filled materials for shoe soles, to be used (5).

Slabstock Molding (Free–Rise Foaming)

This is a high–volume production method of molding urethane foam requiring large initial capital outlays for equipment, a large amount of floor space for the work area, and storage space for the finished products. A slab line may be 200–400 feet (61–122 meters) long. *Flexible slab* production is used to make mattresses, cushions, pillows, shock–absorbing packaging materials, peeled goods for laminated clothing, paint rollers, toys, and sponges. *Rigid slab* is used for panel insulation, flotation, packaging, and numerous other applications (6).

The process involves continuous delivery of the chemical components, at fixed rates, to a mixing head which moves across a conveyor in a pattern of parallel lines or ribbons. The foam is moved continuously by the conveyor, which is on a slight downgrade to keep the liquid mix from being deposited on top of rising foam. In this manner a uniform end product is maintained. The liquid mix is dispersed into a continuous mold, which is usually roll–paper formed into a trough by the conveyor and adjustable sideboards. From the conveyor the foam is moved through a curing oven. Flexible foams are usually moved through crushing rolls, although crushing is not necessary with most of the one–shot urethane formulations. However, more uniform load–bearing properties are obtained if the foam is flexed. Most one–shot foams can be flexed as soon as they are tack–free (about 15–30 minutes after formation). Foams catalyzed with stannous octoate alone are slower–setting and cannot be flexed for at least one hour after pouring. These foams are sometimes held as long as 16–24 hours before flexing (7).

Temperature control of the raw materials used in slab production is extremely important. An entire production run can be ruined if the temperature of one of the materials is allowed to change, causing a pressure change and off–ratio output. The density and other properties of slab foam are dependent upon the temperature of the foam ingredients. Foam density increases as the temperature of the foam intermediates exceeds 90°F (32°C). The temperatures of the ingredients are usually kept between 70° and 90°F (21° and 32°C) (8).

Continuous mixers are usually used with high– or low–pressure machines to produce outputs of 100 pounds (45.4 kilograms) or more per minute. The usual range is 150–200 pounds (60–90.7 kilograms). After the foam has risen it is usually passed under curing lamps, heaters, or low–pressure steam to eliminate the surface tack. Exhaust hoods are positioned to remove any irritants which may have volatized during the foaming process. The slab then passes through a series of auxiliary equipment, which may include cut–off saws, horizontal and vertical trimmers, slicing machines, and hot–wire cutters. A variety of machines for cutting and splitting flexible urethane foam are commercially available. The most common processes for the splitting of slabstock are centrifugal peeling and horizontal table splitting (7).

The width of the finished foam bun is determined by the spacing of the adjustable sideboards which bank the paper through. The dimension may be up to 80 inches (203 centimeters) and, for economic reasons, it is usually a multiple of the desired width of the end product. The height of the bun (finished foam product) is a function of throughput and conveyor speed and may be up to 40 inches (102 centimeters). Interrupted slab operations may be used where it is not economically practical to produce foam on a continuous–slab basis. The resultant slabs or buns may be 60–72 inches (152–183 centimeters) wide by 80–96 inches (203–244 centimeters) long by 12–15 inches (30–38 centimeters) high. "On–off" or intermittent equipment is used. While in a continu–ous–slab operation the first few feet of material can be sacrificed, such is not the case here (6).

SPRAYING

Spray techniques are used for filling molds and panels and for applying foam to plane surfaces. Spraying is particularly useful in applications where large areas are involved, such as tanks or building walls. Spraying is the simplest and least expensive way to produce urethane foam. In addition, spraying equipment is reasonably priced and portable, and foam can be applied without molds or jigs of any kind. In spray applications the ingredient mixing is accomplished by atomization of the materials as they leave the nozzle of the spray gun. Resin viscosities in the range of 500–1500 cP (0.5–1.5 Pa· s) should be maintained to ensure intimate mixing of the foam constituents. When necessary, heat may be applied to the polymer to reduce the viscosity to the desired level (9).

Commercial spray guns are capable of applying a 12–inch (30.5 cm) circle of urethane foam when the gun is held at a distance of 3 feet (0.9 meters) from the target surface. Larger deliveries and different spray configuration are possible with special guns. The spray technique may be used for filling molds (spray pouring) by confining the spray pattern as it leaves the spray gun. This is accomplished by attaching a cone to the gun head. The cone may be from 8 inches (20.3 centimeters) to 2 feet (61 centimeters) in length and is attached with the smaller diameter at the outlet end. This device also serves to minimize catalyst loss into the air and confines the foam discharge to a smaller area. With this method large deliveries of foam material are possible. This technique provides interesting possibilities when filling intricate or large molds. It is relatively inexpensive and highly flexible. Adhesion of the sprayed foam to most surfaces is excellent (9).

Conventional spraying is usually carried out with portable metering equipment, which should be lightweight, inexpensive and compact, so that it can be carried on a small truck or trailer. Most of these units consist of air–driven dual–acting piston pumps which can be readily calibrated and used with the original 5– or 55–gallon (19– or 208–liter) shipping containers. There are two basic types of guns used for spray operations, *internal–mix* or *external–mix*. The internal–mix guns use a number of different methods of mixing. The external–mix guns use high–pressure air to bring the components out of the gun and then, through a swirling action, the air mixes and atomizes the components between the gun and the substrate to be sprayed (7).

To prevent sagging of sprayed urethane foams formulations with fast reaction rates are being used. The faster rates are obtained by using additional catalysts. Additional layers can be sprayed almost immediately. The nominal spray rate is about 4–8 pounds (1.8–3.6 kilograms) per minute (10). Particular attention should be given to providing adequate ventilation during the spraying operation (9). The development which has been most responsible for improvement in the spray process was the (1:1) ratio, low–viscosity, one–shot, totally catalyzed system (11). This system brought about the development of the *airless gun*. Previously atomization was achieved by the introduction of high–pressure air into a conventional mixing chamber. Such spray heads are known as air–assisted (12)(13).

The rise time for airless sprayed urethane foam is about 30 seconds. It can be refoamed for additional thickness after this period. The foam can be walked on after 3–4 minutes and reaches its full properties in 24 hours. The airless spray gun is held about 30 inches (76 centimeters) from the surface and moved steadily over it (12). Dispens–

ing rates of 4–6 pounds (1.8–3.6 kilograms) per minute are generally considered optimum for most spray applications. The surface on which the foam is sprayed must be free of loose scale or grease. The adhesion of urethane foam to steel is essentially equal to the tensile strength of the foam, provided the surface is clean. Aluminum surfaces, on the other hand, do not provide a good bond unless a primer coat, such as vinyl wash, is used prior to spray foaming (14). In contrast to the successful techniques developed for spraying rigid and semi–rigid foams, methods for spraying flexible urethane foams are at a relatively early stage of development and at present, are rarely used commercially. To date only polyester systems have been used. Applications are for carpet underlay and textile backing (7).

Spray–applied urethane foam in buildings must be protected by a non–burning barrier approved by the appropriate building codes. Without such barriers polyurethane foam will spread fire rapidly once ignited, even with fire–retardant grades. With few exceptions, all model codes require that foam plastic insulation be covered by a thermal barrier equal in resistance to 1/2–inch (1.27 centimeters) gypsum board, or be used only in sprinklered buildings. The gypsum board or equivalent is supposed to prevent the foam from reaching a temperature of 325°F (163°C) for a 15–minute period when subjected to the ASTM E 119 time–temperature curve, which averages 1100°F (593°C)(15).

FROTHING

This process was introduced by Du Pont in 1961. The process operates in a two–stage expansion system. The main idea is to introduce into the formulation another more volatile liquid, such as Fluorocarbon–12, so that instead of delivering a liquid blend into a cavity for foaming, a pre–expanded one is delivered. In this manner a froth stream of 8–12 lb/ft^3 (128–192 kg/m^3) density is poured into a cavity where it completes its expansion to a low density of 1.5–2.5 lb/ft^3 (24–40 kg/m^3), which is the density range desired. In the froth stage the stream is quite fluid, resembling shaving cream in appearance, and flows readily. Frothing is particularly suited for void filling or panel manufacture. The advantages of this process are (16)(17)(7):

- Lower mold pressures during foaming, minimizing the need for jigs and making possible the forming of larger panels.

- Lower and more uniform foam densities.

- Ability to lay down an expanding foam without causing its collapse or density change.

- Possibility of cross–section molding without excessive sacrifice in density.

- Lower densities obtainable with low–temperature molding.

- Froth may be screeded (leveled) by methods similar to those used with concrete in the construction industry.

Frothing of the urethane foam–component mix occurs when the volatile liquid (Fluorocarbon–12) is vaporized by a reduction in pressure as the material is discharged from the mixer. By employing a combination of fluorocarbons of different boiling points, such as Fluorocarbon–11 and 12, a two–stage expansion is possible. The vaporization of the low–boiling solvent occurs in the initial stage, with the final foam density resulting from the supplemental expansion of the higher–boiling solvent (Fluorocarbon–11) due to the heat generated by the reactions between the hydroxyl– and isocyanate–containing foam components. With conventional molding techniques the foam must expand 30–40 times in the mold to reach its final volume, while in the frothing process the final expansion in the mold is reduced to only 3–6 times the froth volume (17).

One of the major problems with the frothing process is the lack of fully perfected dispensing equipment. In addition, foam prepared by this method often has many elongated holes approximately 1/8–1/4 inch (0.32–0.64 centimeters) in cross–sectional diameter (7). In molding, when using the frothing process, large vertical pours can be made in one continuous pour and the knit line or hard spots do not pose as big a problem as with a non–froth system. The width of the cavity to be filled is limited by the diameter of the froth streams from the let–down valve. Panels and voids of 2–inch (5.1–centimeter) width with a foam rise up to 8 feet (2.44 meters) have been filled with froth foam. The main problem with the horizontal pouring of froth foam is the distribution of the froth over the mold surface. With current froth systems conveyors and mechanical distribution are necessary with horizontal pouring (18).

Froth may also be *sprayed.* If the equipment is properly designed and adjusted, R–12 pressure reduction may be used, instead of air, to propel the material to the surface being sprayed (16).

Frothing techniques may be used in vertical–pouring applications of rigid urethane foam where the surface–to–volume ratio is high. Examples are found in the insulation of refrigerated railroad cars and trailer vans, large refrigerators, or simple building panels. Because of the fluidity of the product after dispensing and its quick curing, frothing is very well-suited for continuous production of laminated roll stock using flexible skin materials such as foil, corrugated paper, and cardboard. It may be necessary to use some sort of spreader to ensure even spreading of the partially expanded foam. Frothing is particularly useful in insulating railroad freight cars. This technique provides complete wrap-around insulation with no through-wall connections. An average freight car can be insulated by three men in three hours (16).

LAMINATING

One of the largest uses for polyester-type flexible urethane foam is in foam laminates for clothing interlinings. The first step in preparing the foam for clothing lamination is to take slab stock (described above), cut it into roughly square logs, and peel them as one would peel wood log for veneer. Thicknesses of about 8/32 inch (0.64 centimeter) and 3/32 inch (0.24 centimeter) are generally used. The foam can be butt-joined easily by heat to form large continuous rolls of foam (19).

There are basically three methods of producing foam laminates commercially. The first is the *heat-fusion* or *flame-lamination* process, which is most popular in the U.S. In this process controlled naked gas flames are allowed to impinge upon a traveling layer of foam. This produces a layer of molten urethane resin by micromelting the upper portion of the foam (about 1/3 the thickness). The fabric is immediately pressed against the tacky surface and adheres firmly. Good peel strength is obtained. The laminate can be rolled up immediately and prepared for use. No additional adhesive is required. At one time only polyester foams could be utilized for flame laminations, but now polyether types can also be used. To make them susceptible to flame lamination certain organic phosphorus compounds are added. Polyether foams are preferred in soft-clothing interlinings (19).

In most textile applications a cloth lining must be sewn against the foam side. A comparatively new adaptation of flame lamination is now being used to obviate this step. In this method two fabrics form a sandwich with a very thin foam interlining. One fabric is flame-laminated to one side of the foam, followed by flame laminating of another fabric to the other side. After flame laminating so much of the foam is used up that it in reality is only acting as an adhesive. In this way a liner is adhered in place, rather than sewn, with little or no loss in breathability, with gain in warmth, and with practically no gain in weight. Foams as thin as 40 mils (1.0 mm) are being used. Polyester foams are preferable for this process. Polyether foams are difficult to peel in such thin layers. They also have low tear strength and a tendency toward electrostatic build up. This results in the foam clinging to the cutting blade and bunching up. The need for antistatic bar protection is pronounced (19).

The second widely used method for preparing polyurethane foam laminates is the *liquid adhesive*, or *wet process*. Here special adhesives, either in the form of water solutions, or as solutions in organic solvents, are applied to either the fabric or the foam. The equipment is conventional. The water or solvent is evaporated, and the bond may be set by drying or curing at elevated temperature (20). One of the adhesives used for this application is based on acrylic interpolymer latices. This adhesive is used in the manufacture of thermal garments and insulated bags. Carpet underlay can also be made by adhesive laminating (7).

The third method, the so-called *frothing technique*, consists of foaming the material directly onto the base materials, such as films or impervious fabrics. The previously unsolved problem of controlling strike-through when foaming on fabrics has been solved by simply wetting the fabric with ordinary tap water. Almost any fabric can be used, although the technique is more difficult with the more porous fabrics, which may require the use of thickeners. The fabric is usually immersed in water, after which the moisture level is controlled by squeegeeing or heating. After the foam has been poured onto the textile the entire laminate is passed through an oven. This serves the dual purpose of curing the foam and drying out the fabric (20)(21).

STRUCTURAL FOAM PREPARATION

Chapter 3 on Thermoplastic Foams discusses structural foams in general and considers the properties of individual structural-foam types. This discussion is concerned solely with methods of manufacturing these foams.

Structural Foam Molding

There are a number of different systems and machines available for producing structural foams. The most important categories are low-pressure molding and high-pressure molding. *Low-pressure molding* involves the use of nitrogen gas or chemical blowing agents as expansion devices. In low–pressure systems, the molds, which are under very low pressure, are only partially filled with the melt. This results in a "short shot". The melt is then expanded by the nitrogen gas, or by decomposing blowing agents, to fill the mold. With nitrogen–gas molding the resin, in pellet form, is fed into an extruder where it is plasticated and mixed with the nitrogen before it is injected into the mold. With decomposing chemical blowing agents (CBAs) they are either already incorporated into the molding pellet, dry–mixed into the resin by processor, or supplied as agent/resin concentrate. A number of different types of machinery are used in this process. Special equipment is required for the nitrogen process, but with CBAs conventional injection–molding equipment, with minor modifications, can be used. Some machines are modified especially for structural–foam molding. Modifications include oversize injection pumps, large platen areas, and multiple–mold stations (22).

High–pressure molding involves injecting the polymer melt and the blowing agents under higher pressures into the mold cavity to completely fill the mold. The mold then expands or mold inserts are withdrawn to accommodate the foaming action. It is possible to use specially adapted conventional machines in this system. Special machines with accumulators can also be used. Special techniques applicable to structural–foam molding include expansion casting of ABS and other thermoplastic foams, rotational molding, and multi–polymer molding. By using solid chemical blowing agents it is possible to rotationally mold a multi–layer system of solid skin and cellular core. By formulating to proper decomposition temperatures a layer of solid polymer can be laid down near the mold surface to function as a skin. With continuation of heating, expansion then takes place in the rest of the polymer toward the end of the cycle and simultaneously with complete fusion (22).

A third method of producing structural foams is *sandwich molding*, used to produce a foam part with a solid skin and cellular core. This system uses an injection–molding machine with two injection units. The resin from one injection unit is partially injected into a mold cavity. Resin, with incorporated blowing agent, is then injected from the second unit into the same mold. This second shot forces the first polymer to the edges of the mold cavity where it sets to form the solid skin. The second shot, with the blowing agent, then expands to form the cellular core beneath the skin (22).

Structural Foam Extrusion

Structural foams with solid integral skins and cellular cores can be extruded in the form of profiles, pipe, tubing, sheet, etc. Methods in use for extrusion involve conventional single–screw machines handling plastics incorporating chemical blowing agents. Direct gas injection and tandem extrusion set–ups have also been used. Applications include polystyrene and vinyl profiles for use as building trim and moldings, and picture frames (22).

SYNTACTIC FOAM PREPARATION

These foams, treated above in Chapter 2 on Thermosetting Foams, usually consist of tiny hollow spheres of phenolic resin or glass held together by an organic resin binder. By their nature they are always closed–cell materials, with the solid resin matrix being the continuous phase and the hollow spheres the discontinuous phase. The resin matrix or binders are usually phenolic, polyester, or epoxy resins. The uncured resins have the consistency of putty. They are cured or hardened into foams having densities of 10 to 14 lb/ft^3 (160 to 640 kg/m^3). These foams are unique in that they are formed without the use of a blowing agent.

The syntactic foams are made by simply mixing the micro– or macrospheres into a catalyzed resin until the desired consistency is obtained. In most cases the materials are mixed to a putty–like state, or, if a casting material is desired, to a state in which the material can just be cast. The usual ratio of filler to resin is approximately 60% filler by volume (23).

Very low–density syntactic foams can be achieved by using *polymodal packing*, in which different–size hollow spheres form the syntactic foam. Blending two sizes of spheres with a minimum diameter ratio of 10:1 results in more efficient packing, with the smaller spheres filling the spaces between the larger ones. The incorporation of micro– and macrospheres for example, results in a *bimodal system* with a packing efficiency of 75 to 80%, a significant increase from 60 to 65% when either filler is used alone. Such a system has a lower density, replacing a large portion of polyester or epoxy resin (24). Such a reduction in resin–matrix content naturally results in lower compressive strength.

A propeller–type mixer is satisfactory for preparing experimental batches of syntactic foam, but on a commercial scale a kneader–type mixer should be used. With the machine in operation, the hardener is added to the epoxy resin, or the peroxide catalyst and cobalt naphthenate accelerator to the polyester resin. Peroxide and napthenate must be added separately, since they react explosively together. The phenolic spheres are then thoroughly mixed in. The putty–like mass is transferred to a suitable mold, trowelled onto a surface, pressed into cavities, or placed as a core in a sandwich structure. Curing will take place at room temperature, but can be hastened by heating up to 212°F (100°C) (22).

FOAM–IN–PLACE (FOAM–IN–BAG) TECHNIQUES

This is a system used for blocking and bracing of the containers. It is widely used in military applications. It involves the use of polyure–thane foam ingredients being poured around the item to be protected. When the urethane sets the foam acts to protect the item from damage due to rough handling. The item must always be protected by polyethyl–ene bags or film to prevent adhesion of the foam to the item being protected (25).

In one typical method used by the Army a pair of polyethylene bags are placed around the items to be protected. The polyurethane foam ingredients are injected into the bags and, following foam rise and cure, the item is protected from movement within the container until it is removed. The function of the foam is that of blocking and bracing since the foam is rigid. The foam has no cushioning functions. At any time after cure the box may be opened and the "molded" bags removed, exposing the items so protected (25).

Another system used is "encapsulation". In this system the item is completely surrounded by a homogeneous mass of foam. The item is usually protected by a cocoon of plastic film before the foam is applied. The greatest difficulty in this method is removal of the foam after use. Tear strings of wire are frequently used to facilitate this removal (25).

Military standardization documents currently in effect prescribing procedures for foam–in–place (FIP) are listed in Chapter 11. These standards are as follows:

MIL–STD–1191	Foam–in–Place Packaging, Pro–cedures for (1989)
MIL–P–43226A (MI)	Polyether Cushioning Material, Foam–in–Place, Flexible (1991)

MIL–P–83379A (USAF)	Plastic Material, Cellular Poly-urethane, Foam–in–Place, Rigid (3 pounds per cubic foot dens-ity) (1980)
MIL–P–83671 (SA)	Foam–in–Place Packaging Mate-rials, General Specification for (1990)
MIL–F–87075B	Foam–in–Place Packaging Sys-tems, General Specification for (1984)

REFERENCES

1.   Landrock, A.H., "Polyurethane Foams: Technology, Properties and Applications", *PLASTEC Report 37*, (January 1969). Available from NTIS as AD 688 132.

2.   Backus, J.K. and Haag, E.C., Chapter 35, "Urethanes", in *Machine Design*, Reference Issue – Plastics, 38 (14): 117–119 (June 16, 1966).

3.   Dixon, S., "Polyurethane Foam Molding", *British Plastics*, 36 (1): 24–27 (January 1963).

4.   Parkinson, J.C., "Moulding Flexible Polyurethane Foam", *British Plastics*, 27 (3): 146–147 (March 1964).

5.   FOAMS, edition 2, Desk–Top Data Bank® ed. by M.J. Howard, The International Plastics Selector, Inc., San Diego, California, (1980).

6.   Brabandt, H., Jr., "Multi–Component Foams Processing", Part I in Section III, "Manufacturing Equipment for Foamed Plastics", *Handbook of Foamed Plastics*, Ed. by R.J. Bender, Lake Publishing Corp., Libertyville, Ill., (1965), p. 19–38.

7.   Saunders, J.H. and Frisch, K.C., Part II – Technology, in *Polyurethanes — Chemistry and Technology*, Interscience Publishers, New York, (1964).

8.   Anonymous, "Continuous Production of Urethane Foam Building Panels", *Rubber and Plastics Age,* 46 (121): 1283 (November 1965).

9.   E.I. DuPont de Nemours & Co., Inc., Elastomer Chemicals Div., Wilmington, Delaware, "Rigid Urethane Foams — Methods of Application", Bulletin A–4713, (June 1957).

10.   Union Carbide Corp., Chemicals Div., New York, "Rigid Urethane Foam", unnumbered brochure, (1964).

11.   Hersch, P., "How to Buy and Use the Newer Foam Spray Equipment", *Plastics Technology*, 12(12):32–36 (December 1966).

12.   Brooks, W.R., "Advantages of New Airless Atomization Spray Gun Explained", *Journal of Cellular Plastics*, 1(3):419 (July 1966).

13.   Lerner, A., and Coen, R., "Polyurethane Foams", in "How to Buy Equipment for Foam Processing" (Anonymous), *Plastics Technology*, 12(7):97–99 (July 1966).

14.   Bowman, R.A., "Rigid Polyurethane Spray Foam Technology", Technical Papers, Vol. X, 20th Annual Technical Conference, Society of Plastics Engineers, Atlantic City, New Jersey, January 27–30, (1964), Section XXVI, pp. 1–4.

15.   Landrock, A.H., *Handbook of Plastics Flammability and Combustion Toxicology*, Noyes Publications, Park Ridge, NJ, (1983).

16.   Anonymous, "Foamed Organics, A Comprehensive Treatment on Foamed Organics Based on Selected Articles from *Chemical Engineering Process (CEP)* plus unpublished manuscripts on the Subject", A CEP Technical Manual published by the American Institute of Chemical Engineers, New York, (1962).

17   Knox, R.E., "Frothing — A New Method for Producing Urethane Foams", *Chemical Engineering Progress*, 59(10):40–47 (October 1961) (Also in Reference 16 above).

18. Stengard, R.A., "Void Filling with Frothed Rigid Urethane Foam", duPont Brochure, A–19555, (March 29, 1961).

19. Dombrow, B.A., *Polyurethanes*, 2nd Edition, Reinhold, New York (1965).

20. Ehrlich, V.A., "Fabric Foam Laminates", *Modern Textiles*, 44(2): 37–40 (February 1963).

21. Anonymous – "New Technique Foams Urethane on Fabric", *Plastics Technology*, 12(2):11 (February 1966).

22. *Plastics Engineering Handbook of The Society of The Plastics Industry, Inc.*, 4th Edition, ed. J. Frados, Van Nostrand Reinhold, New York (1976). (*Authors Note:* The 5th Edition, edited by M.L. Berins, was published in 1991).

23. Schwartz, S.S. and Goodman, S.H., *Plastic Materials and Processes*, Van Nostrand Reinhold, New York (1982).

24. Wehrenberg, R.H., "Shedding Pounds in Plastics: Microspheres Are Moving", *Materials Engineering*, 88(4):58–63 (October 1978).

25. Landrock, A.H., "Handbook of Plastic Foams", *PLASTEC Report* R52 (February 1985). Available from NTIS as AD A156758.

9

SOURCES OF INFORMATION

Arthur H. Landrock

INTRODUCTION

Considerable useful and up–to–date information on plastic and elastomeric foams is available from journals, manufacturers' bulletins, technical conferences and their published proceedings, seminars and workshops, standardization activities, trade associations, consultants and information centers (such as PLASTEC), in addition to books, many of which have been cited in the previous chapters. Some of these sources will be listed and commented upon briefly in this chapter.

JOURNALS AND OTHER PERIODICALS

Journal of Cellular Plastics

This is a bimonthly publication published by Technomic Publishing Co., Inc. at 851 New Holland Avenue, Box 3535, Lancaster, PA 17604. This journal has been published for 24 years and is a major source of information in foamed plastics technology, covering new developments and applications In addition to articles on product applications and markets there is an index of foamed plastic patents from industrial nations and a section on industry news.

Journal of Thermal Insulation

This is a quarterly publication published since 1977. The journal provides in-depth reports on important developments in thermal insulation, including cellular plastics.

Cellular Polymers—An International Journal

This highly regarded journal is published by the Publications Group, RAPRA Technology Ltd. in the UK. Six issues per volume, one volume per year. The journal can be obtained in the US through Elsevier Science Publishing Co., Inc., Journal Information Center, 655 Avenue of the Americas, New York, N.Y. 10010.

The journal deals with not only every type of cellular polymer, from elastomeric materials to rigid plastics, but also raw materials, manufacturing processes, and application technology. Coverage includes Conference Reports, Conferences and Seminars Listings, Selected Abstracts, and Book Reviews. Averages three to five articles per issue.

Plastics in Building Construction

This 12-page publication is published monthly by Technomic Publishing Co., Inc, 851 New Holland Avenue, Box 3535, Lancaster, PA 17604. The publication covers *Industry News, Product News, Digests,* and usually has a short feature article with references. Because of the subject area, plastic foams are often featured, usually involving thermal insulation.

Urethane Plastics and Products-Marketing-Technology-Applications

This 8- to 12- page publication is published monthly by Technomic Publishing Company, address above. Typical issues have discussions of new processes, applications, new materials, acquisitions, phaseouts, and other news items of interest in both cellular and non-cellular urethanes. Illustrations are sometimes used.

Urethane Abstracts

This monthly publication, averaging about 43 pages per issues, is also published by Technomic Publishing Co., address above. It contains abstracts of articles and patents of interest in both cellular and non–cellular urethanes.

Plastics Focus—An Interpretive News Report

This six–page "newsy" publication is published bi–weekly by the Plastics Connection, Inc, P.O. Box 814, Amherst, MA 01004. It contains news items concerning all aspects of the plastics industry, including cellular plastics. Typical entries include information about mergers, new products, acquisitions, and industry news.

Modern Plastics

This monthly trade magazine is published by McGraw Hill, Inc, 1221 Sixth Avenue, New York, NY 10020. The well–known *Modern Plastics Encyclopedia* is published as a second issue in October of each year and includes considerable technical information on plastic foams. Headings of the magazine currently are *Business & Management*, *Engineering & Technical*, and *Departments*. Just about any subject matter of interest to the plastic foam user is covered under any one of a large number of headings. A typical issue has about 140 pages.

Plastics Technology

This monthly trade magazine, somewhat smaller than *Modern Plastics*, calls itself "The Magazine of Plastics Manufacturing Productivity." It is published by Bill Communications, Inc, 633 Third Avenue, New York, NY 10164–0735. The emphasis is on new developments. *Plastics Technology* also publishes a valuable Manufacturing Handbook and Buyers' Guide in mid–July. This publication is available separately.

Plastics World

This is a trade magazine published for "managers" in plastics. It is published monthly with an extra issue in March, by Cahners Publishing, a Division of Reed Publishing USA. Subscriptions are available at Plastics World, 270 St. Paul St., Denver, CO 80206. Each issue has a "mix" of Business News and Technology News, along with a number of other headings, all aimed at management. Controlled circulation (free) subscriptions are available to qualified individuals.

Materials Engineering

This magazine is published monthly by Penton Publishing, a subsidiary of Pittway Corp, 1100 Superior Avenue, Cleveland, OH 44114. The magazine combines technological developments and news. State-of-the-art articles are frequently published on all materials, including plastic foams. A *Materials Selector* issue is published annually and is available hardbound.

Plastics Engineering

This is a monthly publication issued free to all members of the Society of Plastics Engineers (SPE). It is published by SPE at 14 Fairfield Drive, Brookfield Center, CT 06805. The emphasis is on practical aspects of plastics engineering. Coverage frequently includes additives, materials, design, processing, machinery, and announcements. One section covers new literature. Dates of forthcoming events are also given.

Plastics Design Forum

This magazine is intended to serve "designers of products and components in plastics." It is published bimonthly by Resin Publications Inc, a subsidiary of Edgell Communications Inc. Subscriptions are available from Plastics Design Forum, P.O. Box 6349, Duluth, MN 55806-9933. The magazine provides articles from designers of products and components in plastics. Book reviews are sometimes given.

Plastics Compounding

This magazine is published bimonthly, with an extra issue in June, by Resin Publications, a subsidiary of Edgell Communications Inc. Subscriptions are available from Plastics Compounding, P.O. Box 6350, Duluth, MN 55806. The magazine is intended primarily for use by resin producers, formulators, and compounders. The *Plastics Compounding Redbook*, issued in June of each year, is designed to provide the compounder with detailed information on the additives, modifiers, colorants fillers, reinforcements, and equipment for compounding. *Foam blowing agents* are included.

Machine Design

This trade magazine is published twice monthly, except for November, when there is a single issue. In addition, a Reference Issue is published in each of five months. The special Reference Issue on Materials is issued in mid April. This covers all types of plastics, including foams, in a primer format, with excellent tables and illustrations. The Reference Issues may be purchased separately. Subscriptions are available by contacting Machine Design, 1111 Chester Avenue, Cleveland, OH 44114. Controlled (free) circulation subscriptions are available to qualified individuals. The publisher is Penton/IPC, Inc.

Advanced Materials & Processes (incorporating Metal Progress)

This trade magazine is published monthly, plus bi-imonthly in June, by ASM International, Materials Park, OH 44073. It covers all high-performance engineering materials and contains much "newsy" information. A number of excellent technical articles on non-metals, such as composites and reinforcements, have also been published. ASM International now includes all materials, including cellular, in its scope. A typical issue had four feature articles and the following departments: Global Report, Materials in Action, News at a Glance, Materials Newsletter, Listings of Products, Company Literature, Book Reviews, and a Calender of Events.

Plastics Packaging

This is a new glossy trade magazine published bi-monthly by Edgell Communications, Inc, 7500 Old Oak Blvd, Cleveland, OH 44130. Editorial offices are at 1129 East 17th Avenue, Denver, CO 80218. The emphasis in this magazine is on Design, Manufacture and Performance, as indicated in the subtitle. Based on the subject area, it is quite likely that cellular plastic cushioning materials will be a topic of concern to be covered.

Plastic Trends

This is a small bimonthly trade magazine covering technology with an emphasis on business. Subscriptions are available from Plastic Trends Publishing Inc 128 E. Katella Avenue, Orange, CA 92667.

Design News

This magazine, covering "news for design engineers," is published twice monthly, with one directory issue, in each of the months of March, May, July, September and November, by Cahners Publishing Co., a division of Reed Holdings, Inc., 275 Washington Street, Newton, MA 02168. A typical issue runs over 200 pages and covers many types of engineering materials, including cellular plastics.

SAMPE Journal

This journal is published bimonthly by the Society for the Advancement of Material and Process Engineering (SAMPE), a professional society covering all materials and processes, including plastics and plastic foams. Subscriptions are available from the SAMPE International Business Office, 843 West Glentana, P.O. Box 2459, Covina, CA 91722. A typical issue will have about six feature articles along with a number of features, including activities, news of SAMPE chapters, and materials news.

SAMPE Quarterly

This publication is also published by SAMPE (see above). This publication carries no advertising. A typical issue has 6–8 technical articles with abstracts plus a "materials calendar."

ASTM Journal of Testing and Evaluation

This journal is published bimonthly by the American Society for Testing and Materials (ASTM) at 1916 Race Street, Philadelphia, PA 19103. The editorial objectives are to 1) Publish new technical information derived from the field and laboratory testing, performance, quantitative characterization, and evaluation of materials, products, systems and services, 2) Present new test methods and data and critical evaluation of these methods and data, 3) Report the users' experience with test methods and the results of interlaboratory testing and analysis, 4) Provide the scientific basis for both new and improved ASTM standards, 5) Stimulate new ideas in the fields of testing and evaluation.

The journal also publishes review articles, technical notes, research briefs, and commentary. From time to time papers and discussions on plastic foams are included. This journal is highly scientific and contains no advertising. Some issues have book reviews. Each article published has an abstract and a list of key words.

ASTM Standardization News (SN)

This magazine is published monthly by ASTM (see address above). It is basically a news magazine of all ASTM activities and usually has featured articles on the activities of particular ASTM committees. It also publishes actions on current ASTM Society Ballots and Standards Actions (accepted revisions of standards and new standards).

Fire and Materials—An International Journal

This excellent journal is an international forum for scientific and technological communications on the behavior of all materials and composites in fire in a variety of environments; on fire hazard, e.g. flame, smoke, toxic and hot gases, and heat transfer; on test methods; on the investigation of ignitability, flammability, retardancy and extinguishment; and on the fundamentals of fire. The journal is published quarterly by John Wiley & Sons Ltd, Baffins Lane, Chichester, Sussex, England. U.S. Subscriptions are available through Publication Expediting Services Inc., 200 Meacham Avenue, Elmont, NY 11003.

A typical issue will have four highly scientific articles with many references, abstracts and key–word listings. Occasionally Short Notes, Correspondence, and News and Diary sections are published.

Fire & Flamability Bulletin (formerly *Rubber and Plastics Fire and Flammability*)

This publication considers itself a "monthly monitor of the international fire scene." The bulletin provides a continuous update on flammability news and comprehensive coverage of a number of areas, including:

- legislation and regulations
- standards and codes of practice
- new products, equipment and services
- test procedures and methods
- fire statistics and trends
- research results

- fire–detection and suppression equipment
- recently published literature
- meetings, conferences and exhibitions

Subscriptions are available from Elsevier Science Inc., 655 Avenue of the Americas, New York, N.Y. 10010.

Fire Technology

This small–format journal is a quarterly publication of the National Fire Protection Association (NFPA), Batterymarch Park, Quincy, MA 02269. Each issue contains in–depth materials from leading fire–protection engineers, scientists and researchers. Issues typically contain:
- current worldwide research reports
- wide range of theroretical and applied engineering topics
- book and software reviews
- viewpoints of leading experts in fire technology

High Performance Plastics

This is a new monthly newsletter bridging the gap between polymer research and the practical applications of high performance plastics. Its scope includes reinforced plastics, engineering plastics, cellular plastics, and other polymers produced for heavy–duty or other special applications. Subscriptions are available from Elsevier Science Inc., 655 Avenue of the Americas, New York, N.Y. 10010.

Polymer Testing—An International Journal

This journal is published six issues per volume, one volume per year. It provides a forum for developments in the testing of polymers and polymeric products, including foams (cellular materials). Subscriptions are available from Elsevier Science Inc., 655 Avenue of the Americas, New York, N.Y. 10010.

BOOKS

Plastic Foams, **Kurt C. Frisch and James H. Saunders, editors, Marcel Dekker, New York, NY.**

This two-part series is an excellent source prepared by authors expert in their fields. Part 1 has an overview of the entire field of plastic foams in the introductory chapter, an extensive treatment of the fundamental principals of foam formation of nearly all types in Chapter 2, followed by specific chapters on foam varieties which are predominantly flexible. A chapter on testing of all types is included. Part 2 covers primarily rigid foam types and has a series of chapters on applications of foams.

Part 1, 1972, 464 pp
Part 2, 1973, 704 pp

Plastic Foams: The Physics and Chemistry of Product Performance and Process Technology, **by Calvin J. Benning, Wiley–Interscience, New York, NY, 1969.**

Volume 1:   Chemistry and Physics of Foam Formation, 665 pages.

Volume 2:   Structure, Properties and Applications, 363 pages.

Both rigid and flexible foams are covered in each volume.

Mechanics of Cellular Plastics, **N.C. Hilyard, editor, Elsevier Applied Science Publishers, UK, 1981, 401 pp.**

This book may be obtained from Elsevier Science Inc., 655 Avenue of the Americas, New York, N.Y. 10010. It is also available from Technomic Publishing Co., Lancaster, PA as Technomic Book No. 1906. The book covers theory; composition and structure; analysis; properties; rigid and flexible foams; and applications.

Introduction to Structural Foam, **by Stefan Semerdjiev, SPE Processing Series. Society of Plastics Engineers, Brookfield Center, CT, 1982, 123 pp.**

The author, a Senior Research Scientist in Bulgaria, has written this paper–covered book specifically for SPE. The book covers the characteristics of the structural foam process, the various machinery options needed to process a foamable resin, the important aspects of structural foam part design, and the various applications that have enjoyed commercial success.

Engineering Guide to Structural Foam, **Bruce C. Wendle, editor, Technomic Publishing Co., Lancaster, PA, 1976, 228 pp.**

This useful book has 14 chapters, as follows:

I.   Introduction
II.   Applications of Structural Foam
        A. Common Resins
        B. Engineering Resins
III.   Part Design
IV.   Post Molding Operations
V.   Materials
VI.   Mechanical Properties of Thermoplastic Structural Foams
VII.   Additives and Fillers for Structural Foam
VIII.   Chemical Blowing Agents
IX.   Processing Structural Foam
X.   Tooling
XI.   Machining for Structural Foams
XII.   Auxiliary Equipment for Structural Foam Processing
        A. Temperature Control
        B. Material Handling
XIII.   Planning a Structural Foam Production Facility
XIV.   Future of Structural Foams

Foams—Edition 2, Desk–Top Data Bank, **Michael J. Howard, editor, The International Plastics Selector, Inc., publishers, San Diego, CA, 1980, 845 pp.**

This soft–cover book, which has not been updated, is a compendium of technical property data of polymers used in the foam industry. The data has been compiled, organized and displayed in various ways in order to assist the user in locating candidate materials for foam processing. Each type of foam material is listed with its properties, as reported by the manufacturer, and is presented in a format that allows ease of comparison between one polymer and another.

Handbook of Plastic Foams **by Arthur H. Landrock Plastics Technical Evaluation Center, Picatinny Arsenal, NJ, PLASTEC Rept. R52, February 1985, 86 pp. Available from the National Technical Information Service (NTIS), 5285 Port Royal Road, Springfield, VA 72161, as AD D439 305.**

This brief summary report describes the state of the art of all types of cellular materials, particularly plastics and elastomers. The report is organized in the form of a handbook and has an Introduction and sections on Types of Foams (Cellular Materials), Methods of Manufacture, Foam Properties, Solvent Cementing and Adhesive Bonding of Foams, Methods of Making Foams Conductive, and Applications. There are two Appendices, one on Standardization Documents (Test Methods, Practices and Specifications) and the other on Definitions.

Polyurethanes: Chemistry and Technology, **by J.H. Saunders and K.C. Frisch, Interscience Publishers, New York, NY, 1964.**

These two volumes are an excellent authoritative source on all aspects of polyurethanes, not only foams, although the book is obviously outdated.

Part I. Chemistry, 368 pp
Part II. Technology, 881 pp

Polyurethane Handbook—Chemistry Raw Materials—Processing— Application Properties, **Gunter Oertel, editor, Carl Hanser Verlag, Munich, 1985.**

This book is part of the Kunstoff-Handbook series and has been translated from the German. It has fifteen (15) chapters. Chapters particularly relevant to foams are Chapter 5—Flexible Foams, Chapter 6—Polyurethane Rigid Foams, and Chapter 7—Polyurethane In-tegral Skin Foams. The many authors are internationally known experts in the field of polyurethane technology.

International Progress in Urethanes, **Technomic Publishing Co, Inc., Lancaster, PA.**

This book series presents in–depth reports on the most important new developments in urethane technology. Volumes 2–5 contain papers written only by Japanese specialists. A number of these papers concern rigid and flexible foams. Polyurethanes and polyisocyanurates are widely used in Japan. The papers cover recent developments in chemistry, processing, properties and applications.

Vol. 1	1977 291 pp. international in scope.
Vols. 2–5	all papers by Japanese polyurethane specialists.
Vol. 2	1980, 173 pp.
Vol. 3	1982, 260 pp.
Vol. 4	1985, 141 pp.
Vol. 5	1988, 231 pp.

Flexible Polyurethane Foams—Chemistry and Technology, **by George Woods, Applied Science Publishers, London, U.K., 1982.**

This book describes the chemistry, manufacture and use of the wide range of flexible polyurethane foams, from low–density open–cell upholstery foams to microcellular and reaction– injection–molded and reinforeed–reaction–injection–molded materials. The related effects of varying the raw–material chemistry and the production process and machinery on the properties and service performance of the final product are indicated.

Urethane Foams—Technology and Applications, **by Yale L. Meltzer, Noyes Data Corporation, Park Ridge, N.J., 1971.**

The detailed descriptive information in this paper–covered book is based on U.S. patents since 1965 relating to polyurethane foam manufacture. A few very relevant earlier patents are also included. The book supplies detailed technical information and can be used as a guide to the U.S. patent literature in this field. The Table of Contents is organized in such a way that it can serve as a subject index.

Precautions for the Proper Usage of Polyurethanes, Polyisocyanurates, and Related Materials, **prepared by the Chemical Division of the Upjohn Chemical Company, Kalamazoo, Michigan, 2nd Edition, 1980. Available from Technomic Publishing Company.**

This well known 98 page book was first published in 1974 as Upjohn Technical Bulletin 107. Its purpose is to inform and warn users and potential users of any chemical or any product used in making polyurethane or polyisocyanurate plastics, or any product made from them, of dangers in their misuse.

Developments in Polyurethane—1, **J.M. Buist, editor, Applied Science Publishers, London, U.K., 1978.**

This book has ten (10) chapters written by a number of special-ists, many involving urethane foams.

Modern Plastics Encyclopedia 92, **Mid–October Issue, published by McGraw–Hill, Inc., New York, N.Y., 824 pp. (This hard–bound volume is supplied free to subscribers of *Modern Plastics Magazine* as a second issue in October. It is also available for separate purchase).**

This is an excellent source. The "Textbook" section has articles on:

> Foaming agents
> Polyurethane foam catalysts
> Foam processing

The "Engineering Data Bank" has tabular data on Foaming Agents, and the "Buyer's Guide" has an alphabetical list of companies (and their addresses) involved in plastic foams and foaming agents.

The Wiley Encyclopedia of Packaging Technology, **Marilyn Bakker, editor–in–chief, Wiley–Interscience, New York, N.Y., 1986, 746 pp.**

This monumental source has an informative four–page article on Foam Cushioning covering the following topics:

- Expanded Polystyrene (EPS) Foam
- Expanded Polyethylene (EPE) Foam
- Expanded Polyethylene Copolymer (EPC)

● Expanded Styrene–Acrylonitride (SAN) Copolymer

The book also has a six–page article on Cushioning Design, with excellent illustrations, and a one–page article on Foam, Extruded Polystyrene (sheet).

***International Plastics Handbook*, by H. Saechtling, Hanser Publications, Munich, Vienna, and New York, 1983, 530+ pages.**

This book, sponsored by the Society of Plastics Engineers (SPE) is available in the U.S. from Scientific and Technical Books, MacMillan Publishing Co., Inc, New York, N.Y. Coverage of plastic foams includes the following:

3.1.5	Expanded plastics
4.2.4.3	PVC foams
4.7	Lightweight cellular plastics (summarizing survey)

The book indexes all ASTM, ISO and DIN standards mentioned in the text. It also lists trade names and references them to the text.

***Handbook of Fillers for Plastics*, H.S. Katz and J. Milewski, editors, Van Nostrand Reinhold Co., New York, N.Y., 1987, 480 pp.**

This excellent book, a revision of the 1978 *Handbook of Fillers and Reinforcements*, covers only fillers and has a chapter on hollow microballons for use in syntactic foams. A companion volume by the same authors covering reinforcements does not consider foam applications.

***Plastics Additives*, (2nd Edition), R. Gachter and H. Muller, editors, Hanser Publishers, Munich, Vienna, and New York, 1987, 742 pp.**

This reference book covers seventeen (17) individual types of additives. From the viewpoint of plastic foams, the most important additives considered are blowing agents and fire retardants, both of which are covered in detail in this revision.

Plastic Materials and Processes, **by Seymour S. Schwartz and Sidney H. Goodman, Van Nostrand Reinhold, New York, N.Y., 1982, 965 pp.**

This is an extremely useful book devoting equal coverage to both plastics materials and processes. The book has 23 chapters including one, Chapter 15, on Foam Processes. Other chapters also have useful information on cellular plastics and elastomers. The book is well illustrated and has a detailed table of contents and index.

SPI Plastics Engineering Handbook **(of the Society of the Plastics Industry, Inc.), 5th Edition, Michael L. Berins, editor, Van Nostrand Reinhold, New York, N.Y., 1991, 845 pp.**

Almost every chapter of this well–known book has either been completely rewritten or revised extensively. All aspects of cellular plastics have been covered in Chapter 19, a 67 page chapter.

Encyclopedia of Polymer Science and Engineering, **2nd Edition, J.I. Kroschivitz, Editor–in–Chief, John Wiley & Sons, Inc., New York, N.Y.**

Volume 3, 820 pp, 1986, has an article on *Cellular Polymers*. This article has not been seen by this editor, but a similar article in the 1st edition was very useful.

Handbook of Thermal Insulation Applications, **by Don E. Croy and Douglas A. Dougherty, EMC Engineers, Inc., Denver, Co, for the Naval Civil Engineering Laboratory, Port Hueneme, CA. January 1983, Approved for Public Release. Contractor Report No. CR83005. Approx. 220 pp.**

This compilation is a voluminous work intended to provide current design information in thermal insulation materials and assemblies for building envelopes and mechanical systems. Compiled data includes thermophysical properties of commonly used generic thermal–insulating materials and their application in wall, floor and roof assemblies, door and window assemblies and in mechanical piping, tanks, vessels, equipment, and air–duct installations. Listings of insulation materials are sorted by both manufacturers' trade names and by product descriptions. Covers all kinds of insulating materials, not only foams.

Building Insulation Materials Compilation, **by J.G. Bourne, D.L. Brownell, E.C. Guyer, R.C. Thompson, and R.P. Tye, Dynatech R/D Company, Cambridge, MA, for the Naval Civil Engineering Laboratory, Port Hueneme, CA. September 1979, Approved for Public Release. Contractor Report No. CR 80.001, 137 pp.**

This document is intended to provide a ready source of the thermal characteristics, availability, safety and other pertinent information for the selected types of insulation. The compilation covers all commonly used types of products whose primary purpose is to provide thermal resistance to heat flow through the building envelope. Due to the importance of controlling the migration of moisture to the insulating material, vapor barriers are also included.

Where possible the generic properties of a particular insulating material are presented to allow the designer to draw comparisons between types. In addition, where data are available, the variation of conductivity with, temperature and/or density is given for each generic material.

International Plastics Flammability Handbook—Principles—Regulations Testing and Approval, **by J. Troitzsch, Hanser Publishers, Munich, Vienna, and New York, 1983, 506 pp.**

This book is available in the U.S. from Scientific and Technical Books, Macmillan Publishing Co., Inc, New York, N.Y. It deals with all aspects of plastics flammability, from fundamentals to the detailed descriptions of national and international regulations, standards, test methods, product approval procedures for plastics and plastic components in the various fields of application. Foams are given extensive coverage throughout the book. The tables and illustrations are extremely well prepared.

Cellular Solids—Structure & Properties, **by L.J. Gibson and M.F. Ashby, Pergamon Press, Elmsford, N.Y., 1988, 357 pp.**

This book consists of 12 well–presented chapters and explains the structure and properties of cellular solids and of the ways in which they can be exploited in engineering design. By unifying the modeling of many different types of cellular solids similarities in behavior of these diverse materials are explained by the authors. Case studies are used to provide examples of foam behavior and selection for engineering applications.

Proceedings of the Workshop on Acoustic Attenuation of Materials Systems, **prepared by the National Materials Advisory Board (NMAB) Committee on Structural Application of Syntactic Foam, NMAB–339, 1978, 147 pp.**

This report, available for public sale from the National Technical Information Service (NTIS), was sponsored by the Office of Naval Research.

Basic Principles: Polymeric Foams—Preparation, Processes and Properties, **Technomic Publishing Co., Lancaster, PA**

This two–day seminar was presented in Atlanta, GA May 18–19, 1988 by Technomic Publishing Co., Lancaster, PA. It covered rigid, flexible, reticulated, syntactic, sintered, microporous, bead and rotomolded foams.

Handbook of Plastics Flammability and Combustion Toxicology, **by A.H. Landrock, Noyes Publications, Park Ridge, N.J., 1983, 308 pp.**

This interdisciplinary book covers all aspects of plastics flammability. Useful information relating to cellular plastics is scattered throughout the book. Chapter 5 on Fire Safety Aspects of Plastics has an extensive section on Cellular Polymers (pp 67–84). Considerable attention is given to standard test methods in Chapter 7.

Fire and Cellular Polymers, **J.M. Buist, S.J. Grayson, and W.D. Wooley, editors, Elsevier Science Publishing Co, New York, N.Y., 1986, 320 pp.**

This book, consisting of 16 chapters, is the proceedings of the Conference, "Fire and Cellular Polymers," organized by the Fire and Materials Centre, Queen Mary College and the Fire Research Station, UK, in October 1984. First published in *Cellular Polymers*, the material presented reviews the most important cellular polymers and their applica- tions in different environments where fire hazards exist. Topics include cellular polymers, rigid foam insulation board, polyurethane foams, phenolic foams, polystyrene foams, PVC foams, isocyanate–based rigid foams, flame retardants, smoke– and gas–suppressant systems, legislation, and use in transportation. Fire–related characteristics of these materials and how they can be assessed in light of the latest information and technology are discussed.

Fire and Flammability Series **(Carlos J. Hilado, Editor) The following paper–bound books were published by Technomic Publishing Company, Inc, 851 New Holland Avenue, Lancaster, PA 17604.**

> Volume 8   – Flammability of Cellular Plastics, Part 1, 227 pages.
> Volume 19 – Flammability of Cellular Plastics, Part 2, 154 pages.
> Volume 20 – Flammability of Cellular Plastics, Part 3, 296 pages.

These volumes present a selection of technical articles reprinted from various Technomic journals.

Polymer Foams—Processing and Production Technology, **Fyodov A. Shutov, editor, Technomic Publishing Co., Lancaster, PA, 1991, 474 pp.**

Handbook of Polymeric Foams and Foam Technology, **Klempner, D. and Frisch, K.C., eds., Hanser Publishers, Munich, Vienna, and New York, 1992, 454 pp.**

CONFERENCES, PROCEEDINGS, TECHNICAL BULLETINS, AND TECHNICAL REPORTS

Society of the Plastics Industry (SPI)

This trade association has several divisions holding annual technical and marketing conferences involving cellular plastics. The following listing provides details on conference proceedings and bulletins currently available from The Society of the Plastics Industry, Inc., Literature Sales Department, 1275 K Street, NW, Suite 400. Washington, DC 20005, Telephone (202) 371–5200 or (800) 541–0736.

SPI's Polyurethane Division held its 33rd Annual Technical/Marketing Conference at Orlando, FL in 1990. Attendance was by persons representing research, design, manufacturing, sales and marketing aspects of the polyurethane industry, including polyurethane foams.

Copies of the proceedings are available from Technomic Publishing Co., Inc., Lancaster, PA. Other SPI divisions holding technical conferences include the Expanded Polystyrene Division, the Polyurethane Foam Contractors Division, and the Structural Plastics Division. The 18th Annual Structural Foam Conference was held April 1–4, 1990 in New Orleans, LA. The 19th Conference was held April 14–17, 1991 in Atlanta, GA.

Copies of these conference proceedings and of future proceedings may be obtained from SPI as indicated below.

SPI Publications Available

General: *Facts and Figures of the U.S. Plastics Industry (1990 edition)*, published annually. Features data on resin production, sales, markets, and capacity utilization rates for plastics resins and domestic consumer patterns, The 1990 edition has 129 pages.

Expanded Polystyrene Division (partial listing):

AH–101 EFS 301 *Fire Safety Guidelines for Use of Expanded Polystyrene in Building Construction.*

AH–102 EPS 302 *A Description of Corner Wall Testing and Thermal Barrier Terminology.*

AH 103 EPS 303 *Guidelines for Reporting Thermal Transmission Properties of Polystyrene Foam Insulating Materials, Thermal Conductance, and Transmittance of Built–Up Construction Systems.*

AH–109 *Expanding World of EPS.* Advantages of EPS molded packaging.

AH–111 *Extruded Polystyrene Foam.*

AH–120 *Literature Review of the Combustion Toxicity of Expanded Polystyrene* (Southwest Research Institute, 1986).

AH–122 *Molded EPS Roofing, Wall and Foundation Design Ideas.*

Foam Cup and Container Division

AK–103 *Let's Be Ecological About Polystyrene Foam.*

AK 104 *True, Molded Foam–Cups Do Not Contain CFC's.*

Polyurethane Division (Conference Proceedings)

AX–121 *Exploring New Horizons* (29th Annual Technical/ Marketing Conference, October 15–17, 1986, Toronto, Canada).

AX–123 *Polyurethane World Congress 1987: 50 Years of Poly-urethanes.* Collection of conference papers presented at this congress September 29 October 2, 1987 at Aachen, FRG. 156 presentations.

AX–130 *Polyurethanes 88* (31st Annual Technical/Marketing Conference, October 18–21, 1988, Philadelphia, PA).

Polyurethane Division (other publications)

AX–105 *Expand Your Line ... Expand Your Sales.* Foam Core Mattresses Give the Consumer a Choice (U–114).

AX–106 *Polyurethane and Polyisocyanurate Foam Insulation* (Sweets Brochure).

AX–107 *Fire Safety Guidelines for Use of Rigid Polyurethane Foam in Building Construction* (U–100R2).

AX–108 *An Update Report on Findings of Fire Study of Rigid Cellular Plastics Materials for Wall and Roof Ceiling Insulation* (U–102R).

AX–109 *Large–Scale Corner Wall Tests of Spray–On Coatings Over Rigid Polyurethane Foam Insulation* (U103).

AX–110 *Evolution of the Fire Performance of Carpet Underlay-ment* (1–105).

AX–111 *Fire Safety Guidelines on Flexible Polyurethane Foams Used in Upholstered Furniture and Bedding* (UIO6).

AX–112 *Room–Scale Compartment Corner Tests of Spray–On Coatings Over Rigid Polyurethane Foam Insulation* (U–107).

AX–113 *An Assessment of the Thermal Performance of Rigid Polyurethane Foam Insulation for Use in Building Construction* (U–108).

AX–127 *Flexible Polyurethane Foams And Combustibility* (U–110).

AX–118 *Foam Plastics Insulation. Polyurethane and Urea Formaldehyde – There's a Big Difference.* (U–117).

AX–119 *Guide for the Safe Handling And Use of Polyurethane and Polyisocyanurate Foam Systems* (U–118).

AX–124 *Test Methods for Polyurethane Raw Materials. Collection of methods for analyzing polyurethane raw materials. ASTM and other methods.*

Polyurethane Foam Contractors Division

AY–101 *Sprayed Polyurethane Foam Systems* (Industry Brochure). A 4–page color, sprayed PUF promotional piece with illustrations.

AY–102 *A Guide for Selection of Protective Coatings over Sprayed Polyurethane Foam Roofing–Systems.* Recommended design considerations and guide specifications. PFCD–PCI–1/88.

AY–103 *The Application and Maintenance of Polyurethane Foam Systems for Outdoor Service Vessels Operating Between –30°F and 225°F.* Recommended design considerations and guide specifications (PFCD–G52–10/87.

AY–104 *Spray Polyurethane Foam and Protective Coating Systems for New and Remedial Roofing* – An illustrated specifier's guide in C.S.I. format. PFCDGS1–1/88.

AY–105 *Spray Polyurethane Foam Roofing Buyer's Check List.* What to look for when you have chosen a spray polyurethane foam roofing system. "Prebid and Post–bid."

AY–106 *Introduction to Spray Polyurethane Foam Roofing Systems.* A general overview of a sprayed PUF roofing system.

AY–107 *Spray Polyurethane Foam Blisters.* Their causes, types, prevention and repair.

AY–108 *1989 Directory and Buyer's Guide.*

Structural Plastics Division (formerly Structural Foam Division)

BB–101 *Combustibility Report: Large–Scale Test Program on Electrical and Electronic Equipment and Electronic Equipment and Application Enclosures.*

BB–102 *Electromagnetic Shielding and Physical Property Test Experiments on Structural Foam Plastic Equipment Enclosures.*

BB–103 *Structural Foam.* An examination of the available technology of structural foam.

Structural Plastics Division (Conference Proceedings)

BB–104 9th Annual Structural Foam Conference (1981)

BB-105 10th Annual Structural Foam Conference (1982)
BB-106 11th Annual Structural Foam Conference (1983)
BB-107 12th Annual Structural Foam Conference (1984)
BB-108 13th Annual Structural Foam Conference (1985)
BB-109 14th Annual Structural Foam Conference (1986)
BB-110 15th Annual Structural Foam Conference (1987)
BB-111 16th Annual Structural Plastics Conference (1988)
BB-112 17th Annual Structural Plastics Conference (1989)
BB-113 18th Annual Structural Plastics Conference (1990)

10
TEST METHODS

Arthur H. Landrock

INTRODUCTION

This chapter is aimed at expanding upon the listings of standard test methods presented in the following Chapter 11 on Standardization Documents. The first section of the chapter lists 130 properties of cellular plastics and elastomers and tabulates the standard test methods (U.S., ISO and British Standards) known to be applicable to each. Only number designations are given. The reader will find the titles and complete citations for the standards in Chapter 11. A key to the symbols used is given to indicate the type of cellular materials used (flexible or rigid) and a number of other matters of interest.

The second section is a somewhat detailed discussion of twenty-two (22) foam properties tested by standard test methods. In most cases only one or a few methods are listed and discussed, but, in the case of combustion properties a total of thirteen (13) methods are considered. This emphasis was made partly because of the considerable attention given to this important subject in recent years.

The third section is a brief invited presentation of several non-standardized test methods currently in use. These methods are thermal analysis, analytical pyrolysis, and molecular–weight determination and distribution.

COMPILATION OF STANDARD TEST METHODS

Key to symbols used
* The standard is primarily or entirely concerned with cellular plastics or cellular elastomers.

(F) The standard is concerned only with *flexible* cellular plastics or rubbers, including microcellular rubber.

(R) The standard is concerned only with *rigid* cellular plastics.

(TM) The standard is basically a test method, or involves test methods.

(SP) The standard is an ASTM *standard practice*.

(SPEC) The standard is a *specification* that includes test–method procedures.

(UB) The standard is under ASTM Society Ballot and is expected to be adopted.

Abrasion
ASTM C 421 (R) (TM)
ASTM D 1044 (TM)
ASTM D 1242 (R) (IM)
ASTM D 2632 (F) (TM)
* ASTM D 3489 (F) (TM)
MIL–HDBK–304, Chapt. 6 (TM)

Acoustical Properties
ASTM C 384 (TM)
ASTM C 423 (TM)
ASTM C 522 (TM)

Aging (Accelerated) (Air–Oven) (Dry–Heat) (Thermal)
ASTM D 573 (F) (TM)
ASTM D 1055 (F) (SPEC)
* ASTM D 1056 (F) (SPEC)
* ASTM D 1565 (F) (SPEC)
* ASTM D 1667 (F) (SPEC)
* ASTM D 2126 (F) (TM)
* ASTM D 3574 – Test K (F) (TM)
* ASTM D 3676 (F) (SPEC)
* BS 4443: Method 12 (F) (TM)
ISO 2578 (TM) (Time–Temperature Limits)
* MIL–O–12420 (F) (SPEC)

Aging (Humid)
* ASTM D 2126 (R) (TM)
* BS 4443: Method 11 (F) (TM)
* ISO 2440 (F) (TM)

Aging (Steam Autoclave)
* ASTM D 3574 – Test J (F) (TM)

Air–Flow Resistance (Permeability) (Porosity)
ASTM C 522 (TM)
* ASTM D 3574 – Test G (F) (TM)
* BS 4443: Method 16 (F) (TM)
* BS 4578 (F) (TM)
* ISO 4638 (F) (TM)
* ISO 7231 (F) (TM)

Air Pycnometer
* ASTM D 2856 (R) (TM)
* ISO 4590 (R) (TM)

Apparent Density
See Density

Ash
* ASTM D 3489 (F) (TM) (for Microcellular Urethane)

Bacterial Resistance
ASTM G 22 (SP)

Breaking Load
ASTM C 203 (R) (TM)

Brittleness Temperature
ASTM D 746 (TM)

Bulk Density
See Density

Buoyancy
* ASTM D 3575 (F) (TM)
* MIL–P–12420 (F) (SPEC)

* MIL–P–40619 (R) (SPEC)

Cell Count (open cells)
See Open–Cell Content (count)

Cell Count (closed cells)
See Closed–cell Content (Count)

Cell Size
* ASTM D 3576 (R) (TM)

Chemical Resistance
ASTM D 471 (F) (TM)
ASTM D 543 (TM)
* ASTM D 1056 (F) (SPEC)

Closed–cell Content (Count) (see also open–cell Content (Count))
* BS 4370: Method 10 (R) (TM)
* ISO 4500 (R) (TM)

Combustion Properties (Fire–Related Properties)
ASTM D 1929 (TM)
ASTM D 2843 (TM)
ASTM D 2863 (TM)
* ASTM D 3014 (R) (TM)
* ASTM D 3675 (F) (TM)
* ASTM D 3894 (R) (TM)
ASTM D 4100 (TM)
ASTM E 84 (TM)
ASTM E 119 (TM)
ASTM E 162 (TM)
ASTM E 286 (TM)
ASTM E 662 (TM)
* UL 94 (TM) (Appendix A)
* UL 746A (TM)
* UL 746B (TM)
* BS 4735 (TM)
* BS 5111 (TM)
* BS 5946 (R) (TM) (Phenolic Foam)
ISO 1210 (R) (TM)
ISO 3582 (TM)

ISO/DIS 9772.2 (TM)
ISO/DIS 10351 (TM)
* MIL–F–83671 (F,R, & semi–R) (SPEC)
* MIL–P–12420 (F) (SPEC)
* MIL–P–26514 (R&F) (SPEC)

Compression/Deflection Properties
ASTM C 165 (R) (TM)

Conductance
ASTM D 257 (TM)

Corrosivity
* MIL–P–26514 (R or F) SPEC)

Creep
ASTM C 273 (R) (TM)
ASTM C 480 (R) (TM)
ASTM D 945 (F) (TM)
* ASTM D 2221 (F) (TM)
ASTM D 2990 (TM)
* ASTM D 3575 (F) (TM)
* BS 4443: Method 8 (F) (TM)
* ISO 7616 (R) (TM)
* ISO 7850 (R) (TM)
* MIL–F–83671 (F, R and semi–R) (SPEC)
* MIL–P–26541 (R or F) (SPEC)

Creep Rupture
ASTM D 2990 (TM)
MM–HDBK–304, Chapt. 6 (TM)

Cross–Breaking Strength
* BS 4370: Method 4 (R) (TM)

Cushioning Properties
* ASTM D 1596 (F) (TM)
* ASTM D 3575 (F) (TM)
* ASTM D 4168 (F) (TM)
* MIL–P–12420 (F) (SPEC)
* MIL–P–26514 (R & F) (SPEC)

* MIL–P–83671 (F, R and semi–R) (SPEC)
  MIL–HDBK–304, Chapt. 6 (TM)
* BS 4443: Method 9 (F) (TM)
* ISO 4651 (TM)

Cut–Growth Resistance
ASTM D 1052 (F) (TM)
* ASTM D 3489 (F) (TM)

D–C Resistance
ASTM D 257 (TM)

Deflection Temperature Under Load (DTUL)
See also Heat–Distortion Temperature
ASTM D 648 (TM)

Delamination Resistance (for Cushioning Underlay)
* ASTM D 3676 (F) (SPEC)

Density (and Specific Gravity)
ASTM C 271 (R) (TM)
ASTM C 303 (R) (TM)
ASTM D 792 (TM)
* ASTM D 1565 (F) (SPEC)
* ASTM D 1622 (R) (TM)
* ASTM D 1667 (F) (SPEC)
* ASTM D 2840 (TM)
* ASTM D 3849 (F) (TM)
* ASTM D 3575 (F) (TM)
* MIL–P–26514 (R or F) (SPEC)
  MIL–HDBK–304, Chapt. 6 (TM)
* BS 4370: Method 2 (R) (TM)
* BS 4443: Method 2 (F) (TM)
* ISO 845 (TM)
* ISO/R 1855 (F) (TM)

Dielectric Properties
ASTM D 149 (TM)
ASTM D 150 (TM)
ASTM D 1673 (TM)

Dimensional Stability
* ASTM D 2126 (R) (TM)
* ASTM D 3575 (F) (TM)
* BS 4370: Method 5 (R) (TM)
* ISO 1923 (TM)
* ISO 2796 (R) (TM)

Dimensions, Measurement of (Dimensioning)
* BS 4370: Method 1 (R) (TM)

Dissipation Factor
ASTM D 1673 (TM)

Dustiness
MIL–P–13607 (Append 2) (F) (SPEC)

Elasticity, Bulk Modulus of
ASTM D 2926 (R) (TM)

Electrical Properties
ASTM D 149 (TM)
ASTM D 150 (TM)
ASTM D 257 (TM)
ASTM D 495 (TM)
* ASTM D 1673 (TM)

Elongation at Break (see also Tensile Strength)
ASTM D 638 (TM)
* ASTM D 1623 (R) (TM)
* ASTM D 3575 (F) (TM)
* BS 4443: Methods 3A and 3B (F) (TM)
* ISO 1798 (F) (TM)

Expansion
See Thermal Expansion

Fatigue
ASTM C 394 (R) (TM)
ASTM D 638 (TM)
* ASTM D 1055 (F) (SPEC)
* ASTM D 1568 (F) (SPEC)

* ASTM D 2221 (F) (TM)
* ASTM D 3574 – Tests I_2 and I_3 (F) (TM)
* BS 4443: Method 13 (F) (TM)
* ISO 3385 (F) (TM)

Flammability
See Combustion Properties

Flexibility (of Cushioning Materials)
MIL–HDBK–304, Chapt. 6 (TM)
MIL–P–12420 (F) (SPEC)
MIL–P–13607 (F) (SPEC)

Flexible Cellular Materials Testing
* ASTM D 1565 (F) (TM)
* ASTM D 3574 (F) (TM)
  MIL–P–13607 (F) (SPEC)
  MIL–HDBK–304, Chapt. 6 (TM)
* BS 4443 (F) (TM)

Flexing Test
* ASTM D 1055 (F) (SPEC)
* ASTM–D 1565 (F) (SPEC)

Flexural Creep
ASTM C 480 (R) (TM)
ASTM D 2990 (TM)

Flexural Properties (see also Flexural Creep)
ASTM C 203 (R) (TM)
ASTM C 393 (R) (TM)
ASTM C 480 (R) (TM)
ASTM D 747 (TM)
ASTM D 790 – Method I (TM)
* ASTM D 1565 (F) (SPEC)
* ASTM D 3489 (F) (TM)
* ASTM D 3768 (F) (TM)
* ISO 1209 (R) (TM)
* MIL–P–12420 (F) (SPEC)

Flexural Recovery
>   ASTM D 3768 (F) (TM)

Fluid–Immersion Test
>   * ASTM D 471 (F) (TM)
>   * ASTM D 1056 (F) (SPEC)
>   * BS 4443: Method 10 (F) (TM)

Fragmentation (Dusting and Breakdown) (Friability)
>   ASTM C 367 (R) (TM)
>   * ASTM C 421 (R) (TM)
>   * MIL–HDBK–304, Chapt. 6

Friability
>   See also Fragmentation and Dusting
>   * MIL–P–83671 (R, F and semi–R) (SPEC)

Fungal Resistance
>   ASTM G 21 (SP)
>   * MIL–HDBK–304, Chapt. 6 (TM)

Guarded–Hot Plate Method
>   ASTM C 177 (TM)
>   * BS 4370: Method 7 (R) (TM)

Hardness
>   ASTM C 367 (TM)
>   ASTM C 569 (TM)
>   ASTM D 531 (F) (TM)
>   ASTM D 785 (TM)
>   ASTM D 1415 (F) (TM)
>   ASTM D 2240 (F) (TM)
>   * ASTM D 3489 (F) (TM)
>   * BS 4443: Method 7 (F) (TM)
>   * BS 4578 (F) (TM)
>   * ISO 2439 (F) (TM)

Heat–Distortion Temperature (see also Deflection Temperature Under Load)
>   * BS 4379: Method 11 (R) (TM)

Heat–Flow Meter
      ASTM C 518 (TM)
   * ISO 2581 (R) (TM)

Heat–Release Rate
      ASTM E 906 (TM)

Hydrogen Ion Concentration (pH)
      MIL–HDBK–304, Chapt. 6 (TM)

Hydrolytic Resistance
      ASTM D 3137 (F) (TM)
   * ASTM D 2440 (F) (TM)
   * ASTM D 3489 (F) (TM)
   * MIL–F–83671 (F, R, and semi–R) (SPEC)

Hygrothermal–Exposure Test
      MIL–HDBK–304, Chapt. 6 (TM)

Impact Strength
      ASTM D 256 (TM)
      ASTM D 746 (TM)
      ASTM D 1822 (TM)
   * ASTM D 3489 (F) (TM)
   * ASTM D 3575 (F) (TM)

Indentation–Force Deflection (See also Compression/Deflection Properties)
   * ASTM D 1565 (F) (SPEC)
   * ASTM D 3574 – Tests B_1 and B_2 (F) (TM)
   * MIL–F–83671 (F, R and semi–R) (SPEC)
   * MIL–P–26514 (R or F) (SPEC)

Indentation Hardness
      See Hardness

K–Factor
      See Thermal Conductivity

Lacquer Lifting
   * MIL–P–12420 (F) (SPEC)

Low-Temperature Test (Compression and Deflection)
* ASTM D 1055 (F) (SPEC)
* ASTM D 1056 (F) (SPEC)
* ASTM D 1667 (F) (SPEC)

Maximum-Use Temperature
ASTM C 447 (R) (SP)

Microcellular Urethane Testing
* ASTM D 3489 (F) (TM)

Microspheres, Hollow, Testing
* ASTM D 2840 (R) (TM)

Moisture Content
MIL-HDBK-304, Chapt. 6 (TM)

Nail-and Screw-Holding Properties
ASTM D 1037 (R) (TM)

NBS/NIST Smoke-Density Test
ASTM E 662 (TM)

Nuclear-Magnetic-Resonance Spectroscopy
ASTM D 4273 (TM)

Oil Resistance
* MIL-F-83671 (F, R and semi-R) (SPEC)

Olefin Cellular Plastics, Flexible, Testing
* ASTM D 3575 (F) (TM)

Open-Cell Content (Count)
* ASTM D 2856 (R) (TM)
* BS 4443: Method 4 (F) (TM)
* ISO 4590 (R) (TM)

Outgassing
ASTM E 595 (TM)

Oxidation
* ISO 2440 (F) (TM)

Oxygen–Index Test
ASTM D 2863 (TM)

Permittivity
ASTM D 150 (TM)
* ASTM D 1673 (TM)

Piston–Cylinder Method
* ASTM D 2926 (R) (TM)

Pliability
* MIL–P–26514 (R or F) (SPEC)
* MIL–F–83671 (F) (SPEC)

Polyethylene Cellular Plastics
* ISO 7214

Porosity (see also Open–Cell Content)
* ASTM D 2856 (R) (TM)

Punking Behavior—Phenolic Foam
* BS 5946 (R) (TM)

Radiant–Panel Test
* ASTM D 3657 (F) (TM)
ASTM E 162 (TM)

Rate of Rise, Urethane Foaming Systems
* ASTM D 2237 (TM)

Resilience
ASTM D 945 (F) (TM)
ASTM D 1054 (F) (TM)
* ASTM D 3489 (F) (TM)
* ASTM D 3574 – Test M (F) (TM)

Resistance (Electrical), Resistivity
ASTM D 257 (TM)

Rigid–Cellular–Materials Testing
>   ASTM D 4098 (SP)
> * BS 4370 (R) (TM)
> * MIL–F–83671 (R) (SPEC)

Rupture, Modulus of
>   ASTM C 446 (TM)

Sag, High–Temperature
> * ASTM D 3489 (F) (TM)
> * ASTM D 3769 (F) (TM)

Sandwich Structure, Testing
>   ASTM C 364 (R) (TM)
>   ASTM C 365 (R) (TM)
>   ASTM C 366 (R) (TM)
>   ASTM C 393 (R) (TM)
>   ASTM C 394 (R) (TM)
>   ASTM C 480 (R) (TM)
>   ASTM C 481 (R) (TM)

Screw– and Nail–Holding Properties
>   ASTM D 1037 (R) (TM)
>   ISO 9054 (R) (TM)

Shear Fatigue
>   ASTM C 394 (R) (TM)

Shear Modulus
>   ASTM C 393 (R) (TM)
>   ASTM E 143 (TM)
> * BS 4370: Method 6 (R) (TM)

Shear Strength
>   ASTM C 273 (R) (TM)
>   ASTM D 732 (TM)
>   ASTM D 945 (TM)
> * BS 4370: Method 6 (R) (TM)
> * ISO 1922 (R) (TM)

Shock–Absorption Characteristics (Dynamic Compression)
* ASTM D 1596 (TM)
* ASTM D 4168 (TM)
  MIL–HBDK–304, Chapt. 6 (TM)

Shrinkage (of Thermal Insulation)
  ASTM C 356 (TM)

Small–Corner Test (Flammability)
* ASTM D 3894 (R) (TM)

Smoke Generation (Smoke–Release Rate)
  ASTM D 2843 (Rohm and Haas Test) (TM)
  ASTM D 4100 (Gravimetric Test) (TM)
  ASTM E 662 (NBS/NIST Test) (TM)
  ASTM E 906 (TM)
* BS 5111 (TM)

Solvent Swelling (Solvent Resistance)
  ASTM D 543 (TM)
* BS 4443: Method 10 (F) (TM)

Sound Absorption
  See Acoustical Properties

Specific Gravity
  See Density

Specific Heat
  ASTM C 351 (TM)

Static Properties
* MIL–P–26514 (R or F) (SPEC)

Steiner Tunnel Test
  ASTM E 84 (TM)

Stiffness
  See Flexural Properties

Stress Relaxation
>   ASTM D 1390 (F) (TM)

Surface Flammability (Burning Characteristics)
>   ASTM D 3675 (F) (TM)
>   ASTM E 84 (TM)
>   ASTM E 162 (TM)
>   ASTM E 286 (TM)

Tear Strength
>   ASTM D 624 (TM)
>   ASTM D 3574 – Test F (F) (TM)
>   ASTM D 3575 (F) (TM)
> * BS 4443: Methods 15 and 16 (F) (TM)

Tensile Strength
>   ASTM D 412 (F) (TM)
>   ASTM D 638 (TM)
> * ASTM D 1623 (R) (TM)
>   ASTM D 2707 (R) (TM)
>   ASTM D 2990 (TM)
> * ASTM D 3489 (F) (TM)
> * ASTM D 3574 – Test F (TM)
> * ASTM D 3575 (F) (TM)
> * MIL–F–83671 (F, R and semi–R) (SPEC)
>   MIL–HDBK–304, Chapt. 6
> * BS 4370: Method 9 (R) (TM)
> * BS 4443: Method 3A (F) (TM)
> * ISO 1798 (F) (TM)
> * ISO 1926 (R) (TM)

Tension Test
>   See Tensile Strength

Thermal Aging
>   See Aging (Thermal)

Thermal Conductivity
>   ASTM C 177 (TM)
>   ASTM C 335 (TM)
>   ASTM C 518 (TM)

BS 874 (TM)
* BS 4370: Method 7 (R) (TM)
* ISO 2581 (R) (TM)

Thermal Expansion, Coefficient of Linear
ASTM C 367 (TM)
ASTM D 696 (TM)
* BS 4370: Method 13 (R) (TM)

Thermal Stability
ASTM D 3575 (F) (TM)
* MIL-F-83671 (F, R and semi-R) (SPEC)

Thickness Measurement, Padding Materials
MIL-P-13607 (F) (SPEC)

Thickness Measurement, Sandwich Cores
* ASTM D 366 (R) (TM)
MIL-HDBK- 304, Chapt. 6 (TM)

Upjohn Mini-Corner Test (Flammability)
* ASTM D 3894 (R) (TM)

Urethane Raw Materials, Isocyanates
ASTM D 1638 (TM)
ASTM D 4659 (TM)
ASTM D 4660 (TM)
ASTM D 4661 (TM)
ASTM D 4663 (TM)
ASTM D 4664 (TM)
ASTM D 4665 (TM)
ASTM D 4666 (TM)
ASTM D 4667 (TM)
ASTM D 4876 (TM)
ASTM D 4889 (TM)
* MIL-F-83671 (F, R and semi-R) (SPEC)

Urethane Raw Materials, Polyols
ASTM D 4273 (TM)
ASTM D 4662 (TM)
ASTM D 4668 (TM)

ASTM D 4669 (TM)
ASTM D 4670 (TM)
ASTM D 4671 (TM)
ASTM D 4672 (TM)
ASTM D 4875 (TM)
ASTM D 4890 (TM)

Vibration Transmissibility
MIL–HDBK–304, Chapt. 6 (TM)

Vicat Softening Point
ASTM D 1525

Water Absorption
* ASTM C 272 (R) (TM)
ASTM D 570 (TM)
* ASTM D 1667 (F) (SPEC)
* ASTM D 2842 (F) (TM)
* ASTM D 3575 (F) (TM)
* MIL–F–83671 (F and semi–R) (SPEC)
MIL–P–12420D (F) (SPEC)
* MIL–P–26514 (R or F) (SPEC)
* MIL–P–40619A (R) (SPEC)
* ISO 2896 (R) (TM)

Water–Extract Preparation
* BS 4443: Method 14 (F) (TM)

Water–Vapor–Transmission Rate (WVTR) of Rigid Cellular Plastics
ASTM E 96 (TM)
* MIL–F–83671 (R) (SPEC)
* ISO 1663 (TM)

Weathering, Accelerated Testing (Simulated Service)
ASTM D 750 (F) TM)
ASTM D 756 (SP)
ASTM D 1499 (SP)
ASTM D 1667 (F) (SPEC)
ASTM D 2565 (SP)
ASTM G 23 (SP)
ASTM G 26 (SP)

Weathering, Outdoor
   ASTM D 1435 (SP)

DISCUSSION OF SELECTED TEST METHODS

Abrasiveness (of cushioning materials)

MIL–HDBK–304, Chapt. 6, outlines a method for "abrasive qualities" of cushioning materials, including plastic foams. The cushioning material is attached with pressure–sensitive–two–side–coated tape to a rectangular metal block, which serves as a weight. The foam specimen and weight are placed upright on the test area of an aluminum alloy (No. 1100–H25) sheet with one side of the aluminum sheet in a direction perpendicular to the machine direction of the aluminum sheet with a stroke of approximately 6 inches and a speed of approximately one foot per second. Rubbing is carried out for 30 seconds, or less, if scratches are clearly developed on the aluminum surface. The sheet is then examined for any adverse effect on its surface.

Aging

Most of the published test methods involving aging of cellular polymers are concerned with flexible rubber–like foams. In ASTM D 3574, Test J covers *Steam Autoclave Aging*. This test consists in treating the foam specimen in a low–pressure steam autoclave, drying, and then observing the effect on physical properties. The test is carried out at 105°– 125°C (Procedure J_1 calls for three hours at 105°C and Procedure J_2 calls for five hours at 125°C). After autoclave exposure, the specimens are dried three hours for each 25 mm of thickness at 70°C in a mechanically convected dry–air oven, then removed from the oven and allowed to come to room temperature equilibrium (24°C and 50 percent RH). Testing is then carried out for the prescribed physical property and the results reported as percent change in property. Typically the compression force deflection is obtained before and after exposure. The steam autoclave compression set value is also calculated (1).

ASTM D 3574 – Test K covers dry–heat aging of flexible cellular materials. This test consists in exposing foam samples in an air–circulating oven at 140°C for 22 hours. The specimens are then removed and conditioned for not less than 12 hours at 23°C and 50 percent RH. As with the Steam Autoclave Tests (J_1 and J_2) physical tests are carried out before and after aging. According to Shah (1) tensile properties are usually studied after dry–heat aging.

Air–Flow Resistance

ASTM C 522 covers the measurement of airflow resistance and the related measurements of specific airflow resistance and airflow resistivity of porous materials that can be used for the absorption and attenuation of sound. The method describes how to measure a steady flow of air through a test specimen, how to measure the air–pressure difference across the specimen, and how to measure the volume velocity of airflow through the specimen. The airflow resistance, R, the specific airflow resistance, r, and the airflow resistivity, r_0, may be calculated from the measurements. The apparatus includes a suction generator or positive air supply arranged to draw or force air at a uniform rate through the specimen. A flowmeter is used to measure the volume velocity of airflow through the specimen, and a differential–pressure–measuring device measures the static–pressure difference between the two faces of the specimen with respect to atmosphere.

ASTM C 384 is another method used for acoustical materials. This method is an impedance–tube method. In the test a flexible–foam specimen is placed in a cavity over a vacuum chamber. A specified constant–air–pressure differential is then created. The air–flow value is the rate of flow of air required to maintain this pressure differential. The test is carried out 1) with air flow parallel to foam rise, and 2) with air flow perpendicular to foam rise. Air–flow values are proportional to porosity in flexible foams.

ASTM D 3574 – Test G is another method used for air flow of flexible foams. The test measures the ease with which air passes through a cellular structure. The test consists in placing a flexible–foam core specimen in a cavity over a chamber and creating a specified constant air–pressure differential. The rate of flow of air require to maintain this pressure differential (125 Pa) is the *air–flow value*. The results are reported in cubic decimeters per second (dm^3/sec). Tests are usually carried out in two directions parallel to foam rise and perpendicular to foam rise.

A new air–flow unit based on the ASTM D 3574 – Test G method is now available from Fluid Data, 1201 North Velasco, Angleton, TX 77515 (409–849–2344). This unit, Fluid Data's Model FPI Foam Porosity Instrument, is reported to be easy to read, computer–compatible, and more portable than its predecessors. It uses a digital electronic display instead of a rotameter, has automatic calibration, and prints out consecutively numbered results. An RS–232 serial port allows data to be transferred from the unit and stored in a computer. This unit is illustrated in Figure 10.1.

Figure 10.1: Fluid Data Model FPI Foam Porosity Instrument (used with permission of Fluid Data).

Cell Size

ASTM D 3576 is the method usually used for determination of cell size of rigid cellular plastics. In this method the number of cell–wall intersections in a specified distance are counted. The foam specimen is cut to less than monocellular thickness in a slicer such as a Hobart Model 610, available from the Hobart Corporation, Troy, Ohio, or a small hobby plane. The shadowgraph is then projected on a screen by the use of a cell–size scale slide assembly (Figure 10.2) and projector. The average chord length is then obtained by counting the cells or cell–wall intersections and converting this value to average cell size by mathematical derivation. A conventional 35–mm slide projector may be used. The cell–size scale slide assembly (Figure 10.2) consists of two pieces of slide glass hinged by tape along one edge, between which a calibrated scale 30 mm in length printed on a this plastic sheet is placed. The calibrated glass slide described in the 1977 issue of ASTM D 3576 is no longer available. In the forthcoming updated revision clear acetate stickers with the calibrated reference lines will be available from ASTM Headquarters. These stickers can be affixed to 2 in. x 2 in. glass slides to provide the slide assembly.

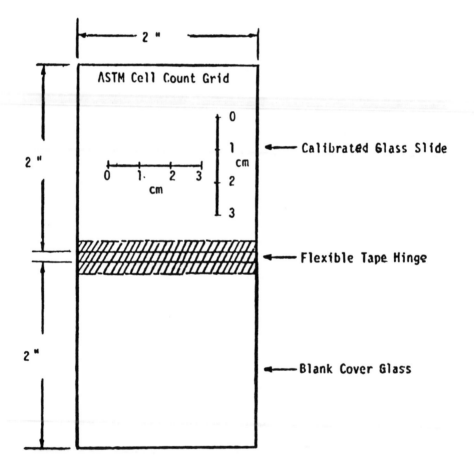

Figure 10.2: Cell–Size Scale Slide Assembly Used in ASTM D 3576 (Copyright ASTM, 1916 Race Street, Philadelphia, PA 19103. Reprinted, with permission.)

ASTM Test Methods D 2842 on water absorption and D 2856 on open–cell content require knowledge of surface cell volume, which uses cell–size values in the calculation.

A new video "contrast enhancement" method has been developed by Dow Chemical USA scientists (2). This method is claimed to be an improvement over the ASTM D 3576 "shadowgraph" method and to provide results more in line with those obtained by the reference micro- scopic method. The new method determines the cell size of a bulk foam sample having only me of its surfaces accurately cut. In eliminating the necessity for a thin slice, the most tedious and time–consuming part of the sample–preparation process has been eliminated. The hypotenuse side of a 45°–90°–45° prism is coated with a viscous wetting agent and pressed lightly against the cut foam surface. Strong light enters through one of the other sides of the prism and a video camera with a microscope objective magnifies and receives a polarized–light image of the film surface. Diamond–scribed reference lines on the contact surface of the prism provides the basis for sizing on the video screen. The lighting, prism and contact wetting greatly enhance the contrast between cell membranes and open volume, with almost nothing below the contrast plane being visible.

The instrument described has been used mostly on polystyrene and polyethylene foams in cell sizes ranging from 0.1 to 2.0 mm. Figure 10.3 shows a schematic of the instrument. A study made with 14 different samples of polystyrene foam ranging from 0.17 to 1.61 m (microscopic method) showed a correlation coefficient of 0.9952 for the ASTM D 3576 shadowgraph method vs. 0.9964 for the video method.

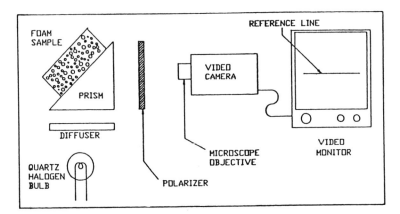

Figure 10.3: Schematic of Dow Video Cell–Size Instrument (Ref. 2) (Courtesy of Dow Chemical USA).

Combustion Properties (Flammability) (Smoke Evolution)

Combustion properties of interest in cellular plastics and elastomers include ease of ignition (ignitibility), support of combustion (oxygen index), relative extent and time of burning, surface flammability, flame spread, smoke evolution properties, and rate of heat release. The following test methods are either concerned solely with cellular plastics, or are used for both cellular and solid plastics.

ASTM D 2843 for Smoke Density. This method was developed for the original Rohm and Haas XP2 apparatus. The method measures the loss of light transmission when a standard specimen is burned in a test chamber (Figure 10.4) under standard conditions. Smoke density is measured by a photoelectric cell across a 12–inch light path. High smoke–producing compositions will yield smoke–density ratings of 90% or higher, while law–smoke generators will give values in the range of 30–40%. The test chamber is fitted with a white backing sheet and a glazed front to permit visual observation of the degree to which an EXIT sign in the backing sheet becomes obscured by the smoke. The percentage light absorption is determined at 15–second intervals up to a total of four minutes. The *maximum smoke density* and the *smoke density rating* are then determined. Figure 10.4 shows the test chamber.

Figure 10.4: ASTM D 2843 Smoke–density chamber (Courtesy of Rohm and Haas Co.).

ASTM D 4100 – Gravimetric Determination of Smoke Particulates: This method is a laboratory test procedure for the gravimetric determination of smoke particulate matter produced from the combustion or pyrolysis of plastic materials, including rigid and flexible polyurethane foams. The method applies to standard–size samples of plastic materials in a slab configuration. The test is conducted in the laboratory under dynamic conditions in a flaming mode of combustion. Percent char and the total amount of plastic burned in a designated time period are determined. The apparatus used is the Arapahoe Chamber (Arapahoe Chemicals, Inc., Boulder, CO), or its equivalent. This consists of a combustion chamber, ignition source, chimney, particulate–filtration system, and a decharring roll mill. A laboratory analytical balance and propane and nitrogen gas source are also required.

In testing, a specimen and glass–fiber filter are weighed separately on an analytical balance. The specimen and filter are positioned in a laboratory test chamber with a high–capacity vacuum source and filtration system. Air is drawn through the test chamber at a prescribed rate. A propane–fueled micro–Bunsen burner is ignited. The burner is applied to the specimen and smoke particulates from the flaming combustion of the specimen impinge upon the filter. The sample is allowed to burn for 30 seconds, unless greater burn times are needed to generate the minimum smoke particulates required for accuracy. The ignition flame is then extinguished and the air flow allowed to continue for 30 additional seconds to collect any particulates in the chimney. On completion of the burns, the specimen and filter are removed from the instrument and weighed on an analytical balance. Char is mechanically or manually removed from the burned specimen and the sample again weighed. Smoke particulate and char mass are calculated by difference and expressed as percentages of the total amount of sample burned.

ASTM E 662 NBS (NIST) Smoke Density Test: This method covers determination of the specific optical density of smoke generated by solid materials and assemblies mounted in a vertical position in thicknesses up to one inch. Measurement is made of the attenuation of a light beam by smoke (suspended solid or liquid particles) accumulating within a closed chamber as a result of nonflowing pyrolytic decomposition and flaming combustion. Results are expressed in terms of *specific optical density*, which is derived from a geometrical factor, and the *measured optical density*, a measurement characteristic of the concentration of smoke. The method uses an electrically heated radiant-energy source for the nonflamming conditions. A six–inch burner is used for flaming conditions, A photometric system with a vertical light path is used to measure the varying light transmission as smoke accumulates. A typical chamber made by Newport Scientific, Inc., Jessup, MD (301–498–6700) is shown in Figure 10.5. This company has an excellent technical bulletin describing the unit.

Figure 10.5: NBS (NIST) Smoke Density Chamber (ASTM E 662). This apparatus is also used in NFPA 258 and ASTM F 814. (Courtesy of Newport Scientific, Inc., Jessup, MD.)

ASTM D 2863 Oxygen Index Test: This is a procedure for determining the relative flammability of a plastic material by measuring the minimum concentration of oxygen in a flowing mixture of oxygen and nitrogen that will just support flaming combustion. The gas mixture, based on an initial concentration determined by experience with similar materials, is caused to flow upwards in a test column. The vertically positioned specimen, supported in a frame, if necessary, is exposed to an ignition flame at its top.   If ignition does not occur at the starting concentration, the oxygen concentration is increased until ignition occurs. The volume percentage of oxygen used is reported as the oxygen index or limiting oxygen index (LOI).   In general, LOI's lower than 20.8 (the volume % of O_2 in normal air) are obtained with plastics easy to burn. Phenolic foam, which is difficult to burn, has an LOI of about 30–40, while non–fire–retarded rigid polyurethane foam (1.6–2.0 lb/ft³) has an LOI of 18–21.4.   Landrock (3) has discussed a number of aspects of oxygen index and tabulated LOI data on a number of plastics, including some foams.

ASTM D 3014 Flame Height, Time of Burning, and Loss of Weight of *Rigid Thermoset Cellular Plastics*, in a Vertical Position (Butler Chimney Test). This is a small–scale laboratory screening test for comparing relative extent of burning (flame height, time of burning and loss of weight). The specimen is mounted in a vertical chimney with a glass front and ignited with a Bunsen burner for ten seconds. The height and duration of flame and the weight percent retained by the specimen are recorded.

ASTM D 3675 Surface Flammability of Flexible Cellular Materials Using a Radiant Heat Energy Source: This method may be used on cellular elastomeric materials such as flexible polyurethane foam and neoprene foam.   It employs a radiant panel heat source consisting of a 300 by 460–mm (12 by 18 in.) panel in front of which an inclined 150 by 460–m (6 by 18 in.) specimen of the material is placed. The orientation of the specimen is such that ignition is forced near its upper edge, and the flame front progresses downward. Factors derived from the rate of progress of the flame front and heat liberated by the material under test are combined to provide a *flame spread index*.   The method was developed to test cellular elastomeric materials which could not be tested by ASTM E 162.

ASTM D 3894 Small Corner Test: This method, based on the Upjohn Mini–Corner Test, is a small–scale laboratory procedure used in screening rigid cellular plastics to determine their performance in a Factory Mutual Full–Scale Corner Wall Test. The cellular plastic specimen, consisting of two sides (walls) and a ceiling assembled to form a covered corner, is ignited with a gas burner.   The flame spread, temperature rise, and damage are recorded.   A corner provides a critical surface geometry for evaluating the fire response of material surfaces.   It incorporates three adjacent surfaces providing a combined heat flux that includes the conductive, convective and radiative responses of any specific burning material.

ASTM D 3574 Methods of Testing Flexible Cellular Materials.
Although not included in the body of this standard, an Appendix lists
U.S. Government and ASTM tests covering combustibility of flexible
urethane foam used in cushioning. These include DOT and FAA tests not
listed in this book.

ASTM E 84 Steiner Tunnel Test. This test, which uses very large
samples (20 ft × 20 1/4 in.) is referenced in all model building codes for
evaluating flame spread and smoke emission of foam plastic insulation.
The test apparatus consists of a chamber or tunnel 25 ft. long and 17 3/4
× 17 5/8 in. in cross section, one end of which contains two gas burners.
The test specimen is exposed to the gas flame for ten minutes, while the
maximum extent of the flame spread and the temperature down the tunnel
are observed through windows. Smoke evolution can also be measured
by use of a photoelectric cell. The flame spread and smoke evolution are
reported in an arbitrary scale for which asbestos and red oak have values
of 0 and 100, respectively. More highly fire–retardant materials have
ratings of 0–25 by this method.

ASTM E 162 Radiant–Panel Test. This test is intended solely for
research and development. It employs a radiant–heat source kept at
1238°F (670°C). A specimen 6 in. × 18 in. is supported in front of the
panel, with the longer dimension inclined 30° from the vertical. A pilot
burner is used to ignite the top of the specimen so that the flame front
progresses downward along the underside exposed to the radiant panel.
The temperature rise above the base level of 356° – 446°F (180° –
230°C), as recorded by stack thermocouples, is used as a measure of heat
release. A smoke–sampling device which collects smoke particles on
glass–fiber filter paper is used to measure smoke density. The *flame–
spread index* is calculated as a product of the flame spread and heat
evolution factors obtained from the test data. Highly fire–retardant
materials give flame–spread index values of 25 or less, while standard red
oak yields an index of 100, and highly flammable materials may exhibit
index values as high as 1000.

For cellular elastomers and cellular plastics, rigid or flexible, the
back and sides of the test specimen must be wrapped with aluminum foil
and asbestos cement board used as backing. Other special requirements
for cellular materials are also in effect.

ASTM E 286 Eight–Foot Tunnel Test. This test, a smaller
version of the Steiner Tunnel Test (ASTM E 84), covers the measurement
of surface flame spread of materials capable of being mounted and
supported within a 13.75 in. (349–mm) x 8 ft. (2.44–m) test frame. The
test also includes techniques for measuring the smoke density and heat

produced. The test is intended or use in research and development only.

ASTM E 906 Heat and Visible–Smoke–Release Rate Test (based on Ohio State University Release Rate Apparatus). In this test the specimen is injected into an environmental chamber through which a constant flow of air passes. A radiant–heat source is used to produce the desired heat flux in the specimen, which way be tested horizontally or vertically. Combustion may be initiated by a number of different methods. The changes in temperature and optical density of the gas leaving the chamber are monitored, and from the resultant data the heat–release rate and visible smoke–release rate are calculated. Toxic–gas release and oxygen consumption rates may also be measured.

UL 94 Appendix A—Horizontal Burning Test for Classifying Foamed Materials 94HBF, 94 RF–1 or 94 BF–2. This test is in an Appendix and is *not* a part of UL 94. It is published for informational purposes only. The apparatus consists of a test chamber, enclosure or laboratory hood, Bunsen or Tirell burner, laboratory ring stands, and a supply of technical–grade methane gas, wire cloth, stop watch, dry absorbent surgical cotton, desiccator, conditioning room, and a full–draft circulating oven kept at 70°C (158°F). Detailed directions are given for classifying the foamed materials after observing their burning characteristics.

BS 5946 Punking Behavior of Phenol–Formaldehyde Foam. Punking is found in some phenol–formaldehyde foams. It consists of a slow combustion (smoldering) initiated by localized application of a source of heat. Propagation of the combustion front continues without further outside application of heat until the foam is reduced to a carbonaceous char. The test method is primarily for the purposes of monitoring the consistency of production of phenolic foam. Its use gives an indication of the punking or non–punking behavior of the foam and is not to be considered as an overall indication of its potential fire hazard. In the procedure test cubes of 120–mm sides are cut out and placed in wire cages of 125–mm sides. The test cages with foam are then exposed to a Bunsen flame under standard conditions and the temperature within the foam measured by any appropriate means. Punking foam shows a continuing risc in tcmpcraturc when temperature is plotted against time.

Compression/Deflection Properties

Compression Force Deflection Test: ASTM D 3574 – Test C. This test consists in measuring the force necessary to produce a 50% compression over the entire top area of a flexible foam specimen. The

50% compression deflection value is reported in pascals. Prior to testing at the 50% value the specimen is preflexed twice to 75–80% of its original thickness, then allowed to rest for ten minutes.

Constant–Deflection–Compression–Set Test: ASTM D 3574 – Test D. This test consists in deflecting the flexible foam specimen under specified conditions of time and temperature and noting the effect on the thickness of the specimen. Ordinarily the specimen is deflected to 50%, 75%, or 90% of its thickness. The entire assembly is then placed in a mechanically convected air oven at 70°C (158°F) and 5% RH for 22 hours. Following this exposure the specimen is then removed from the apparatus and the recovered thickness measured. The constant–deflection compression set, expressed as a percentage of the original thickness, is then calculated.

Indentation–Force–Deflection (IFD) (ILD) Test (to Specified Deflection): ASTM D 3574 – Test B_1. This test consists in measuring the force necessary to produce 25% and 65% or other designated indentations in the flexible foam product. The results are reported in newtons required for 25% and 65% indentations. These figures are known as the 25% and 65% IFD (or ILD) values, respectively.

Indentation–Force–Deflection (IFD) (ILD) Test (to Specified Force): ASTM D 3574 – Test B_2. This test consists in measuring the Indentation–Residual–Deflection Force of a flexible foam. The measurements are known as IRDF values. The force deflection is determined by measuring the thickness of a foam pad under fixed forces of 4.5N, 110N, and 220N on a 323-cm^3 circular indentor foot. This test is useful in determining how thick the padding should be under the average person sitting on a seat cushion.

Compressive Properties of Rigid Cellular Plastics: ASTM D 3574 – Test A (for flexible cellular materials). This test is similar to ASTM D 1622, except that it uses the terms *section density* and *interior density* for apparent overall density and apparent core density, respectively.

MIL–HDBK–304 – Chapter 6: This handbook on Package Cushioning Design has a simple procedure for determining the density in lb/ft^3 (pcf) of any type of plastic foam cushioning materials. The foam is conditioned to standard conditions, such as 73°F (22.8°C) and 50% RH and then the dimensions (length, width and thickness) are measured accurately to within 0.01 inch. The density is then calculated according to the following formula:

$$D = \frac{1728W}{LWT}$$

Where   D = density in lb/ft^3
         W = weight in lb
         L = length in inches
         W = width in inches
         T = thickness in inches

Fatigue

ASTM D 3574 on flexible foams has three procedures for measuring fatigue, as follows:

Test I$_1$ Static Force Loss at Constant Deflection
Test I$_2$ Dynamic Fatigue by the Roller Shear at Constant Force
Test I$_3$ Dynamic–Fatigue Test by Constant–Force Pounding.

The purpose of Test I is to determine: (1) a loss of force support, (2) a loss of thickness, and (3) structural breakdown by visual examination. This procedure tests the specimens statistically at a 75% constant deflection brought about by two parallel plates. Both 25% and 65% IFD deflections are used in the test initially. Testing is carried out at 75% deflection held for 22 hours at 23°C (73.4°F) and 50% RH.

Test I$_2$ fatigues the specimen dynamically at a constant force, deflecting the material both vertically and laterally. The test may be carried out by using 8,000 cycles for approximately 5 hours (Procedure A), or 20,000 cycles for approximately 12 hours (Procedure B). Specifications are given for the roller assembly. The test is conducted at a frequency of 0.45 +/– 0.5 Hz. A cycle is a complete forward and reverse stroke. The stroke length is 300 +/– 10 mm. The IRDF is determined initially. Within one hour after the test is completed the final thickness is measured and the percent loss is thickness determined.

Test I$_3$ determines: (1) the loss of force support at 40% IFD (indentation force deflection), (2) loss in thickness, and (3) structural breakdown as determined by visual inspection. This procedure describes tests that deflect the material by a flat–horizontal indentation exerting a force of 750 +/– 20 N in the test specimen. Test I$_3$ is similar to ISO 3385 and BS 4443: Method 13 and is used primarily to determine the loss in thickness and in hardness of flexible cellular materials intended for use in load– bearing upholstery. The test involves repeated indentation of a specimen by an indentor smaller in area than the test piece. The flexible cellular materials tested are usually latex and polyether urethanes.

Fragmentation (Friability) (Dusting)

ASTM C 421 on block–type thermal insulation describes a simple test for rigid materials, including foams, in which the specimen is tumbled in a small oak test box along with small oak cubes. The mass loss is then reported. MIL–HDBK–304 – Chapter 6 on package cushioning design has a procedure for testing cushioning materials. The materials are placed in a wire basket, which is then placed in a paint pail (can) and agitated in a paint shaker. A sample of the airborne dust is then withdrawn from the container and the number of dust particles within a fractional sample are counted. Fragmentation is determined by weighing the fragments of the specimen that fall to the bottom of the pail during agitation. This value is converted to a percentage figure.

Flexibility (of Cushioning Materials)

MIL–HDBK–304 – Chapter 6 has a simple procedure in which strips of cellular cushioning materials are wrapped around a cylinder or mandrel under standard conditions through a total angle of 180°. The specimens are then examined for failure, such as cracking, delamination, surface spalling, etc.

Flexural Properties

ASTM D 1565, a specification, outlines a test method for *dynamic flexing* of flexible vinyl cellular materials. This test uses a flexing machine which oscillates at 1 Hz. A minimum of 250,000 flexes are applied. After alternate compression and relaxation the effect on the structure and thickness of the foam is observed. The percentage loss of thickness is reported. *Flexural modulus* of microcellular urethane is described in ASTM D 3489. This method uses the general procedure in ASTM D 790, Method I. ASTM D 3768 outlines a procedure for determining *flexural recovery* of microcellular urethanes. The method is used to indicate the ability of a material to recover after a 180° bend around a 12.7–mm (0.5 in.) diameter mandrel at room temperature.

Flexural properties of *rigid* cellular plastics may be determined by ISO/DIS 1209, which specifies two methods using three–point loading. Method A may be used to determine either the load for a specified deformation, or the load at break. Method B way be used to determine the load at break and the flexural strength. These methods are useful only when no significant crushing is observed.

Fungal Resistance

MIL–HDBK–304, Chapter 6 on package cushioning materials has a procedure for evaluating the fungal resistance of cushioning materials. A total of five test organisms are used to inoculate a sterile agar layer in various types of glass containers containing the specimens, along with a viability control of sterile filter paper. Directions are given for incubation of the specimen containers at 28° to 30°C (82.4 to 86°F) and at least 85% RH. At the end of the 14–day incubation period the specimens are examined to determine the extent of fungal growth.

ASTM G 21 is a standard practice somewhat similar to the procedure given in MIL–HDBK–304. The ASTM practice uses a 21–day incubation period, however, and one of the five test organisms is different.

Hydrolytic Stability

ASTM D 3489 on testing microcellular urethane calls out ASTM D 3137 as the method prescribed for testing hydrolytic stability. In this method test specimens are exposed to the influence of humid environments under defined conditions of temperature, humidity and time (85°C or 185°F for 96 hours) for the purpose of measuring the resulting hydrolytic degradation by noting the change in tensile strength after exposure over distilled water.

Impact Strength (Brittle Strength)

ASTM D 3489 in microcellular urethane specifies the use of ASTM D 256, Method A (Izod Test) or Method B (Charpy Test) at -30°C (-22°F). Other temperatures may also be used.

Open–Cell Content

ASTM D 2856 is the method used to determine the open–cell content of rigid cellular plastics by use of the *air–pycnometer*. This method is used where *porosity* of the cellular plastic has a direct bearing on the end use involved. For example, in thermal insulation, a high percentage of closed cells is essential to prevent the escape of gases and thereby promote low thermal conductivity. Also, in flotation applications high closed–cell contents generally prevent water absorption.

This method is based on Boyle's Law, which states that a decrease in volume of a confined gas results in a proportionate increase in pressure. The apparatus used in the procedure consists of two equal-volume cylinders with a specimen chamber provided in one of the cylinders for insertion of the cellular material. Pistons in both cylinders permit volume changes. The same increased pressure is obtained by decreasing the volumes when a cellular material is present in the specimen chamber. The volume change for the specimen cylinder is smaller than for the empty chamber because of the presence of the sample. The extent of this difference is a measure of closed cells and solid polymer present. Open-cell content is determined by difference. This method uses three Procedures, as follows:

Procedure A—Testing cell dimensions to correct for cells cut by sample preparation.

Procedure B—Correcting for cells opened in sample preparation by cutting the test samples into eight smaller samples.

Procedure C—Determination of open-cell content without corecting for surface cells opened by cutting.

ASTM D 1940, which was formerly used for open-cell content, has been discontinued.

Resilience (Ball-Rebound Test)

This test, described in ASTM D 3574 – Test H, consists in dropping a steel ball (16-m diameter, 16.3 g-weight) through a vertical clear plastic tube graduated with a series of circles to represent the rebound heights directly in percentages of the original height. The ball is dropped onto a foam specimen and the height of the rebound noted. The drop height is 500 m. The foam specimens must be cut at least 50 mm thick and have an area no less than 100 by 100 mm. The results are determined as a percentage of the 500-mm drop length. Higher drop heights indicate more "lively" foams.

Tear Resistance (Tear Strength)

In ASTM D 3574 – Test F, tests are carried out using a power-driven apparatus which will indicate the final load at which rupture of a block-type test specimen takes place. An automatic machine may be used to draw the actual curve, or a style or scale may be used with an indicator that remains at the point of maximum force after rupture. The specimen is a block free of skin and densification lines. A 38-mm (1.5-in.) cut is made on one side of the specimen of dimensions 1.5 in. (38.1 mm) × 1 in. (25.4 mm) × 1 in. (25.4 mm). The results are reported in newtons per meter (N/m). This test is similar to BS 4443: Method 15.

Tension Test

ASTM D 3574 – Test E for *flexible foams* determines the effect of the application of a tensile force on the foam. Measurements are made, using a power–driven testing machine, of tensile stress, tensile strength, and ultimate elongation. The samples are cut from flat–sheet material 12.5 mm thick in a dumbbell configuration. The tensile strength is reported in kilopascals, the tensile stress in kilopascals at a predetermined elongation, and the ultimate elongation in percent.

ASTM D 1623 is used for tensile and tensile adhesion properties of *rigid cellular plastics*. The procedure is essentially similar to that of ASTM D 3574 described above. Three types of specimens are used – Type A, where adequate material is available, type B, where only smaller specimens are available, as in sandwich panels, and Type C, which covers tensile adhesion properties of a cellular plastic to a substrate, as in a sandwich panel, or the bonding strength of a cellular plastic to a single substrate.

Water Absorption

There are two ASTM methods used for rigid cellular plastics. ASTM C 272 covers water absorption of core materials for structural sandwich constructions. In the test a 3 x 3 in. (7.62 x 7.62 cm) specimen is cut from a sandwich panel and its edges smoothed. After conditioning and weighing, the specimen is immersed completely in water at 23°C (75.4°F) for either 2 or 24 hours. Samples that float are held under water with a net. After the immersion period the specimen is removed, dried and weighed. Water absorption is reported as water gained/cm^3 of specimen, or as a percentage weight gain.

ASTM D 2842 covers water absorption of rigid cellular plastics. This method measures the change in buoyant force resulting from immersion under a 5.1–cm (2–in.) head of water for 46 hours. Longer periods are sometimes used. The procedure is rather complicated and entails measuring average cell size. The principle is as follows: the buoyant force of an object less dense than water is equal to the weight of the volume of water it displaces when submerged, less the dry weight of the object. Water absorbed into the object lowers the buoyant force by reducing the volume of water actually displaced. By knowing the volume and initial dry weight of the object the initial buoyant force can be calculated. The final buoyant force at the end of the immersion period is measured with an underwater weighing assembly. The difference between the initial and final buoyant force is the weight of water absorbed, which can be expressed in terms of water absorbed per unit of specimen volume.

Water-Vapor Transmission

In the U.S. ASTM E 96 is most commonly used to determine WVT (the passage of water vapor through materials). The method is a gravimetric one, based on a gain or loss in weight of a test dish holding the specimen. There are two basic methods used. In the Desiccant Method the test specimen, which may be a foam sheet, is sealed to the open mouth of a test dish containing a desiccant, and the assembly placed in a controlled atmosphere, usually 100°F (38°C) and 90% RH. Weighings are made on an analytical balance until a constant rate of gain is obtained. In the Water Method the test dish contains water and is placed in an atmosphere of low relative humidity, usually 0% (anhydrous $CaCl_2$) as with the Desiccant Method, until a constant rate of loss in weight is obtained.

ISO 1663 is a method specifically designed for cellular plastics. Basically it is similar in principle to ASTM E 96. However, a tall beaker or dish is used instead of a shallow dish to accommodate thick foam specimens, and anhydrous $CaCl_2$ is used in the beaker or dish, and any one of three saturated salt solutions are placed in the bottom of a desiccator to provide varying relative humidities. Potassium nitrate is used at 38°C 100°F) to provide 88.5% RH.

Both of these methods may be used to calculate permeability, permeance, and water vapor transmission rate (WVTR). ASTM C 355, formerly used for rigid foams, has been discontinued and replaced by ASTM E 96.

SPECIAL NON-STANDARDIZED TEST METHODS[*]

In addition to the conventional test procedures discussed above there are a number of emerging characterization techniques that have not yet attained the status of standardized test methods. These methods are generally used in the research laboratory, but they can also be very useful in practical problems with plastic foams, particularly in assessing thermal stability and in quality control.

---

[*] This discussion was written by John A. Brown, Ph.D., John Brown Associates, Inc., P. O. Box 145, Berkeley Heights, NJ 07422.

Thermal Analysis

Thermal analysis involves observation of the usually very delicate response of a sample to controlled heat stimuli. The elements of thermal–analysis techniques have been known since 1887 when Le Chatelier used an elementary form of differential thermal analysis to study clays (4), but wide application did not come until the introduction of convenient instrumentation by du Pont, Perkin–Elmer, Mettler and other sources in the 1960's. Currently, instrumentation and procedures are commercially available for DTA, DSC, TGA, TMA, and a number of so–called hyphenated methods. Several methods are currently under study by ASTM committees for consideration as to their suitability for adoption as ASTM standards.

Differential Thermal Analysis (DTA). DTA and the closely related Differential Scanning Calorimetry (DSC) are both methods of detecting and measuring endothermic events such as melting, and exothermic events, such as polymerization, or certain phase transitions. The methods consist of burying a sensitive thermocouple in the sample and measuring either temperature changes (DTA) or heat flow (DSC) in the sample relative to that in inert control materials as the temperature of the system is scanned upward or downward. DTA is an excellent means of measuring melting points, phase–transition temperatures, onsets of reaction, and glass–transition temperatures.

The model curve shown in Figure 10.6 illustrates the kind of data obtained on polymers (5):

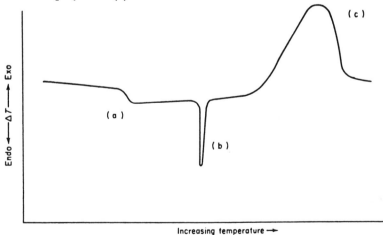

Figure 10.6. Hypothetical DTA trace for a polymer tested in air (5). (a) glass–transition temperature (b) Melting point (c) Exothermic decomposition.

Figure 10.7 summarizes the major polymer properties that can be measured or explored by DTA/DSC analysis (6).

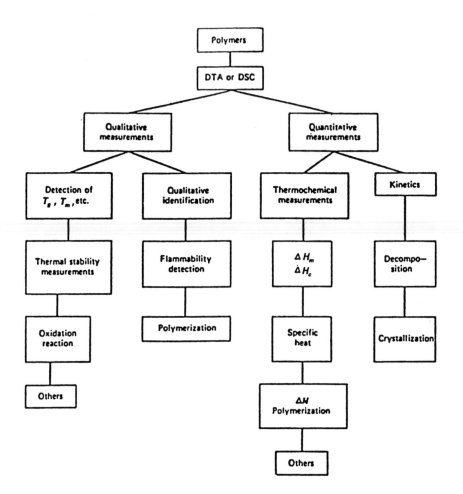

Figure 10.7. Applications of DTA and DSC to polymers (6).

Thermogravimetric Analysis (TGA). TGA follows the loss in weight of a sample as it is heated, and measures the amount of volatiles versus temperature. Carried to pyrolysis temperature, in nitrogen or in oxygen, it can be used to determine the percent of organic matrix versus inorganic fillers, as well as percent moisture or solvents.

Figures 10.8 and 10.9 are data traces showing TGS weight–loss curves for various polymers, including polyurethane foam.

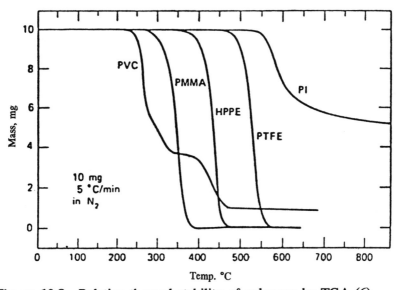

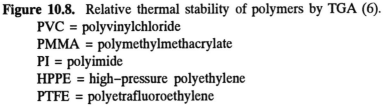

Figure 10.8. Relative thermal stability of polymers by TGA (6).

    PVC = polyvinylchloride

    PMMA = polymethylmethacrylate

    PI = polyimide

    HPPE = high-pressure polyethylene

    PTFE = polyetrafluoroethylene

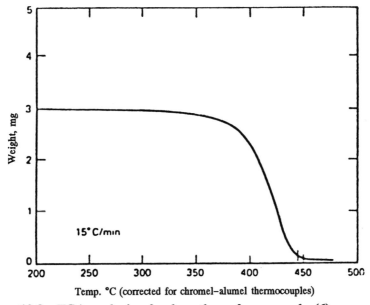

Figure 10.9. TGA analysis of polyurethane foam sample (6).

Thermal Stability. The above techniques (DTA, DSC and TGA) provide measures of thermal stability of polymers. TGA provides a direct measure by observing the temperature(s) at which actual weight loss occurs. DTA/DSC can be used to determine safe storage – and use– temperatures of materials that decompose exothermally by finding the temperature at which exothermic decomposition occurs. These techniques are used extensively in the study of reactive monomers, catalysts and explosives.

A particularly useful application is the oxidation stability test in which the thermal activity of a specimen is followed in oxygen until a sudden large exotherm occurs. If the sample contains an oxidation inhibitor the onset of the exotherm will be delayed, compared to that of an uninhibited sample. The amount of this delay is a measure of the amount of oxidation inhibitor remaining.

Thermal Mechanical Analysis (TMA). TMA follows some particular mechanical property – hardness, modulus, or other – as the temperature of the specimen is scanned. A particularly useful version is TMA by penetration, in which a weighted probe is placed on the sample and the temperature scanned in a slow, linear fashion. The resulting displacement–versus–temperature curve reveals the coefficient of expansion as the sample warms up, and a sharp change in the slope of the curve signals the glass–transition temperature. If the sample melts or softens, the probe sinks into it at the melting or softening temperature, which is revealed as a sudden dip in the curve.

Analytical Pyrolysis

Analytical pyrolysis consists of heating a sample, preferably suddenly, in a vacuum, and analyzing the vapors produced. It is very much an art, but identification of the vapors – olefins such as ethylene, vinyl chloride, MCN, aldehydes, etc. – provides obvious clues to the composition of the material from which they are broken down. Analysis of the vapors is usually by gas chromatography (GC) or infrared (IR) spectrophotometry.

Molecular Weight and Molecular–Weight Distribution

If a polymer is soluble in some solvent, High Performance Liquid Chromatography (HPLC) offers a powerful means of determining both molecular weight and molecular–weight distribution. In this technique a dilute solution of the polymer is passed through a chromatographic column packed with a porous filler that sorts the solute molecules by physical size and allows them to elute from the column in size order. The resulting chromatogram is a plot of molecular weight versus weight percent, and can be interpreted in terms of polymer properties such as toughness, tackiness and hardness. In addition, plasticizers and stabilizers can often be seen as peaks on the low–molecular–weight end of the chromatogram. Figure 10.10 is a model chromatogram illustrating the kind of data obtainable.

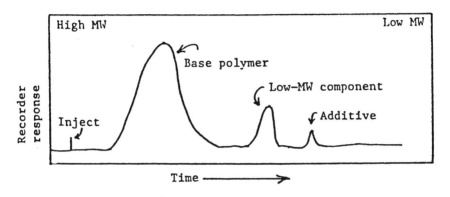

Figure 10.10.   Model chromatogram obtained by high–performance liquid chromatography (HPLC) (7)(8)(9)(10).

REFERENCES

1.   Shah, V., *Handbook of Plastics Testing Technology*, Wiley Inter-science, New York (1984).

2.   Wineland, S.H. and Bartz, A.M., "An Instrument for Measuring the Cell Size of Polystyrene and Polyethylene Foams." *Journal of Cellular Plastics*, 22(2): 122–138 (March/April 1986).

3.   Landrock, A.H., *Handbook of Plastics Flammability and Combus-tion Technology*, Noyes Publications, Park Ridge, NJ (1983).

4.   Le Chatelier, H., *Compt. Rend. Hebe, Seanc. Acad. Sci., Paris, 104*, 144 (1887).

5.   Pope, M.J. and Judd, M.D., *Differential Thermal Analysis*, Heyden, London (1977).

6.   Wendlandt, W.W., *Thermal Methods of Analysis*, Wiley, New York (1974).

7.   Turi, E.A., editor, *Thermal Characterization of Polymeric Materials*, Academic Press, New York (1981).

8.   Waters Associates, Inc., Milford, PA, Product literature and applications, Undated.

9.   Du Pont Company, Wilmington, DE, Product literature and applications, Undated.

10.   Perkin–Elmer Corporation, Norwalk, CT, Product literature and applications, Undated.

11

STANDARDIZATION DOCUMENTS

Arthur H. Landrock

INTRODUCTION

Chapter 10 discussed test methods of interest in cellular plastics and related materials from a general point of view and listed industry, government, British and International (ISO) standards in a number of relevant subject areas. This chapter will list published specifications (139) and test methods (116) used in the United States, in addition to British Standards (28) and ISO International Standards (40). United States industry standards inlcude ASTM Test Methods (116), ASTM Practices, Guides, Definitions, Terminologies and Abbreviations (24), ASTM Specifications (23), SAE–AMS Specifications (25), and Underwriters Laboratories Standards (1). U.S. Government standards covered include Military Specifications (67), Military Standards (5), Federal Standards (2), Military Handbooks (6), and Federal Specifications (24). Most of these standards are undergoing frequent revision and unused standards are constantly being withdrawn. A total of 361 standards are covered.

In the case of ASTM Standards the responsible ASTM Committee, such as D–20 in Plastics, is listed when known, along with each entry. Other ASTM Committees listed are as follows:

C–16	on Thermal Insulation
C–24	on Building Seals and Sealants

D-4   on Road Paving Materials
D-9   on Electrical Insulating Materials
D-10   on Packaging
D-11   on Rubber
E-5   on Fire Standards
E-10   on Nuclear Technology
E-21   on Space Simulation and Applications of
       Space Technology
E-28   on Mechanical Testing
E-33   on Aerospace and Aircraft
F-17   on Plastic Piping Systems
G-3   on Durability of Nonmetallic Materials

Of these fifteen (15) committees the greatest representation is as follows:

D-20   on Plastics (53 standards)
D-11   on Rubber (30 standards)
C-16   on Thermal Insulation (24 standards)
F-7   on Aerospace and Aircraft (12 standards)

With ASTM standards in the following entries the year of issue is given after the number designation, as ASTM D 1673–79 (1989). The year of reapproval without substantial change, except for minor editorial change, is given in parentheses, as (1989) in the example. In the case of Aerospace Material Specifications (AMS) the letters A, B, C etc. following the AMS number designation, as AMS 3193B–85, indicate revisions. AMS 3193 would be the first issue, 3193A the first revision, 3193B the second revision, etc. The 85 indicates that 1985 was the year of issue. Some AMS documents have supplementary detail specifications, such as AMS 3568–83. AMS 3568/1–83, AMS 3568/2–83, and AMS 3568/3–83 are the detail specifications. The year of issue is frequently not given.

Many ASTM and AMS standards have been adopted by the U.S. Department of Defense (DOD) for inclusion in the Department of Defense Index of Specifications and Standards (DODISS). As such, they can be used in military procurement. Such documents are so indicated by the notation (DOD Adopted). Documents not so indicted may be DOD Adopted at a later date, since such possible acceptance is constantly being reviewed. Federal Supply Classification (FSC) numbers and Preparing Activity codes are given for all DOD–Adopted ASTM and

AMS standards and for all Military and Federal standardization documents (specifications, standards, handbooks) and DOD Adopted industry standards when known.

In the case of Military and Federal Specifications, Standards and Handbooks revisions are indicated by the letters A, B, C, etc., as with AMS documents. A number in parentheses following the document number indicates an Amendment. An Interim Amendment is indicated by the designation Int Amend 1, Int Amend 2, etc. Such entries as (GL), (AR), (MI), etc., following the document number or Interim Amendment number indicate a Limited Coordination document. Such documents have been prepared by a specific Military or Federal agency solely for its own use, although it may be used by other agencies. A list of symbols are, in any event, also listed to designate the Preparing Activity given on the last page of all military or Federal standardization documents. In all cases for the document listed the date given is the data of latest action on the document. If there is an Amendment, the date is that of the Amendment. In some cases the Amendment lists new Types or Classes, or both, or modifies the requirements. In most cases, however, the bulk of the changes are editorial, covering spelling and similar changes, although technical changes are also made. The number of pages are indicated by p or pp.

In all Federal and Military standardization documents the Preparing Activity Code (Table 11.1) and Federal Supply Classes, Federal Supply Groups, and Standardization Areas (Table 11.2) are given. When a Qualified Products List (QPL) of preapproved products, materials and supplies is available for Federal and Military Specificatons that fact is indicated on the last line of the entry by the notation (QPL). QPLs are documents containing lists of Federally tested and recognized commercial products known to meet the requirements of the various Types, Classes and Grades in their corresponding specifications.

Table 11.1: Preparing Activity Codes for Military and Federal Standardization Documents Listed in This Chapter and Recognizant Agencies for Industry Standards (ASTM and SAE–AMS)

AR	Army Armament, Munitions and Chemical Command
AS	Naval Air Systems Command
AT	Army Tank–Automotive Command
Com–NBS	Dept. of Commerce – National Bureau of Standards (now NIST)

(continued)

Table 11.1: (continued)

DM	Defense Logistics Agency (DepSo)
DS	Defense Nuclear Agency
EC	Space and Naval Warfare Systems Command
FSS	Federal Supply Service
GL	Army Natick Research, Development and Engineering Center
GSA	General Services Administration
ME	(Army) Belvoir Research, Development and Engineering Center
MI	Army Missile Command
MR	(Army) Materials Technology Laboratory
NU	Naval Clothing and Textile Research Facility
OS	Naval Sea Systems Command (Ordnance Sytstems)
PA	Symbol formerly used for Picatinny Arsenal (now AR)
SA	Naval Supply Systems Command
SH	Naval Sea Systems Command (Ship Systems)
SM	AMC Packaging, Storage & Containerization Center (Army)
YD	Naval Facilities Engineering Command
11	Air Force Aeronautical Systems Division (AFSC)
20	Air Force Wright Aeronautical Laboratory
69	Air Force Packaging Evaluation Activity (HQ AFLC)
82	Air Force San Antonio Air Logistics Center (AFLC)
99	Air Force Cataloging and Standardization Center (AFLC) (Battle Creek)

"Limited Coordination" documents are those which have been coordinated by one Service only. Such documents have their Preparing Activity codes given in parentheses immediately following the designation, as MIL–J–822698B(SA). Air Force standards, however, frequently list (USAF) following the designation, rather than the specific agency code.

In this chapter the author has provided brief descriptions of the contents of some of the standards listed, but space did not permit such coverage on all standards. Additional information on industry and government specifications and standards and how they are used may be found in References 1 and 2.

The reader will note that in Chapter 10 a number of ASTM test methods were listed that were not particularly applicable to cellular plastics or elastomers. In such cases no asterisk (*) was used. The same situation prevails in this chapter except that the asterisks were not used. Such general methods, however, have been used in evaluating the properties of cellular materials, often in spite of statements in the scope of such methods that they are not intended for use in such applications.

Table 11.2: Standardization Areas, Federal Supply Classes (FSCs) and Groups Listed in This Chapter

Standardization Areas

CMPS	Composite Technology
FACR	Facilities Engineering and Design Requirements
MISC	Miscellaneous
PACK	Packaging, Preservation and Transportability

Federal Supply Classes

1055	Launchers, Grenade, Rocket and Pyrotechnic
1195	Miscellaneous Nuclear Ordnance
1240	Optical Sighting and Ranging Equipment
1320	Ammunition, Over 120 nim
1338	Guided Missiles, Special Vehicle Propulsion Units, Solid Fuel and Chemical
1660	Aircraft Air Conditioning, Heating and Pressurizing Equipment
2040	Marine Hardware and Hull Items
2590	Miscellaneous Vehicular Components
3540	Wrapping and Packaging Machinery
4220	Marine Lifesaving and Diving Equipment
5340	Miscellaneous Hardware
5410	Prefabricated and Portable Buildings
5510	Lumber and Related Basic Wood Materials
5640	Wallboard, Building Paper and Thermal Insulation Materials
5680	Miscellaneous Construction Materials
5970	Electrical Insulators and Insulating Materials
6145	Wire and Cable, Electrical
7110	Office Furniture
7210	Household Furnishings
7220	Floor Coverings
7330	Kitchen Hand Tools and Utensils

(continued)

Table 11.2: (continued)

7510	Office Supplies
8030	Preservative and Sealing Compounds
8040	Adhesives
8115	Boxes, Cartons and Crates
8135	Packaging and Packing Bulk Materials
8140	Ammunition and Nuclear Ordnance Boxes and Specialized Containers
8305	Textile Fabrics
8415	Clothing, Special Purpose
8430	Footwear, Mens'
9320	Rubber Fabricated Materials
9330	Plastics Fabricated Materials
9340	Glass Fabricated Materials

Federal Supply Groups

56GP	Construction and Building Materials
93CP	Nonmetallic Fabricated Materials

INDUSTRY STANDARDS

American Society for Testing and Materials (ASTM) (See Reference 3)

ASTM Specifications (See Reference 3)

ASTM C 509-90 Standard Specification for *Cellular Elastomeric Preformed Gasket and Sealing Material*, 4 pp (Comm C-24)

ASTM C 534-88 Standard Specification for *Preformed Flexible Elastomeric Cellular Thermal Insulation in Sheet and Tubular Form*, 3 pp (DOD Adopted) (FSC 5640) (YD) (Comm C-16)

ASTM C 552-88 Standard Specification for *Cellular Glass Thermal Insulation*, 4 pp (DOD Adopted) (FSC 5640) (YD) (Comm C-16)

ASTM C 578-87a Standard Specification for *Preformed, Cellular Polystyrene Thermal Insulation*, 7 pp (DOD Adopted)* (FSC 5640) (YD) (Comm C-16)

ASTM C 591-85 Standard Specification for *Unfaced Preformed Rigid Cellular Polyurethane Thermal Insulation*, 4 pp (DOD Adopted) (FSC 5640) (YD) (Comm C-16)

ASTM C 725-88 Standard Specification for *Mineral Fiber and Roof Insulation Board*, 2 pp (DOD Adopted)** (FSC 5640) (Comm C-16)

ASTM C 984-83 Standard Specification for *Perlite Board and Rigid Cellular Polyurethane Composite Roof Installation*, 4 pp (Comm C-16)

ASTM C 1029-90 Standard Specification for *Spray-Applied Polyurethane Thermal Insulation*, 4 pp (Comm C-16)

ASTM C 1050-91 Standard Specification for *Rigid Cellular Polystyrene Cellulosic Fiber Composite Roof Insulation*, 3 pp (Comm C-16)

This specification covers the composition and physical properties of insulating boards composed of wood-fiber insulation board laminated to rigid cellular polystyrene insulation boards, flat or tapered, used principally above structural roof decks for thermal insulation as well as for a base for roofing in building construction.

ASTM C 1126-89 Standard Specification for *Faced or Unfaced Rigid Cellular Phenolic Thermal Insulation*, 4 pp (Comm C-16)

This specification applies to faced or unfaced, rigid cellular phenolic thermal insulation in either board or tubular form. Materials covered are used as roof insulation, sheathing or rigid board for non-load bearing, building material applications, and pipe insulation between -40° and 125°C (-40° and 257°F).

---

*   Replaces Federal Specification HH-I-524C.

**   Replaces Federal Specification HH-I-526.

ASTM D 1055–90 Standard Specification for *Flexible Cellular Materials – Latex Foam*, 6 pp (DOD Adopted) (FSC 9320) (MR) (Comm D–11)

ASTM D 1056–85 Standard Specification for *Flexible Cellular Materials Sponge or Expanded Rubber*, 14 pp (DOD Adopted) (FSC 9320) (MR) (Comm–11)

ASTM D 1565–81 A Standard Specification for *Flexible Cellular Materials Vinyl—Chloride Polymers and Copolymers (Open–Cell Foam)*, 9 pp (DOD Adopted) (FSC 9320) (MR) (Comm D–11)

This specification is identical to SAE J15.

ASTM D 1667–76 (1986) Standard Specification for *Flexible Cellular Materials—Vinyl Chloride Polymers and Copolymers (Closed–Cell Foam)*, 6 pp (DOD Adopted) (FSC 9320) (YD) (Comm D–11)

ASTM D 1752–84 Standard Specification for *Preformed Sponge Rubber and Cork Expansion Joint Filler for Concrete Paving and Structural Construction*, 2 pp (DOD Adopted) (FSC 5610) (YD) (Comm D–4)

ASTM D 1786–90 Standard Specification for *Toluene Diisocyanate*, 2 pp (Comm D–20)

ASTM D 3204–86 Standard Specification for *Prefoamed Cellular Plastic Joint Fillers for Relieving Pressure*, 2 pp (Comm D–4)

Type I is closed–cell polyethylene and Type II is open–cell polyurethane.

ASTM D 3453–80 (1990) Standard Specification for *Flexible Cellular Materials — Urethane for Furniture and Automotive Cushioning, Bedding and Similar Applications*, 4 pp (DOD Adopted) (FSC 9320) (MR) (Comm D–11)

ASTM D 3490–80 (1985) Standard Specification for *Flexible Cellular Materials — Bonded Urethane Foams*, 3 pp (Comm D–11)

ASTM D 3676–78 (1989) Standard Specification for *Rubber Cellular Cushion Used for Carpet or Rug Underlay*, 3 pp (Comm D–11)

ASTM D 3770–79 (1985) Standard Specification for *Flexible Cellular Materials – High Resilience Polyurethane Foam (HR)*, 2 pp (Comm D–11)

ASTM D 3851–84 Standard Specification for *Urethane Microcellular Shoe Soling Materials*, 3 pp (Comm D–11)

ASTM F 891–90 Standard Specification for *Coextruded Poly(Vinyl Chloride) Plastic Pipe with a Cellular Core*, 8 pp (Comm F–17)

ASTM Test Methods (See Reference 3)

ASTM C 165–91 Standard Test Method for *Measuring Compressive Properties of Thermal Insulation*, 5 pp (DOD Adopted) (Area MISC) (84) (Comm C–16)

ASTM C 177–85 Standard Test Methods for *Steady–State Heat Flux Measurements and Thermal Transmission Properties by Means of the Guarded–Hot–Plate Apparatus*, 16 pp (DOD Adopted) (Area PACK) (AS) (Comm C–16)

ASTM C 203–91 Standard Test Methods for *Breaking Load and Flexural Properties of Block–Type Thermal Insulation*, 6 pp (DOD Adopted) (Comm C–16)

ASTM C 271–61 (1988) Standard Test Method for *Density of Core Materials for Structural Sandwich Constructions*, 2 pp (Comm F–7)

ASTM C 272–53 (1980) Standard Test Method for *Water Absorption of Core Materials for Structural Sandwich Constructions*, 3 pp (Comm F–7)

Includes cellular materials and honeycomb.

ASTM C 273–61 (1988) Standard Method of *Shear Test in Flatwise Plane of Flat Sandwich Constructions or Sandwich Cores*, 3 pp (DOD Adopted) (FSC Area FACR) (YD) (Comm F–7)

ASTM C 297–61 (1988) Standard Method of *Tension Test of Flat Sandwich Construction in Flatwise Plane*, 3 pp (Comm F–7)

ASTM C 335–89 Standard Test Method for *Steady–State Heat Transfer Properties of Horizontal Pipe Insulation*, 10 pp (Comm C–16)

This method replaces ASTM C 691–84.

ASTM C 351 Standard Test Method for *Mean Specific Heat of Thermal Insulation*, 5 pp (Comm C–16)

This procedure uses the classical methods of mixtures, consisting essentially of adding a known mass of material at a known high temperature to a known mass of water at a known low temperature, and determining the equilibrium temperature that results. The heat absorbed by the water and its containing vessel can be calculated and this value equated to the expression for the heat given up by the hot material. From this equation the unknown specific heat can be calculated.

ASTM C 356–87 Standard Test Method for *Linear Shrinkage of Pre–formed High–Temperature Thermal Insulation Subjected to Soaking Heat*, 3 pp (Comm C–16)

ASTM C 364–61 (1988) Standard Test Method for *Edgewise Compressive Strength of Flat Sandwich Constructions*, 3 pp (Comm F–7)

ASTM C 365–57 (1988) Standard Test Method for *Flatwise Compressive Strength of Sandwich Cores*, 4 pp (DOD Adopted) (Comm F–7) (Area FACR) (Navy YD)

ASTM C 366–57 (1988) Standard Methods for *Measurement of Thickness of Sandwich Cores*, 5 pp (DOD Adopted) (FSC 5680) (AS) (Comm F–7)

ASTM C 384–90a) Standard Test Method for *Impedance and Absorption of Acoustical Materials by the Impedance Tube Method*, 11 pp (Comm E–33)

ASTM C 393–62 (1988) Standard Method for *Flexure Test of Flat Sandwich Constructions*, 3 pp (Comm F–7)

ASTM C 394–62 (1988) Standard Method of Test for *Shear Fatigue of Sandwich Core Materials*, 2 pp (Comm F–7)

ASTM C 411–82 (1987) Standard Test Method for *Hot–Surface Performance of High–Temperature Thermal Insulation*, 5 pp (Comm C–16)

ASTM C 421–88 Standard Test Method for *Tumbling Friability of Preformed Block–Type Thermal Insulation*, 2 pp (Comm C–16)

   Covers the determination of the mass loss of preformed block–type thermal insulation as a result of a combination of abrasion and impact produced by a laboratory tumbling mechanism.

ASTM C 423–90a Standard Test Method for *Sound Absorption and Sound Absorption Coefficients by the Reverberation Room Method*, 7 pp (DOD Adopted) (FSC 5640) (YD) (Comm E–33)

   Covers the measurement of sound absorption in a diffuse reverberant sound field by measuring the decay rate.

ASTM C 446–88 Standard Test Method for *Breaking Load and Calculated Modulus of Rupture of Preformed Insulation for Pipes*, 2 pp (Comm C–16)

ASTM C 480–62 (1988) Standard Test Method for *Flexure–Creep of Sandwich Constructions*, 2 pp (Comm F–7)

ASTM C 481–62 (1988) Standard Test Method for *Laboratory Aging of Sandwich Constructions*, 3 pp (DOD Adopted) (FSC FACR) (YD) (Comm F–7)

ASTM C 518–85 Standard Test Method for *Steady–State Heat Flux Measurements and Thermal Transmission Properties by Means of the Heat Flow Meter Apparatus*, 17 pp (Comm C16)

   This is a secondary or comparative method. Use ASTM C 177 for measuring the thermal resistance of standard specimens. Use this method only for materials with densities lower than 900 kg/m^3.

ASTM C 522–87 Standard Test Method for *Airflow Resistance Of Acoustical Materials*, 5 pp (Comm E–33)

   Airflow resistance is a property that determines sound–absorption and sound–transmitting properties.

ASTM C 1166–91 Standard Test Method for *Flame Propagation of Dense and Cellular Elastomeric Gaskets And Accessories*

Covers a laboratory procedure for determining flame–propagation characteristics of a dense or cellular elastomeric gasket, or an accessory, when exposed to heat and flame, with no significance being attached to such matters as fuel contribution, rate of flame spread, smoke developed, or the nature and temperature of the products of combustion.

ASTM D 149–90 Standard Test Method for *Dielectric Breakdown Voltage and Dielectric Strength of Solid Electrical Insulating Materials at Commercial Power Frequencies*, 11 pp (DOD Adopted) (FSC 9330) (MR) (Comm D–9)

These tests are usually carried out at 60 Hz, although they may be carried out over the range of 25 to 399 Hz.

ASTM D 150–87 Standard Test Methods for *A–C Loss Characteristics and Permittivity (Dielectric Constant) of Solid Electrical Insulating Materials*, 19 pp (DOD Adopted) (FSC 9330) (MR) (Comm D–9)

ASTM D 543–87 Standard Test Method for *Resistance of Plastics to Chemical Reagents*, 7 pp (DOD Adopted) (FSC 9330) (MR) (Comm D–20)

ASTM D 570–81 (1988) Standard Test Method for *Water Absorption of Plastics*, 3 pp (DOD Adopted) (FSC 9330) (MR) (Comm D–20)

ASTM D 573–88 Standard Test Method for *Rubber Deterioration in an Air Oven*, 5 pp (DOD Adopted) (FSC 9320) (MR) (Comm D–11)

ASTM D 575–88 Standard Test Methods for *Rubber Properties in Compression*, 3 pp (Comm D–11)

Covers two procedures for determining the compression–deflection characteristics of rubber compounds other than those usually classified as hard rubber or sponge rubber.

ASTM D 624–86 Standard Test Method for *Rubber Property — Tear Resistance*, 5 pp (DOD Adopted) (FSC 9320) (MR) (Comm D–11)

Covers vulcanized rubber, including flexible foams.

ASTM D 638–89 Standard Test Method for *Tensile Properties of Plastics*, 12 pp (DOD Adopted) (FSC 9330) (MR) (Comm D–20)

This method may be used on specimens up to 14 mm (0.55 in.). There is a metric version, ASTM D 638M–89.

ASTM D 695–90 Standard Test Method for *Compressive Properties of Rigid Plastics*, 7 pp (DOD Adopted) (FSC 9330) (MR) (Comm D–20)

There is a metric version, ASTM D 695M–90.

ASTM D 696–79 (1988) Standard Test Method for *Coefficient of Linear Thermal Expansion of Plastics*, 4 pp (DOD Adopted) (FSC 9330) (MR) Comm D–20)

Uses a vitreous silica dilatometer to measure the coefficient of linear thermal expansion of plastics which are not distorted by the thrust of the dilatometer on the specimen.

ASTM D 732–90 Standard Test Method for *Shear Strength of Plastics by Punch Tool*, 3 pp (DOD Adopted) (FSC 9330) (MR) (Comm D–20)

This test covers the punch–type of shear test and is intended for use in determining the shear strength of test specimens of organic plastics in the form of sheets and molded disks in thicknesses from 0.127 to 12.7 mm (0.050 to 0.500 in.).

ASTM D 746–79 (1987) Standard Test Method for *Brittleness Tempera-ture of Plastics and Elastomers by Impact*, 7 pp (DOD Adopted) (FSC 9330) (Comm D–20)

ASTM D 750–85 Standard Test Method for *Rubber Deterioration in Carbon–Arc Weathering Apparatus*, 3 pp (DOD Adopted) (FSC 9320) (MR) (Comm D–11)

ASTM D 785–89 Standard Test Method for *Rockwell Hardness of Plastics and Electrical Insulating Materials*, 5 pp (DOD Adopted) (FSC 9320) (MR) (Comm D–11)

ASTM D 790-90 Standard Test Methods for *Flexural Properties of Unreinforced and Reinforced Plastics and Electrical Insulating Materials*, 44 pp (DOD Adopted) (FSC 9330) (MR) (Comm D-20)

Applicable to rigid and semi-rigid materials that break or fail in the outer fibers. Method A is a three-point-loading system and Method B is a four-point-loading system. ASTM D 790M-86 is a metric version.

ASTM D 792-86 Standard Test Methods for *Specific Gravity (Relative Density and Density of Plastics by Displacement)*, 6 pp (DOD Adopted) (FSC 9330) (MR) (Comm D-20)

ASTM D 945-87 Standard Test Methods for *Rubber Properties in Compression or Shear (Mechanical Oscillograph)*, 15 pp (DOD Adopted) (FSC 9320) (MR) (Comm D 11)

ASTM D 1037-89 Standard Methods of *Evaluating the Properties of Wood-Base Fiber and Particle Panel Materials*, 30 pp (DOD Adopted) (FSC 5510) (YD) (Comm D-7)

This standard consists of 22 test methods, some of which are used in evaluating high-density rigid polyurethane foams for furniture applications. These methods are concerned with fastener-holding properties: Lateral Nail Resistance Test, Nail Withdrawal Test, Nail-Head Pull-through Test, and Direct Screw Withdrawal Test.

ASTM D 1044-90 Standard Test Method for *Resistance of Transparent Plastics to Surface Abrasion*, 4pp (DOD Adopted) (FSC 9330) (MR) (Comm D-20)

Uses a Taber Abraser, sometimes to test plastic foams.

ASTM D 1052-85 Standard Test Method for *Measuring Rubber Deterioration—Cut Growth Using Ross Flexing Apparatus*, 4 pp (Comm D-11)

ASTM D 1054-87 Standard Test Method for *Rubber Property — Resilience Using a Rebound Pendulum*, 6 pp (DOD Adopted) (FSC 9320) (MR) (Comm D-11)

ASTM D 1229–87 Standard Test Method for *Rubber Property — Compression Set at Low Temperatures*, 3pp (DOD Adopted) (FSC 0320) (MR) (Comm D–11)

ASTM D 1242–87 Standard Test Methods for *Resistance of Plastic Materials to Abrasion*, 7 pp (DOD Adopted) (FSC 9330) (MR) (Comm D–20)

ASTM D 1390–76 Standard Test Method for *Rubber Property — Stress Relaxation in Compression*, 4 pp (Comm D–11)

ASTM D 1415–88 Standard Test Method for *Rubber Property — International Hardness*, 4 pp (DOD Adopted) (FSC 9320) (MR) (Comm D–11)

ASTM D 1525–87 Standard Test Method for *Vicat Softening Temperature of Plastics*, 3 pp (DOD Adopted) (FSC 9330) (MR) (Comm D–20)

Covers the determination of the temperature at which a specified needle penetration occurs when specimens are subjected to specified test conditions. The method is useful for many thermoplastic materials. The specimen and needle are heated at either of two permissable rates. The temperature at which the needle has penetrated to a depth of 1 mm is the Vicat softening temperature.

ASTM D 1596–91 Standard Test Method for *Shock Absorbing Characteristics of Package Cushioning Materials*, 6 pp (Comm D–10)

ASTM D 1621–73 (1979) Standard Test Method for *Compressive Properties of Rigid Cellular Plastics*, 4 pp (DOD Adopted) (FSC 9330) (MR) (Comm D–20)

Has two procedures. Procedure A uses crosshead motion for determining compressive properties, and Procedure B uses strain-- measuring devices mounted on the specimen.

ASTM D 1622–88 Standard Test Method for *Apparent Density of Rigid Cellular Plastics*, 3 pp (DOD Adopted) (FSC 9330) (Comm D–20)

Covers apparent *overall* density and apparent *core* density. A proposed revision will clarify the potential air–buoyancy effect on fresh closed–cell foam.

ASTM D 1623–78 Standard Test Method for *Tensile and Tensile Adhesion Properties of Rigid Cellular Plastics*, 9 pp (DOD Adopted) (FSC 9330) (YD) (Comm D–20)

This is a comparison test for ASTM D 638 on Tensile Properties of Plastics. It uses three types of specimens, Type A, where adequate sample material is available, Type B where smaller specimens are available, and Type C, which covers the tensile–adhesion properties of a cellular plastic to a substrate, as in a sandwich panel, or the bonding strength of a cellular plastic to a single substrate.

ASTM D 1630–83 Standard Test Method for *Rubber Property —Abrasion Resistance (NBS Abrader)*, 4 pp (DOD Adopted) (FSC 9320) (MR) (Comm D–11)

This method is recommended for use on soles and heels of footware.

ASTM D 1638–74 Standard Methods of Testing *Urethane Foam Isocyanate Raw Materials*, 20 pp (DOD Adopted) (Comm D–20)

These methods cover procedures for testing the isocyanate raw materials used in preparing urethane foams. The isocyanates covered are TDI (purified) and modified or crude isocyanates derived from TDI or crude methylene–bis–(4–phenylisocyanate). The procedures covered are:

TDI (purified)
Assay, Isomer Content, Total Chlorine, Hydrolyzable Chlorine, Acidity, Freezing Point, Specific Gravity, and Color

Crude or Modified Isocyanates
Acidity, Amine Equivalent, and Brookfield Viscosity

ASTM D 1673–79 (1989) Standard Test Methods for *Relative Permittivity and Dissipation Factor of Expanded Cellular Plastics Used for Electrical Insulation*, 5 pp (Comm D–9)

Covers flat sheets or slabs of expanded cellular plastics, both rigid and flexible, at frequencies from about 60 Hz to 100 Hz.

ASTM D 1822–89 Standard Test Method for *Tensile–Impact Energy to Break Plastics and Electrical Insulating Materials*, 9 pp (Comm D–20) There is a metric version, ASTM D 1822M–89

ASTM D 1929–91a Standard Test Method for *Ignition Properties of Plastics*, 6 pp (DOD Adopted) (FSC 9330) (MR) (Comm D–20)

This is called the Setchkin Technique and covers the laboratory determination of the self–ignition and flash–ignition temperatures of plastics using a hot–air–ignition furnace as the heat source.

ASTM D 2126–87 Standard Test Method for *Response of Rigid Cellular Plastics to Thermal and Humid Aging*, 2 pp (DOD Adopted) (FSC 9330) (MR) (Comm D–20)

This method provides ten conditions for the thermal and humid aging of rigid cellular plastics.

ASTM D 2221–68 (1984) Standard Test Method for *Creep Properties of Package Cushioning Materials*, 5 pp (DOD Adopted) (Area Pack) (AS) (Comm D–10)

Covers materials in bulk, sheet, or molded form. Measures the change in thickness of a loaded cushion with time.

ASTM D 2240–86 Standard Test Method for *Rubber Property — Durometer Hardness*, 5 pp (DOD Adopted) (FSC 9320) (MR) (Comm D–11)

ASTM D 2632–88 Standard Test Method for *Rubber Property — Resilience by Vertical Rebound*, 3 pp (DOD Adopted) (FSC 9320), (MR) (Comm D–11)

The scope of this method states that it is not applicable to the testing of cellular rubbers, but it has been used for that purpose.

ASTM D 2842–69 (1990) Standard Test Method for *Water Absorption of Rigid Cellular Plastics*, 9 pp (DOD Adopted) (Comm D–20)

Covers the determination of water absorption by measuring the change in buoyant force resulting from immersion under a 5.1–cm (2–in) head of water for 96 hours.

ASTM D 2843–88 Standard Test Method for *Density of Smoke from the Burning or Decomposition of Plastics*, 8 pp (DOD Adopted) (FSC 9330) (MR) (Comm D–20)

In this test, developed from the original Rohm and Haas XP–2 apparatus, measurements are made in terms of the loss of light transmission through a collected volume of smoke produced under controlled, standardized conditions. The flame and smoke can be observed during the test. Smoke density is measured by a photoelectric cell across a 12–inch light path.

ASTM D 2856–87 Standard Test Method for *Open Cell Content of Rigid Cellular Plastics by the Air Pycnometer*, 7 pp (DOD Adopted) (FSC 9330) (YD) (Comm D–20)

ASTM D 2863–87 Standard Test Method for *Measuring the Minimum Oxygen Concentration to Support Candle–Like Combustion of Plastics (Oxygen Index)*, (Comm D–20)

Describes a procedure for measuring the minimum concentration of oxygen in a flowing mixture of oxygen and nitrogen that will just barely support flaming combustion of a material initially at room temperature. Covers various forms of plastics, including films and cellular plastics.

ASTM D 2990–77 (1982) Standard Test Methods for *Tensile, Compressive, and Flexural Creep and Creep Rupture of Plastics*, 11 pp (DOD Adopted) (FSC 9330) (MR) (Comm D–20)

ASTM D 3014–89 Standard Test Method for *Flame Height, Time of Burning, and Loss of Weight of Rigid Thermoset Cellular Plastics in a Vertical Position*, 4 pp (DOD Adopted) (FSC 9330) (MR) (Comm D–20)

This method is known as the Butler Chimney Method.

ASTM D 3137–81 (1987) Standard Test Method for *Rubber Property —
Hydrolytic Stability*, 3 pp (DOD Adopted) (FSC 9320) (MR) (Comm
D–11)

Covers the determination of the ability of rubber to withstand the
environmental effects of high humidity and temperature. The effects are
determined by noting the change in tensile strength after exposure over
distilled water.

ASTM D 3489–85 Standard Methods of Testing *Rubber — Microcellular
Urethane*, 5 pp (Comm D–11)

Covers density, tensile properties, tear, hardness, compression set,
compression deflection, resilience, surface and core abrasion, heat aging,
hydrolytic resistance, cut–growth resistance, impact strength, flexural
modulus, ash, flexural recovery, and high–temperature sag.

ASTM D 3574–86 Standard Methods of *Testing Flexible Cellular
Materials — Slab, Bonded and Molded Urethane Foams*, 22 pp (DOD
Adopted) (FSC 9320) (MR) (Comm D–11)

Covers density, indentation–force deflection, compressionforce
deflection, constant deflection, compression set, tension, tear resistance,
air–flow, resilience (ball rebound), static–force loss at constant deflection,
dynamic fatigue, steam autoclave aging, and dry–heat aging.

ASTM D 3575–91 Standard Test Methods for *Flexible Cellular Materials
Made from Olefin Polymers*, 8 pp (Comm D–11)

Covers olefin polymers or blends of olefin polymers with other
polymers. A table lists 31 properties considered.

ASTM D 3576–77 Standard Test Method for *Cell Size of Rigid Cellular
Plastics*, 4 pp (Comm D–20)

Measures the number of cells or cell–wall intersections in a
specified distance and then converts these figures to average cell size by
mathematical derivation.

ASTM D 3675–89 Standard Method for *Surface Flammability of Flexible Cellular Materials Using a Radiant Heat Energy Source*, 7 pp (Comm D–20)

ASTM D 3768–85 Standard Method for Testing *Microcellular Urethanes — Flexural Recovery*, 3 pp (Comm D–11)

Describes a method to indicate the ability of a material to recover after a 180° bend around a 12.7–mm (0.5–in.) diameter mandrel at room temperature.

ASTM D 3769–85 Standard Method for Testing *Microcellular Urethanes — High–Temperature Sag*, 3 pp (Comm D–11)

Indicates the deformation tendency of microcellular materials that may occur in painting operations during assembly.

ASTM D 3894–88 Standard Method for *Evaluation of Fire Response of Rigid Cellular Plastics Using a Small Corner Configuration*, 8 pp (Comm D–20)

Describes a procedure based on the Upjohn Mini–Corner Test.

ASTM D 4100–82 (1989) Standard Test Method for *Gravimetric Determination of Smoke Particulates from Combustion of Plastic Materials*, 8 pp (Comm D–20)

Covers plastic foams as well as solid plastics. Data on rigid and flexible polyurethane foams are given in the section on Precision and Accuracy.

ASTM D 4168–88 Standard Test Methods for *Transmitted Shock Characteristics of Foam–in–Place Cushioning Materials*, 5 pp (Comm D–10)

Method A is a free–fall package–drop test. Method B is a shock test.

ASTM D 4273–83 (1988) Standard Methods for Testing *Polyurethane Raw Materials: Determination of Primary Hydroxyl Content of Polyether Polyols*, 9 pp (Comm D–20)

ASTM D 4274–88 Standard Method for Testing *Polyurethane Polyol Raw Materials: Determination Of Hydroxyl Number of Polyols*, 9 pp (Comm D–20)

ASTM D 4659–87 Standard Test Methods for *Polyurethane Isocyanate Raw Materials: Determination of Specific Gravity*, 3 pp (Comm D–20)

ASTM D 4660–90 Standard Test Method for *Polyurethane Raw Materials: Determination of the Isomer Content of Toluenediisocyanate*, 4 pp (Comm D–20)

ASTM D 4661–87 Standard Test Methods for *Polyurethane Raw Materials: Determination of Total Chlorine in Isocyanates*, 5 pp (Comm D–20)

ASTM D 4662–87 Standard Test Methods for *Polyurethane Raw Materials: Determination of Acid and Alkalinity Numbers of Polyols*, 3 pp (Comm D–20)

ASTM D 4663–87 Standard Test Method for *Polyurethane Raw Materials: Determination of Hydrolyzable Chlorine of Isocyanates*, 2 pp (Comm D–20)

ASTM D 4664–87 Standard Test Method for *Polyurethane Raw Materials: Determination of Freezing Point of Toluene Diisocyanate Mixtures*, 2 pp (Comm D–20)

ASTM D 4665–87 Standard Test Method for *Polyurethane Raw Materials: Determination of Assay of Isocyanates*, 3 pp (Comm D–20)

ASTM D 4666–87 Standard Test Method for *Polyurethane Raw Materials: Determination of Amine Equivalent of Crude or Modified Isocyanates*, 3 pp (Comm D–20)

ASTM D 4667–87 Standard Test Method for *Polyurethane Raw Materials: Determination of Acidity in Toluene Diisocyanate*, 2 pp (Comm D–20)

ASTM D 4668–87 Standard Test Method for *Polyurethane Raw Materials: Determination of Sodium and Potassim in Polyols*, 4 pp (Comm D–20)

ASTM D 4669–87 Standard Test Method for *Polyurethane Raw Materials: Determination of Specific Gravity of Polyols*, 2 pp (Comm D–20)

ASTM D 4670–87 Standard Test Method for *Urethane Raw Materials: Determination of Suspanded Matter in Polyols*, 1 p (Comm D–20)

ASTM D 4671–87 Standard Test Method for *Polyurethane Raw Materials: Determination of Unsaturation of Polyols*, 3 pp (Comm D–20)

ASTM D 4672–87 Standard Test Method for *Polyurethane Raw Materials: Determination of Water Content of Polyols*, 4 pp (Comm D–20)

ASTM D 4875–88 Standard Test Methods for *Polyurethane Raw Materials: Determination of the Polymerized Ethylene Oxide Content of Polyester Polyols*, 5 pp (Comm D–20)

ASTM D 4876–88 Standard Test Method for *Polyurethane Raw Materials: Determination of Crude or Modified Isocyanates*, 2 pp (Comm D–20)

ASTM D 4878–88 Standard Test Method for *Polyurethane Raw Materials: Determination of Viscosity of Polyols*, 3 pp (Comm D–20)

These methods determine the viscosities of polyols in the range from 10 to 10,000 cP at 25°C and 50°C. Test Method A (Brookfield Viscosity) also applies to more viscous samples soluble in n–butyl acetate. Test Method B is the Cannon Fenske method.

ASTM D 4889–88 Standard Test Method for *Polyurethane Raw Materials: Determination of Viscosity of Crue or Modified Isocyantes*, 2 pp (Comm D–20)

ASTM D 4890–88 Standard Test Method for *Polyurethane Raw Materials: Determination of Gardner and Alpha Color of Polyols*, 3 pp (Comm D–20)

ASTM D 4986–89 Standard Test Method for *Horizontal Burning Characteristics of Cellular Polymeric Materials*, 4 pp (Comm D–11)

This method describes a small–scale apparatus test procedure for comparing the relative rate of burning and the extent and time of burning of cellular polymeric materials.

ASTM D 5113–90 Standard Test Method for *Determining Adhesive Attack on Rigid Cellular Polystyrene Foam*, 2 pp (Comm D–14)

This test method is useful in determining the effect of using adhesives on rigid cellular polystyrene (RCPS) foam in building construction. The method covers a practical means of measuring the degree of foam cavitation damage when an adhesive is used to bond these foam substrates.

ASTM D 5132–90 Standard Test Method for *Horizontal Burning Rate of Flexible Cellular and Rubber Materials Used in Occupant Compartments of Motor Vehicles*, 4 pp (Comm D–11)

ASTM E 84–89a Standard Test Method for *Surface Burning Character- istics of Building Materials*, 15 pp (DOD Adopted) (FSC 56 GP) (YD) (Comm E–5)

This is called the Steiner Tunnel Test. Very large samples are required.

ASTM E–90 Standard Test Method for *Water Vapor Transmission of Materials*, 8 pp (Comm C–16)

Describes two cup methods, one a Desiccant Method and the other a Water Method. Periodic weighings are made in both methods.

ASTM E–119–88 Standard Methods of *Fire Tests of Building Construc- tion and Materials*, 21 pp (DOD Adopted) (FSC FACR) (YD) (Comm E–5)

This test, similar to UL 263 and NFPA 251, is used for testing walls, floors, ceiling, roofs, etc.

ASTM E 143–87 Standard Test Method for *Shear Modulus at Room Temperature*, 6 pp (Comm E–28)

ASTM E 162–90 Standard Test Method for *Surface Flammability of Materials Using a Radiant Heat Energy Source*, 8 pp (DOD Adopted) (FSC 9330) (MR) (Comm E–5)

This is called the Radiant Panel Test. A Flame–Spread Index is calculated as a product of the flame–spread and heat–evolution factors. Smoke density is also obtained.

ASTM E 286–85 Standard Test Method for *Surface Flammability of Materials Using an 8–ft (2.44–m) Tunnel Furnace*, 9 pp (Comm E–5)

Covers the measurement of surface flame spread of building materials capable of being mounted and supported within a 14–inch by 8–foot (2.44–m) test frame. The test also includes techniques for measuring the smoke density and heat produced. It is intended for R&D purposes only.

ASTM E 595–90 Standard Test Method for *Total Mass Loss and Collected Volatile Condensable Materials from Outgassing in a Vacuum Environment*, 8 pp (DOD Adopted) (FSC 9330) (MR) (Comm E–21)

ASTM E 662–83 Standard Test Method for *Specific Optical Density of Smoke Generated by Solid Materials*, 29 pp (Comm E–5)

This method is basically the NBS Smoke Density Chamber Test. It covers the determination of the specific optical density of smoke generated by solid materials and assemblies mounted in a vertical position in thickness up to 1 inch (2.54 cm). It measures attenuation of a light beam by smoke accumulating within a closed chamber due to nonflaming pyrolytic decomposition and flaming combustion.

ASTM E 906–83 Standard Test Method for *Heat and Visible Smoke Release Rates for Materials and Products*, 22 pp (Comm E–5)

This method is based on the Ohio State University Release Rate Apparatus. The specimen is injected into the environmental chamber through which a constant flow of air passes. The exposure of the specimen is determined by a radiant–heat source adjusted to produce the desired total flux on the specimen, which may be tested horizontally or vertically. Combustion may be initiated by non–piloted ignition, piloted ignition of evolved gases, or by point ignition of the surface. The changes in temperature and optical density of the gas leaving the chamber are monitored, and from the resultant data the release rate of heat and visible smoke are calculated. This apparatus is also used to measure the rate of toxic gas release and consumption.

ASTM Practices, Definitions, Abbreviations, Guides, Classifications, etc. (See Reference 3)

ASTM C 168–90 Standard Terminology Relating to *Thermal Insulating Materials*, 6 pp (Comm C–16)

ASTM C 274–68 (1988) Standard Definitions of Terms Relating to *Structural Sandwich Constructions*, 1 p (Comm F–7)

ASTM C 390–79 (1989) Standard Criteria for *Sampling and Acceptance of Preformed Thermal Insulation Lots*, 3 pp (DOD Adopted) Comm C–16)

ASTM C 447–85 Standard Practice for *Estimating the Maximum Use Temperature of Thermal Insulations*, 4 pp (Comm C–16)

Covers Loose–fill, blanket, board, and preformed pipe insulations.

ASTM C 945–81 (1987) Standard Practice for *Design Considerations and Spray Application of a Rigid Cellular Polyurethane Insulation System on Outdoor Service Vessels*, 11 pp (Comm C–16)

ASTM C 950–81 (1987) Standard Practice for *Repair of a Rigid Cellular Polyurethane Insulation System on Outdoor Service Vessels*, 2 pp (Comm C–16)

ASTM D 756–78 (1983) Standard Practice for *Determination of Weight and Shape Changes of Plastics Under Accelerated Service Conditions*, 6 pp (DOD Adopted) (FSC 9330) (MR) (Comm D–20)

ASTM D 883–90 Standard Definitions of Terms Relating to *Plastics*, 13 pp, (DOD Adopted) (FSC 9330) (MR) (Comm D–20)

ASTM D 1435–85 Standard Practice for *Outdoor Weathering of Plastics*, 7 pp, (DOD Adopted) (FSC 9330) (MR) (Comm D–20)

ASTM D 1490–84 Standard Practice for *Operating Light– and Water–Exposure Apparatus (Carbon–Arc Type) for Exposure of Plastics*, 4 pp (DOD Adopted) (FSC 9330) (MR) (Comm D–20)

ASTM D–1566–91 Standard Terminology Relating to *Rubber*, 9 pp (Comm D–11)

ASTM D 1600–91 Standard Terminology Relating to *Abbreviations, Acronyms, and Codes Relating to Plastics*, 8 pp, (DOD Adopted) (FSC 9330) (MR) (Comm D–20)

ASTM D 2565–89 Standard Practice for *Operating Xenon Arc–Type (Water–Cooled) Light–Exposure Apparatus With and Without Water For Exposure of Plastics*, 4 pp (Comm D–20)

ASTM D 2841–83 (1988) Standard Practice for *Sampling Hollow Micro–spheres*, 3 pp (Comm D–20)

This practice covers the procedure for obtaining representative samples of hollow microspheres of the type used for syntactic foam buo–yancy materials. The procedure consists of procuring representative samples by the use of "spike" or "thief" samplers which can be inserted all the way to the bottom of the container, thus sampling the entire vertical distance.

ASTM D 3748–83 Standard Practice for *Evaluating High–Density Rigid Cellular Thermoplastics*, 2 pp (Comm D–20)

This practice provides nine appropriate testing methods and a specific data–reporting procedure for the subject materials.

ASTM D 4098–82 (1987) Standard Practice for *Evaluating High–Density Rigid Cellular Thermosets*, 2 pp (Comm D–20)

This practice describes the basic test procedures for determination of physical properties and reporting of data necessary to evaluate high–density rigid cellular thermosets. Eleven ASTM test methods are referred to.

ASTM D 5140–90 Standard Guide for *Testing Polyurethane "Poured in Place" Thermal Break Materials*, 2 pp (Comm D–20)

Thermal–break materials are solid or cellular materials, or combinations of materials, of low thermal transmission placed between components of high thermal transmission in order to reduce the heat flow across the assembly.

ASTM D 3951–90 Standard Practice for *Commercial Packaging*, 2 pp (DOD Adopted) (Comm D–10)

This practice establishes minimum requirements for packaging of supplies and equipment, exclusive of ammunition, explosives, or hazardous materials as covered in Title 49 of the Code of Federal Regulations. Cushioning is covered in general terms.

ASTM D 4098–82 (1987) Standard Practice for *Evaluating High–Density Rigid Cellular* Thermosets, 2 pp (Comm D–20)

This practice describes the basic test procedures for determination of physical properties and reporting of data necessary to evaluate the subject materials.

ASTM E 176–91a Standard Terminology Relating to *Fire Standards*, 5 pp (Comm E–5)

This standard contains terms, related definitions, and descriptions of terms used, or likely to be used, in fire standards and fire–risk standards.

ASTM G 21–70 (1985) Standard Practice for *Determining Resistence of Synthetic Polymeric Materials to Fungi*, 5 pp (DOD Adopted) (FSC 9330) (MR) (Comm G–3)

ASTM G 22–76 (1985) Standard Practice for *Determining Resistance of Plastics to Bacteria*, 4 pp (Com G–3)

ASTM G 23–89 Standard Practice for *Operating Light–Exposure Apparatus (Carbon–Arc Type) with and without Water for Exposure of Nonmetallic Materials*, 8 pp (DOD Adopted) (FSC 9330) (MR) (Comm G–3)

This laboratory weathering procedure has four methods.

ASTM G 26–90 Standard Practice for *Operating Light–Exposure Apparatus (Xenon–Arc Type) with and without Water for Exposure of Nonmetallic Materials*, 9 pp (Comm G–3)

Society of Automotive Engineers (SAE) Aerospace Material Specifications (AMS) (See Reference 4)

AMS 3193–B–85 *Silicone Rubber Sponge, Closed Cell, Medium, Extreme Low Temperature*, 6 pp (DOD Adopted) (FSC 9320) (AS)

Covers temperature range from −100° to +125°C (−170° to +400°F).

AMS 3194B–85 *Silicone Rubber Sponge, Closed Cell, Firm, Extreme Low Temperature*, 6 pp (DOD Adopted) (FSC 9320) (AS)

AMS 3195D–85 *Silicone Rubber Sponge, Closed Cell, Medium*, 6 pp (DOD Adopted) (FSC 9320) (AS)

AMS 3196E–89 *Silicone Rubber Sponge, Closed Cell, Firm*, 6 pp (DOD Adopted) (FSC 9320) (AS)

AMS 3197J–84 *Sponge, Chloroprene (CR) Rubber, Soft*, 7 pp (DOD Adopted) (FSC 9320) (AS)

AMS 3198J–84 *Sponge, Chloroprene (CR) Rubber, Medium*, 7 pp (DOD Adopted) (FSC 9320) (AS)

AMS 3199J–84 *Sponge, Chloroprene (CR) Rubber, Firm*, 7 pp (DOD Adopted) (FSC 9320) (AS)

AMS 3568–83 *Foam Sheet, Polyether Urethane (EU) Elastomer, Shock Absorbing*, 12 pp (DOD Adopted) (FSC 9320) (AS)

AMS 3568/1–83 *Foam, Polyether Urethane (EU) Elastomer, Shock Absorbing, 5 lb/cu ft (72 kg/m³) Density*, 2 pp (DOD Adopted) (FSC 9330) (AS)

AMS 3568/2–83 *Foam, Polyether Urethane (EU) Elastomer, Shock Absorbing, 15 lb/cu ft (240 kg/m³) Density*, 2 pp (DOD Adopted) (FSC 9330) (AS)

AMS 3568/3–83 *Foam, Polyether Urethane (EU) Elastomer, Shock Absorbing, 20 lb/cu ft (320 kg/m³) Density*, 2 pp (DOD Adopted) (FSC 9330) (AS)

AMS 3569A–88 *Foam, Flexible Polyurethane (PUR), Open Pore, Polyvinylchloride Coated*, 9 pp

Covers a reticulated flexible polyurethane foam sheet with hydrolytic stability.

AMS 3570D–89 *Foam, Flexible Polyurethane, Open Cell, Medium Flexibility*, 7 pp (DOD Adopted) (FSC 9330) (AS)

Used for interior padding, cushioning and vibration isolation.

AMS 3572A–84 *Polyurethane, Foam–in–Place, Rigid*, 9 pp

Used for FIP applications such as electronic encapsulating or miscellaneous packaging for use from $-70°$ to $+105°C$ ($-95°$ to $+225°F$).

AMS 3574–85 *Polyurethane, Foam–in–Place, Rigid, Closed Cell, for Helmet Liners*, 7 pp

AMS 3635C–84 *Plastic Sheet and Strip, Modified Vinyl, Foamed, Closed Cell*, 7 pp (DOD Adopted) (FSC 9330) (AS)

This material is primarily for parts such as headrest pads and arm rests, wherever an impact–resistant, slow–recovery, energy–absorbing product is required.

AMS 3699–83 *Resin System, Epoxy, Carbon Microballoon Filled*, 135°C (375°F) Cure, 7 pp

Covers a two-part epoxy-resin system in the form of a bisphenol "A" epoxy resin filled with fumed silica and carbon microspheres and a separate aromatic diamine curing agent.

AMS 3709B-87 *Syntactic Foam Tiles*, 8 pp (DOD Adopted) (FSC 9330) (AS)

Covers dielectrically loaded tiles with a polyimide resin matrix for long-term applications to 315°C (600°F).

AMS 3725-84 *Core, Polyurethane Foam (Polyether), Rigid, Cellular*, 7 pp

Covers both polyurethane and polyisocyanurate foam, board or block, for use from −55° to 70°C (−65° to 160°F). Application is primarily for sandwich construction.

AMS 3725/1-84 *Core, Polyurethane Foam (Polyether), 2.0 lb per cu ft (32 kg/m³) Density*, 2 pp

AMS 3725/2-84 Core, *Polyurethane Foam (Polyether), 4.0 lb per cu ft (64 kg/m³) Density*, 2 pp

AMS 3751B-90 *Microspheres, Hollow Glass*, 7 pp (DOD-Adopted) (FSC 9340) (AS)

Applications are for use as lightweight filler materials for syntactic foam shapes or for parts for dielectric applications.

AMS 3753-83 *Microspheres, Carbon, Hollow*, 7 pp

Applications are for use as filler material on syntactic foam shapes.

AMS 3912A-85 *Radomes, Foam Sandwich, Polyimide/Quartz Construction*, 15 pp (DOD Adopted) (FSC 5680) (AS)

Covers material and process requirements for fabricating sandwich radomes with polyimide-resin-impregnated quartz cloth shells and polyimide-resin syntactic foam cores. Application is primarily as a radar–transparent structure for use as an electromagnetic window up to 315°C (600°F).

AMS 3913A–89 *Radome, Foam Sandwich, Hot Melt, Addition–Type Polyimide*, 13 pp

Covers material and process requirements for fabrication sandwich radomes with hot–melt, addition–reaction polyimide–resin–impregnated quartz cloth shells and polyimide–resin syntactic foam cores. Application is primarily as a radar–transparent structure for use as an electromagnetic window up to 230°C (450°F).

Underwriters Laboratories (UL) (See Reference 5)

UL 94–80 Tests for *Flammability of Plastic Materials for Parts in Devices and Appliances*, 3rd Edition, revised to September 17, 1990, 55 pp (DOD Adopted) (FSC 9330) (MR)

This standard includes an Appendix A — *Horizontal Burning Test for Classifying Foamed Materials 94HBF, 94HF–1, or 94HF–2*. This Appendix is not part of the standard, but is included for information purposes only. The requirements of the standard do not cover foamed plastics for use as materials for building construction or finishing. ASTM D 4986–89 on *Horizontal Burning Characteristics of Cellular Polymeric Materials* was developed by ASTM Committee[11] on Rubber as a similar procedure. A similar method is also being developed as an ISO Standard. Copies of UL 94 may be obtained from UL Publications Stock, 333 Pfingsted Road, Northbrook, IL 60062–2096.

MILITARY SPECIFICATIONS (See Reference 6)

MIL–M–910E(3) *Mats, Floor, Standing*, 17 November 1981, 5 pp amend + 7 pp base spec (FSC 7220) (SH)

The mat is a smooth unicellular base bonded to a solid rubber or plastic top covering.

MIL–C–3133C Notice 1 *Cellular Elastomeric Materials, Molded or Fabricated Parts*, 17 May 1984, 3 pp Notice + 8 pp base spec (FSC 9320) (MR)

Notice 1 specifies the use of ASTM D 1055, D 1056, D 1565, D 1667, D 3453, and D 3574. Notice 2 on 9 Aug 1990 validates this specification.

MIL–B–4792D(1) (USAF) *Bumper, Rubber, Duplex Round*, 30 June 1978, 1 p amend + 5 pp base spec (FSC 1730) (82)

Notice 1 on 31 Oct 1989 validates this specification.

MIL–R–5001A(4) *Rubber Cellular Sheet, Molded and Hand Built Shapes: Latex Foam*, 19 Sept 1974, 2 pp amend +19 pp base spec (FSC 9320) (AS)

Int. Amend 5 was issued 1 April 1991, 4 pp (SA).

MIL–R–6130C *Rubber, Cellular, Chemically Blown*, 18 July 1980, 16 pp (FSC 9320) (AS)

Int. Amend 1 was issued 1 April 1991, 3 pp (SA).

MIL–C–8087C (ASG) *Core Material, Foamed–in–Place, Urethane Type*, 24 April 1968, 15 pp (FSC 9320) (AS)

MIL–C–10870G *Food Container, Insulated, With Inserts*, 30 Sept 1982, 16 pp (FSC 7330) (GL)

Has foamed–in–place insulation.

MIL–P–12420D *Plastic Material, Cellular, Elastomeric*, 13 March 1991, 19 pp (FSC 9330) (GL)

This specification covers expanded unicellular elastomeric plastic material in sheet form intended for use in shock–absorbent containers, impermeable extreme–cold–weather jackets, and decorative insulations. The material specified is a blend of vinyl and a butadiene–acrylonitrile rubber or other thermosetting elastomeric material. Recycled material use is encouraged.

MIL–P–13607B(2)(AR) *Padding Materials, Resilient (for Packaging of Ammunition)*, 10 April 1989, 4 p amend + 17 pp base spec (FSC 8140) (AR)

This performance specification covers resilient padding material for military packaging and packing applications inside of water–resistant ammunition containers. There are two types, Type I – Flexible and Type II – Non–flexible.

MIL–P–14401B(AT) *Pads, Cushioning: Personnel Protection, Vehicular,* 6 May 1980, 11 pp (FSC 2590) (AT)

These pads are for cushioning interior surfaces of military vehicles to protect personnel from injury. There are two types, Type I – Facial Contact, and Type II – Head and Body Contact, with five classes, including closed–cell expanded rubber and cellular vinyl elastomer.

MIL–P–14511C(1) *Insulation Sheet, Cellular, Plastic, Thermal,* 16 February 1988, 2 p amend + 12 pp base spec (FSC 5640) (AT)

MIL–H–15280J *Plastic Material, Unicellular (Sheets and Tubes),* 19 December 1988, 18 pp (FSC 5640) (SH)

This specification covers chemically expanded unicellular elastomeric plastic foam for thermal insulation. It is used as the inherently buoyant foam in MIL–L–18045F. Has a Qualified Products List (QPL).

MIL–I–16562A(1)(OS) *Insulation, Synthetic, Rubber–Like, Chemically Expanded, Cellular (Sheet Form),* 14 January 1977, 2 p amend + 21 pp base spec (FSC 1055) (OS)

MIL–C–17435C *Notice 3 Cushioning Material, Fibrous Glass,* 24 March 1987, 1 p notice + 11 pp base spec (FSC 8135) (OS)

Notice 2 reinstates this specification, which had been cancelled. Notice 3 validates it for use in acquisition.

MIL–L–18045F(1) *Life Preservers, Vest, Inherently Buoyant,* 30 March 1988, 2 pp amend + 11 pp base spec (FSC 4220) (SH)

This specification covers life preservers which use buoyant foam pads made from MIL–H–15280H foam.

MIL–C–18345A Notice 1 *Core Material, Cellular Cellulose Acetate*, 13 September 1988, 1 p Notice + 3 pp base spec (FSC 9330) (AR)

For use in fabrication of structural sandwich parts. The Notice validates the specification written in 1962.

MIL–M–18351F(SH) *Mattresses and Mattress Ticks, Berth, Synthetic Cellular Rubber, Naval Shipboard*, 22 April 1981, 19 pp (FSC 7210) (SH)

Covers low–smoke polychloroprene foam rubber berth mattresses for shipboard use.

MIL–P–19644C *Plastic Molding Material (Polystyrene Foam, Expanded Bead)*, 10 July 1970, 15 pp (FSC 9330) (OS). Int amend 1 was issued 1 April 91, 3 pp (SA)

MIL–R–20092L(1) *Rubber or Plastic Sheets and Assembled and Molded Shapes, Synthetic, Foam or Sponge, Open Cell*, 25 February 1988, 1 p amend + 17 pp base spec (FSC 9320) (SH)

This Navy specification covers inherently combustion–retardant cellular synthetic rubber latex or plastic foams for mattresses, cushioning and packaging.

MIL–B–21408G *Boots, Safety*, 17 January 1984, 40 pp (FSC 8430) (NU)

The cushion in the boots may use a wool felt, a foamed PVC (4 to 6 lb/ft^3), or latex foam rubber. This specification was validated 21 August 1989.

MIL–C–21850C(NU) *Coveralls, Catapult, Crewmen's*, 13 December 1979, 29 pp (FSC 8415) (NU)

Uses a neoprene foam of closed–cell construction as a component. Density is 10 lb/ft 3 (160 kg/m^3) minimum.

MIL–P–21929C *Plastic Material, Cellular Polyurethane, Foam–in–Place, Rigid (2 Pounds per Cubic Foot)*, 15 January 1991, 20 pp (FSC 9330) (SH)

This Navy specification covers a single class of foam, nominal density 2.0 lb/ft^3 rigid unicellular polyurethane foam, and the materials required for preparation by the foam–in–place technique. Requirements cover density, compressive strength, volume change after heat aging, humidity aging, compressive set, unicellularity (% open cells, max.), oil resistance, and fire resistance.

MIL–C–23806A(2)(EC) *Cable, Radio Frequency, Coaxial, Semirigid, Foam Dielectric, General Specification*, 4 April 1984, 1 p amend + 10 pp base spec (FSC 6145) (EC)

MIL–C–23806/1C(EC) *Cable, Radio Frequency, Coaxial, Foam Dielectric, 1/2 inch, 50 and 75 ohm, (RG–231/U, RG–331/U, RG–334/U and RG–335/U*, 20 February 1990, 3 pp (FSC 6145) (EC)

MIL–C–23806/2C(EC) *Cable, Radio Frequency, Coaxial, Semirigid, Foam* Dielectric, 7/8 inch, 50 and 75 ohm, (RG–332/U, RG–333/U, RG–336/U and *RG–306A/U*, 20 February 1990, 3 pp (FSC 6145) (EC)

MIL–C–23806/3C(EC) *Cable, Radio Frequency, Coaxial, Semirigid, Foam Dielectric, 3/4 inch, 50 ohm, Jacketed (RG–360/U)*, 20 January 1990, 4 pp (FSC 6145) (EC)

MIL–S–24154A(2) *Syntactic Buoyancy Material for High Hydrostatic Pressures*, 6 February 1991, 3 pp amend 2 + 17 pp base spec (FSC 9330) (SH)

Revision A of this specification was published 10 March 1957. There are two types, Type I for 4,5000 psig hydrostatic pressure and Type II for 10,000 psig. The QPL was cancelled on 31 March 1986. The syntactic buoyancy material consists of a low–density filler such as hollow–glass microspheres in a resin matrix such as epoxy resin.

MIL–S–24167A (SHIPS) *Syntactic Material, Pour–in–Place, Structural Void Filling*, 6 December 1972, 8 pp (FSC 9330) (SH)

MIL–I–24172A(SH) *Insulation, Cellular Polyurethane, Rigid, Preformed and Foam–in–Place*, 19 November 1986, 10 pp (FSC 5640) (SH)

MIL–A–24179A NOTICE 1 (SHIPS) *Adhesive, Flexible Unicellular —* *Plastic Thermal Insulation*, 22 June 1987, 1 p Notice + 2 pp amend + 11 pp base spec (FSC 8040) (SH)

This specification covers high–initial–strength, heat– and water–resistant, contact–type adhesives for bonding flexible unicellular plastic thermal insulation to itself and to metal surfaces. There are three Types and two Classes. Notice 1 validates the specification.

MIL–P–24249(1)(SHIPS) *Plastic Material, Cellular Polyurethane, Rigid, Void Filler, Foam–in–Place, Large Scale and Installation of,* 6 November 1967, 3 pp amend + 13 pp base spec (FSC 9330) (SH) (QPL)

MIL–I–24703 *Insulation, Pipe, Polyphosphazene, Sheet and Tubular,* 5 April 1988, 19 pp (FSC 5640) (SH)

This specification, prepared by the Navy, covers polyphosphazene elastomeric foam material for thermal insulation on piping, in either sheet or tubing form.   Polyphosphazene foam has excellent fire–retardant properties and is suitable for use in the range −20 to 180°F (−29 to 82.2°C) in tubular form (Form T). Form S covers sheet form.

MIL–T–24708(SH) *Thermal/Acoustic Insulation Barrier Material: Polyimide Foam*, 29 July 1988, 10 pp (FSC 5640) (SH)

Covers lightweight, flexible polyimide foam as a layered thermal–insulation–barrier material developed for use within a specific frequency range on submarines.   Currently there is only one Type, with three Classes, 1, 2, and 3, with single, double and triple layers, respectively. The layers of polyimide foam are bonded to a polyphosphazene/barium sulfate sheet.

MIL–S–25392B(ASG) *Sandwich Construction, Plastic Resin, Glass Fabric Base, Laminated Facings and Urethane Foamed–in–Place Core, for Structural Applications,* 8 May 1968, 12 pp (FSC 9330) (AS)

MIL–P–26514F *Polyurethane Foam, Rigid or Flexible, for Packaging,* 18 December 1987, 24 pp (FSC 8135) (69)

Covers prefoamed polyurethane foams, both rigid and flexible, for packaging applications.   The foams described are intended for use as cushioning and blocking/bracing in packages to protect equipment and items therein from damage by shocks or impacts incurred during shipment and handling. There are two Types, I and III for standard and antistatic foams. Type II in earlier revisions covered foam–in–place. There are three Grades, A, B, and C, each with varying static–stress curves. Class 1 covers rigid foam and Class 2 flexible foam. Foam–in–place foam is now covered in MIL–F–83571(1), which see.

MIL–C–26861C *Cushioning Material, Resilient Type, General*, 9 June 1988, 22 pp (FSG 8135) (69)

This specification, developed by the Air Force, is a performance specification. It covers resilient materials, including foams used for cushioning in the form of rolls, sheets, special diecut pads, or molded forms, as specified, to be used within packages to protect equipment from shocks or impacts incurred during shipment and handling. There are seven (7) Classes, based on loading range, and five (5) Grades, based on peak acceleration.

MIL–S–27332B(USAF) *Seat Cushion Insert, Polyurethane Foam, Plastic, General Specification for*, 26 October 1983, 10 pp (FSC 1660) (82)

MIL–F–29248(YD) *Fenders, Marine, Foam Filled, Netless*, 21 November 1986, 14 pp (FSC 2040) (YD)

Covers a high–energy–absorption elastomeric marine–fender system for protection of ships, harbor craft, wharves, and piers from damage between the interface of vessel to vessel, or vessel to pier. The specification calls for a fender cushion completely filled with a closed–cell crosslinked polyethylene plastic foam.

MIL–C–38226B(USAF) *Containers, Polyurethane, Rigid or Elastic for Packaging Small Engines*, 9 March 1987, 12 pp (FSC 8115) (69)

Covers general requirements for polyurethane shipping containers for packaging small engines not exceeding 500 pounds.

MIL–P–39500(2)(OS) *Polyurethane Foam Kit, Rigid*, 1 June 1982, 3 pp amend + 7 pp base spec (FSC 1195) (AR)

Covers a rigid polyurethane foam kit for use in certain nuclear ordnance, explosive ordnance disposal operations. The system is a three-component liquid foam-in-place type for use in immobilizing materials and objects.

MIL-P-40619A *Plastic Material, Cellular, Polystyrene (for Buoyancy Applications)*, 9 December 1968, 8 pp (FSC 9330) (SH) was cancelled by Notice 1 on 4 January 1988, but was then reinstated by Notice 2 on 15 September 1988. Furthermore, the Army Missile Command (MI) has assumed Preparing Activity responsibility in lieu of the Naval Sea Systems Command (Ship Systems) (SH)

MIL-P-43226A(MI) *Polyether Cushioning Material, Foam-in-Place, Flexible*, 28 June 1991, 10 pp (FSC 8135) (MI)

Covers a polyether-type polyurethane foam. There are 3 Types covering densities of 2-3, 3-4 and 4-5 lb/ft^3.

MIL-C-43858B(GL) *Cloth, Laminated, Nylon Tricot, Pblyurethane Foam Laminate, Chemical Protective and Flame Resistant*, 16 January 1986, 25 pp (FSC 8305) (GL)

Covers three types of tricot knit nylon cloth laminated to a polyester-base polyurethane foam and impregnated with an activated-carbon mixture.

MIL-R-46089B(MR) *Rubber Sponge, Silicone Closed Cell*, 12 February 1981, 7 pp (FSC 9320) (MR)

MIL-C-46111C(MR) *Plastic Foam, Polyurethane (For Use in Aircraft)*, 29 September 1978, 10 pp (FSC 9330) (MR)

This specification covers Army aircraft. It was validated by Notice 1, 3 August 1989.

DOD-T-46151A(2) *Tape, Adhesive, Rubber (Metric)*, 18 March 1983, 1 p amend + 15 pp base spec (FSC 9320) (MR)

Covers polyurethane foam pressure-sensitive adhesive tape used for weather stripping, dust sealing, damping, thermal insulation, sound absorption, and packaging. The specification was validated 6 July 1988.

MIL–F–46194 *Foam, Rigid, Structural, Closed Cell Polymethacrylimide (PMI)*, 18 July 1988, 10 pp (AREA CMPS) (MR)

This specification covers a dimensionally stable, closed–cell rigid polymethacrylimide (PMI) foam. These foams are comparable in density to some types of honeycomb core. They may be viable candidates for structural aerospace applications whenever a core or filler is needed, and on nonstructural applications, including formed (hat section) configurations. Type I is for processing up to 250°F (121°C) with Classes ranging from 2.0–4.7 lb/ft^3 (32–75 kg/m^3). Type II is for processing up to 350°F after post–curing, with Classes ranging from 3.2–18.7 lb/ft^3 (51–300 kg/m^3). A water–absorption requirement and prescribed test method is included. A commercial foam meeting this specification is ROHACELL PMI Rigid Foam made by Rohm Tech Inc., Malden, MA.

MIL–H–46354A(2)(AR) *Headrest, Optical Instrument*, 7 May 1986, 3 pp amend + 12 pp base spec (FSC 1140) (AR) (QPL)

Covers headrests, browrests, and crash pads for use with Fire Control Instrumenta. Pads are cellular elastomeric pads with fungus and oil resistance.

MIL–T–46586C(AR) *Tube, Igniter for Charge, Propelling, 175 mm, M86A2 (Cellular Polyurethane)*, 13 June 1978, 8 pp (FSC 1320) (AR)

MIL–S–46897B(MI) *Sealing Compound, Polyurethane Foam*, 5 October 1990, 7 pp (FSC 8030) (MI)

Covers one type of polyurethane–foam sealing compound used as a nozzle seal in rocket motors. There are two density types, 4.5 lb/ft^3 and 7, 5 lb/ft^3. The use of FREON, chlorotrifluorocarbons, or other com–pounds which are restricted because of their effects on the ozone layer are forbidden.

MIL–P–47099A(MI) *Polyurethane Foam, Polyether Type, Rigid, for Packaging and Encapsulation of Electronic Components*, 31 March 1989, 10 pp (FSC 5970) (MI). (This specification was validated 6 September 1990).

There are three Types, as follows:

Type I	2 lb/ft^3
Type II	2 to 3 lb/ft^3
Type III	4 lb/ft^3

MIL–F–47222(MI) Notice 2 *Foam, Polyurethane, Rigid*, June 24, 1985, 1 p notice + 9 pp base spec (FSC 9330) (MI)

Notice 2 reinstates this specification which was issued originally on 12 July 1974.

MIL–F–52236(ME) *Filter Element, Air, Diver's Polyurethane Foam*, 21 August 1962, 4 pp (FSC 4220) (ME)

Used in divers' supply line in military diving equipment. The foam is a skeletal (reticular) polyurethane foam.

MIL–P–60312C(2)(AR) *Parts, Molded, Plastic Foam, Polystyrene (For Use with Ammunition)*, 5 December 1986, 4 pp amend + 14 pp base spec (FSC 8140) (AR)

Covers an antistatic expanded–bead–type polystyrene foam.

MIL–T–60394A(3)(AR) *Tape, Pressure–Sensitive Film Foam, Double Coated* (For Use with Ammunition), 24 July 1988, 5 pp amend + 16 pp base spec (FSC 1375) (AR)

The tape has a high–density open– or closed–cell foam plastic backing. The application is for demolition charges.

MIL–S–63060(AR) *Spacer for Charge, Propelling, 175 mm, M124 (Cellular Polyurethane)*, 31 March 1976, 10 pp (FSC 1320) (AR)

This specification covers a cellular polyurethane spacer with a density of 13.3 ± 2.5 lb/ft^3. There is a burning characteristic requirement. The specification was validated by Notice 1, 1 November 1990.

MIL–C–63169(AR) NOTICE 1 *Cushion, Forward, M753 8 Inch Projectile*, 18 April 1988, 1 p notice + 22 pp base spec (FSC NUOR) (AR)

Notice 1 states that the specification is currently valid for use in acquisition. The original date of the base specification was 25 June 1982. The cushions are intended for use in assembly of rocket motors for M753, 8–inch projectiles. Type I covers soft–grade silicone sponge rubber and Type II medium–grade silicone sponge rubber.

MIL–C–70473(2)(AR) *Cushion Components for Metal Container for Cartridge 120 mm Tank Ammunition*, 23 May 1988, 4 pp amend + 14 pp base spec (FSC 8140) (AR)

Specifies the use of polyethylene foam for cushioning 120–mm tank ammunition.

MIL–F–81254(WP) *Foam, Urethane*, 15 April 1965, 6 pp (FSC 1338) (OS)

Covers one type of inert–gas–expanded isocyanate urethane foam, closed–cell, for use in rocket motors.

MIL–M–81288(1)(AS) *Mounting Bases, Flexible Plastic Foam*, 15 July 1968, 1 p amend + 17 pp base spec

The mounting bases are intended for use with electronic equipment to absorb shock and vibration energy. The foam must be in accordance with MIL–F–81334.

MIL–F–81334B(AS) *Foam, Plastic, Flexible, Open Cell, Polyester Type, Polyurethane*, 5 July 1977, 16 pp (FSC 5340) (AS)

Interim Amendment – 1 (SA) was issued 1 April 1991. It covers safety and health requirements.

MIL–B–83054B(2) (USAF) *Baffle and Inerting Material, Aircraft Fuel Tank*, 7 May 1984, 3 pp amend + 25 pp base spec (FDC 9330) (11)

The foam is a reticulated polyurethane foam for explosive suppression in aircraft fuel tanks and bay areas. There are five Types.

MIL–P–83379A (USAF) *Plastic Material, Cellular Polyurethane, Foam–in–Place, Rigid (3 pounds per cubic foot density)*, 18 March 1980, 18 pp (FSC 9330) (11)

Notice 1 (30 July 1987) validated the specification. This is the Air Force foam–in–place specification. It covers the materials required and the preparation of the formulation used in the process.

MIL–F–83670 (USAF) *Foam–in–Place Packaging, Procedures for Use*, cancelled 13 April 1986 and replaced by MIL–F–45216A, issued 15 July 1984.

MIL–F–83671 (SA) *Foam–in–Plate Packaging Materials, General Specification for*, INT AMEND 2, 30 March 1990, 2 pp amend + 29 pp base spec (FSC 8135) (SA)

With Amendment 1 this specification supersedes the Type II (Foam–in–Place) requirements in MIL–P–26514F described above. The requirement is for flame–resistant polyurethane foams furnished in two component foam–in–place systems. The specification also supersedes MIL–P–21029B. The basic specification and Amendment 1 are Air Force (60) documents, while Interim Amendment 2 is a Navy (SA) document. There are 3 Classes, Class 1 – Rigid, Class 2 – Flexible, and Class 3 – Semi–Rigid. Static–stress curves are given for Classes 2 and 3.

MIL–F–87075B *Foam–in–Place Packaging Systems, General Specification for*, 26 October 1984, 19 pp (FSC 3540) (SA)

Covers two Types, three Classes and two Kinds using chemicals conforming to MIL–F–83171 and procedures using MIL–F–45216.

MIL–F–87090 (SA) *Foam, Combustion Retardant, for Cushioning Supply Items Aboard Navy Ships*, 5 November 1981, 7 pp (FSC 9330) (SA)

These flexible foams include, but are not limited to, polyimide types.

MILITARY STANDARDS (See Reference 6)

MIL–STD–293 *Visual Inspection Guide for Cellular Rubber Items*, 18 October 1956, 63 pp (93GP) (SH) Validated by Notice 1, 16 June 1988.

Covers only sponge rubber items, including chemically blown or expanded–rubber items having either open or closed cells. Also includes cellular products made from chemically or mechanically foamed latices or liquid elastomers. Provides word descriptions and photographs of defects.

MIL–STD–401B *Sandwich Constructions and Core Materials; General Test Methods*, 26 September 1967, 35 pp (FSC 5680) (AS)

MIL–STD–670B Notice 1 *Classification System And Tests for Cellular Elastomeric Materials*, 14 April 1986, 3 pp Notice + 33 pp base spec (FSC 9320) (MR)

Establishes a system for designating cellular elastomeric materials based on natural, synthetic, or reclaimed rubber, or rubber–like materials, alone or in combination. Cellular ebonite (hard rubber) and rigid cellular plastics are not included. Revision B was issued 30 January 1968. Notice 1 makes this standard inactive for new design, which would thereafter refer to the applicable portions of ASTM D 1055, D 1056, D 1505, D 1667 and D 3574. Cross reference tables are given.

MIL–STD–1186A *Cushioning, Anchoring, Bracing, Blocking and Water–proofing with Appropriate Test Methods*, 12 March 1981, 29 pp (Area PACK) (ME)

Cellular polymeric materials used for cushioning are covered in this standard, which was validated by Notice 1, 9 January 1990.

MIL–STD–1191 *Foam–in–Place Packaging, Prodedures for*, 31 December 1989, pp (Area PACK) (Army–SM)

This standard replaces Military Specification MIL–F–45216A, dated 15 July 1984, which had the same title.

MILITARY HANDBOOKS (See Reference 6)

MIL–H–139 (MU) *Plastics, Processing of*, 30 January 1967, 136 pp (FSC 9330) (AR)

Has a brief section on *foaming* and *cellular laminates*.

MIL–HDBK–149B NOTICE 1 *Rubber*, 24 July 1987, 422 pp (FSC 9320) (MR)

This is a comprehensive volume containing much useful information on many aspects of rubber and rubber–like products. There is a short section on cellular rubber, including properties of eight (8) foam rubbers.   Notice 1, 24 July 1987, incorporated two (2) page changes.

MIL–HDBK–304B NOTICE 1 *Package Cushioning Design*, 28 November 1988, 3 pp Notice + 536 pp basic hdbk) (Area PACK) (69)

Provides basic information on cushioning materials and their uses; Coverage is exhaustive on all types of cushioning materials. The Notice lists revisions to the basic handbook.

MTL–HDBK–699B (MR) *A Guide to the Specifications for Flexible Rubber Products*, 28 February 1989, 123 pp (FSC 9320) (MR)

Provides information in tabular form on all known Federal, Military and nationally recognized technical society specifications and standards for those flexible rubber products of interest to the DOD. Pages 17 through 28 cover cellular materials.

MIL–HDBK–700A *Plastics*, 17 March 1975, 291 pp (FSC 9330) (MR)

This is a general handbook covering all aspects of plastics technology. Chapter 7 — *Cellular (Foamed) Plastics* covers 16 pages and is useful, although not up to date. This handbook is undergoing revision.

MIL–HDBK–772 *Military Packaging Engineering*, 30 March 1981, 378 pp (Area PACK) (SM). Notice 1, 30 December 1988, validates the handbook for use in acquisition.

Section 5.6.2 on Cushioning (10 pages) has useful elementary information on cushioning materials, including cellular plastics and elastomers.

FEDERAL SPECIFICATIONS (See Reference 6)

ZZ-P-75B(1) *Pad, Typewriter, Sponge Rubber*, 16 February 1972, 1 p amend + 6 pp base spec (FSC 7510) (GSA-FSS)

ZZ-M-91E Int Amend 1 *Mattress, Bed, Latex Foam*, 20 Mar 1968, 3 pp amend + 14 pp base spec (FSC 7210) (GSA-FSS)

AA-C-00275E (GSA-FSS) *Chairs, Rotary and Straight, Aluminum, Office*, March 1976, 37 pp (FSC 7110) (GSA-FSS)

Latex or urethane foam is used on cushions or pads in this specification.

L-P-386C *Plastic Material, Cellular Urethane (Flexible)*, 26 August 1977, 8 pp (FSC 9330) (GSA-FSS)

Application is for upholstery, general seating, mattresses and other uses. There are a number of Types, Classes and Grades.

L-S-00626D *Sponges, Synthetic*, 15 January 1970, 7 pp (FSC 7920) (GSA-FSS)

Covers sponges made of regenerated cellulose or synthetic plastic and used for general cleaning operations.

ZZ-C-766D (GSA-FSS) *Cushion, Chair and Stool*, 14 September 1977, 8 pp (FSC 7210) (GSA-FSS)

Urethane foam is used for the cushioning material.

PPP-C-795C *Cushioning Material, Packaging (Flexible Cellular, Plastic Film for Long Shipping Cycle Applications)*, 12 September 1989, 14 pp (FSC 8135) (GSA-FSS)

Covers a plastic film "bubble" composite, *not* a plastic foam. Material consists of plastic film with uniformly distributed closed cells. Interim Amendmend – 1 (5A) was issued 1 April 1991. It covers health and safety requirements.

ZZ–C–00811B (COM–NBS) *Cushion, Carpet and Rug, Cellular Rubber*, 2 January 1963, 9 pp (FSC 7220) (GSA–FSS)

PPP–C–843D *Cushioning Material, Cellulosic*, 22 July 1977, 9 pp (FSC 8135) (GL)

This material is intended for cushioning and for absorbent packaging for fragile cartons of liquids which might break in transit.

PPP–C–85OD(5) *Cushioning Material, Polystyrene Expanded, Resilient (For Packaging Uses)*, 3 September 1976, 3 pp amend + 6 pp base spec (FSC 8135) (69)

PPP–C–1120B *Cushioning Material, Uncompressed Bound Fiber for Packaging*, 15 April 1982, 15 pp (FSC 8135) (69)

Not on foams, but listed for comparison as a cushioning material.

ZZ–P–001235A *Pillow, Bed (Flaked Urethane)*, 13 January 1976, 7 pp (FSC 7210) (GSA–FSS)

PPP–C–1266B Int Amend 1 (DM) *Container, Thermal, Shipping, for Medical Material Requiring Controlled Temperature Ranges*, 2 pp amend + 7 pp base spec (FSC 8115) (DM)

The insulation material is polyurethane foam, MIL–P–76514, Type II, Class 1.

L–C–001369(GSA–FSS) *Cushion, Carpet–and–Rug, Bonded Urethane*, 10 December 1969, 8 pp (FSC 7220) (GSA–FSS)

PPP–C–1683A *Cushioning Material, Expanded Polystyrene Loose–Fill Bulk (For Packaging Application)*, 5 December 1988, 18 pp (FSC 8135) (69)

PPP–C–1752D *Cushioning Material, Packaging, Polyethylene Foam*, 26 December 1989, 20 pp (FSC 8135) (GSA–FSS)

PPP–C–1797A *Cushioning Material, Resilient, Low Density, Unicellular, Polypropylene Foam*, September 1982, 13 pp (FSC 8135) (AS)

PPP–C–1842A(3) *Cushioning Material, Plastic, Open Cell (for Packaging Applications)*, 17 August 1977, 2 pp amend + 13 pp base spec (FSC 8135) (GL)

The cushioning material is made of one, or a composite of two or more sheets of transparent plastic films formed into a network of uniformly distributed open cells.

HH–1–1972/GEN(1) *Insulation Board, Thermal, Faced, Polyurethane or Polyisocyanurate*, 3 October 1985, 2 pp amend + 12 base spec (FSC 5640) (YD)

Has a number of "specification sheets", as follows. Specification Sheet 6 was cancelled. Validated 7 February 1991.

HH–1–1972/1 *Insulation Board, Polyurethane or Polyisocyanurate, Faced with Aluminum Foil on Both Sides of the Foam*, 12 August 1981, 3 pp (FSC 5640) (YD)

Validated 7 February 1991.

HH–I–1972/2(1) *Insulation Board, Thermal, Polyurethane or Polyisocyanurate, Faced with Asphalt/Organic Felt, Asphalt/Asbestos Felt or Asphalt/Glass Felt on Both Sides of the Foam*, 3 October 1985, 2 pp amend + 3 pp base spec (FSC 5640) (YD)

Validated 7 February 1991.

HH–I–1972/3 *Insulation Board, Thermal, Polyurethane or Polyisocyanurate, Faced with Perlite Insulation Board on One Side and Asphalt/Organic Felt or Asphalt/Glass Fiber Felt on the Other Side of the Foam*, 12 August 1981, 3 pp (FSC 5640) (YD)

Validated 7 February 1991.

HH–I–1972/4 *Insulation Board, Thermal, Polyurethane or Polyisocyanurate, Faced with Gypsum Board on One Side and Aluminum Foil or Asphalt/Organic Felt on the Other Side of the Foam*, 12 August 1981, 3 pp (FSC 5640) (YD)

Validated 7 February 1991.

HH–I–1972/5 *Insulation Board, Thermal, Polyurethane or Polyisocya-nurate, Faced with Perlite Board on Both Sides of the Foam*, 12 August 1981, 2 pp (FSC 5640) (YD)

Federal Standards (See Reference 6)

FED–STD–406A *Plastics: Methods of Testing*

This well–known standard, covering a large number of test methods, was finally cancelled on April 5, 1990. The "A" revision was issued January 4, 1982. The cancellation provides a cross–reference table to test methods now covered by ASTM methods and lists those methods cancelled without replacement.

FED–STD–601 Change Notice 7 *Rubber Sampling, and Testing*, 17 August 1976, 405 pp, base standard 12 April 1955 (FSC 9320) (MR)

Has a number of test methods on cellular rubber for which ASTM methods are substituted.

British Standards (See Reference 7)

BS 874:1973 Methods for *Determining Thermal Insulating–Properties with Definitions of Thermal Insulating Terms* (incorporates Amendments 1, 2 and 3 to January 1987), 48 pp

BS 3157:1960 Specification for *Latex Foam Rubber Components for Transport Seating*, 17 pp

BS 3379:1975 Specification for *Flexible Urethane Foam for Loadbearing Applications* (incorporates Amendment 1, April 1978), 16 pp

BS 3837: Specification for *Expanded Polystyrene Boards*

Part 1:1986 Boards manufactured from expandable beads, 10 pp

BS 3869:1965 Specification for *Rigid Expanded Polyvinyl Chloride for Thermal Insulation Purposes and Building Applitations* (incorporates Amendment 1, 21 February 1986 and Amendment 2, 30 April 1975), 18 pp

BS 3927:1986 Specifications for *Rigid Phenolic Foam (PF) for Thermal Insulation in the Form of Slabs and Profiled Sections Applications,* 12 pp

BS 4021:1971 Specification for *Flexible Polyurethane Foam Sheeting for Use in Laminates,* 3 pp

BS 4023:1975 Specification for *Flexible Cellular PVC Sheeting,* 13 pp

BS 4370:1968 Methods of Test for *Rigid Cellular Materials*

Part 1:1988. Methods 1–5, incl. Amendment 1, December 1972, 13 pp

Method 1 – Dimensions
Method 2 – Apparent density
Method 3 – Compression
Method 4 – Cross–breaking strength
Method 5 – Dimensional stability

Part 2:1973. Methods 6–10, 30 pp

Method 6 – Shear strength and shear modulus
Method 7 – Thermal conductivity (guarded hot plate)
Method 8 – Water vapor transmission
Method 9 – Tensile strength
Method 10 – Volume percent closed cells

Part 3:1988. Methods 12–13, 13 pp

Method 12 – Friability
Method 13 – Coefficient of linear thermal expansion at low temperatures

BS 4443:1988 Methods of Test for *Flexibility Cellular Materials*

Part 1:1988. Methods 1–6, 17 pp

Methods 1A, 1B and 1C	– Dimensions
Method 2	– Apparent density
Methods 3A and 3B	– Tensile strength and elongation at break
Method 4	– Cell count

Methods 5A and 5B   – Compression stress–
                       strain characteristics
Methods 6A and 6B   – Compression set

Part 2:1988, 8 pp

Method 7 – Indentation hardness tests

Part 3:1989. Methods 8 and 9, 12 pp

Method 8 – Creep
Method 9 – Dynamic cushioning performance

Part 4:1976. Methods 10–12, 8 pp

Method 10 – Solvent swelling
Method 11 – Humidity aging at an elevated temperature
Method 12 – Heat aging

Part 5:1980, 9 pp

Method 13 – Dynamic fatigue by constant load pounding
(same as ISO 3385–1975)

Part 6:1980. Methods 14–16, 9 pp

Method 14 – Preparation of water extract
Method 15 – Tear strength
Method 16 – Air flow value

Part 7:1983, 8 pp

Method 17 – Tear strength of flexible cellular material
with an integral skin

BS 4541:1970 Specification for *Polyurethane Interior Foam Cores for Domestic Mattresses for Adults*, 9 pp

BS 4578:1970 Methods of test for *Hardness of, and Air flow Through, Infants' Pillows, including Amendment 1*, March 1972, 11 pp

BS 4735:1974 Laboratory Method of Test for *Assessment of the Horizontal Burning Characteristics of Specimens No Larger Than 150 mm x 50 mm (Nominal) of Cellular Plastics and Cellular Rubber Materials when Subjected to a Small Flame*, 13 pp

BS 4840:1985 Specification for *Rigid Polyurethan (PUR) Foam in Slab Form*

> Part 1:1985.   Specification for *PUR Foam for Use in Transport Containers and Insulated Vehicle Bodies*, 11 pp

> Part 2:1985.   Specification for *PUR Foam for Use in Refrigerator Cabinets, Cold Rooms and Stores*, 12 pp

BS 4841 Specification for *Rigid Urethane Foam for Building Applications*

> Part 1:1975.   *Laminated Board for General Purposes*, 10 pp

> Part 2:1975.   *Laminated Board for Use as a Wall and Ceiling Insulation*, 12 pp

> Part 3:1975.   *Two Types of Laminated Board (Roofboards) with Auto-Adhesively Bonded Reinforcing Facings for Use as Roofboard Thermal Insulation for Built-Up Roofs*, 17 pp

BS 5111 Laboratory Methods of Test for *Determination of Smoke Generation Characteristics of Cellular Plastics and Cellular Rubber Materials*

> Part 1:1974.   Method for *Testing a 25 mm Cube Test Specimen of Low Density Material (up to 130 kg/m³) to Continuous Flaming Conditions*, 14 pp

Uses a photoelectric cell to determine smoke obscuration.

BS 5131:1975 *Methods of Test for Footwear and Footwear Materials Introduction*: 1975

Part 2 – *Soilings*

Section 2.6:1979   *Split Tear Strength of Cellular Solings,*
                        5 pp

Section 2.10:1980   *Measurement of the Heat Shrinkage of*
                        Cellular Solings, 2 pp

BS 5223 Specification for *Hospital Bedding*

Part 2:1979. *Flexible Polyurethane Foam Mattresses*, 9 pp

Part 3:1976. *Flexible Polyurethane Foam Pillows*, 9 pp

BS 5241 Specification for *Rigid Polyurethane (PUR) and Polyisocyanu‐*
rate (PIR) Foam When Dispensed or Sprayed on a Construction Site

Part 1:1989. *Sprayed Foam Thermal Insulation Applied Exter‐*
              nally, 9 pp

BS 5340:1976 Specification for *Flexible Polyurethane Foam Pillows for*
Domestic Use, 5 pp

BS 5608:1986 Specification for *Performed Rigid Polyurethane (PUR) and*
Polyisocyanurate (PIR) Foam for Thermal Insulation of Pipework and
Equipment, 9 pp

BS 5617:1985 Specification for *Urea–Formaldehyde (UF) Foam Systems*
Suitable for Thermal Insulation of Cavity Walls with Masonry or
Concrete Inner and Outer Leaves, 17 pp

BS 5618:1985 Code of Practice for *Thermal Insulation of Cavity Walls*
(with Masonry or Concrete Inner and Outer Leaves) by Filling with
Urea–Formaldehyde (UF) Foam Systems, 59 pp

BS 5852 *Fire Tests for Furniture*

Part 1:1979. Methods of Test for the *Ignitability by Smokers'*
              Materials of Upholstered Composites for Seating,
              11 pp

Part 2:1982. Methods of Test for the *Ignitability of Upholstered Composites by Flaming Sources*, 19 pp

BS 5946:1980 Method of Test for *Detetmination of the Punking Behaviour of Phenol–Formaldehyde Foam*, 9 pp

BS 6203:1982 *Fire Characteristics and Fire Performance of Expanded Polystyrene (EPS) Used in Building Applications*, 25 pp

BS 6586:1985 *Rigid Polyurethane (PUR) Foam Produced by the Press Injection Method*

Part 1:1985. Specification for *PUR Foam for Insulated Panels for Transport Containers and Insulated Vehicle Bodies*, 12 pp

ISO STANDARDS (See Reference 7)

ISO 472–88 *Plastics — Vocabulary*, Second Edition – 1988, 154 pp

The Second Edition has 11 terms concerning *cellular plastics* defined in English and French.

ISO 844–78 *Cellular Plastics — Compression Test for Rigid Materials*, 4 pp,

Amendment Slip, 1979, 6 pp

This standard specifies a method for determining a) the compressive strength and corresponding relative deformation or b) the compression stress at 10% relative deformation, of rigid cellular plastics.

ISO 845–88 *Cellular Plastics and Rubbers Determination of Apparent (Bulk) Density* – 1988, Second Edition, 6 pp

This standard covers the apparent core density of rigid cellular plastics and the bulk density of semi–rigid and flexible cellular plastics and rubbers.

ISO 1043 – Part 1–87 *Plastics —Symbols —Part 1: Basic Polymers and Their Special Characteristics*, First Edition — 1987, 8 pp

ISO 1043 – Part 2–88 *Plastics — Symbols — Part 3: Fillers and Reinforcing Materials*, First Edition — 1988, 4 pp

ISO 1043 – Part 3–88 *Plastics —Symbols —Part 3: Plasticizers*, First Edition — 1988, 3 pp

ISO 1209 – Part 1–90 *Cellular Plastics, Rigid Flexural Tests —Part 1: Bending Test*, First Edition, 6 pp

ISO 1209 – Part 2–90 *Cellular Plastics, Rigid - Flexural Tests —Part 2: Determination of Flexural Properties*, First Edition, 6 pp

ISO/DIS 1210.2 *Plastics - Determination of the Burning Behaviour of Horizontal and Vertical Specimens in Contact with a Small–Flame Ignition Source*

This Draft International Standard, circulated in May 1990, specifies a small–scale laboratory screening procedure for comparing the relative burning behaviour of vertically or horizontally oriented plastic specimens exposed to a low–energy ignition. It covers rigid solid or cellular plastics having an apparent density of not less than 250 kg/m^3, determined in accordance with ISO 845.

ISO 1663–81 *Cellular Plastics — Determination of Water Vapour Transmission Rate of Rigid Materials*, 4 pp

Covers water vapor transmission rate (WVTR), water vapor permeance, and water vapor permeability.

ISO 1798–83 *Polymeric Materials, Cellular Flexible —Determination of Tensile Strength and Elongation at Break*, 2 pp

ISO 1856–80 *Polymeric Materials, Cellular Flexible —Determination of Compression Set*, Second Edition, 5 pp (Erratum 1981)

Describes three methods, A, B, and C, for latex and polyurethane foams of thicknesses greater than 2 mm.

ISO 1922–81 *Cellular Plastics – Determination of Shear Strength of Rigid Materials*, Second Edition, 6 pp

Also covers shear modulus.

ISO 1923–81 *Cellular Plastics and Rubber — Determination of Linear Dimensions*, 3 pp

Covers sheets, blocks or test specimens, flexible and rigid.

ISO 1926–79 *Cellular Plastics — Determination of Tensile Properties of Rigid Materials*, 4 pp

ISO 2439–80 *Polymeric Materials, Cellular Flexible — Determination of Hardness (Indentation Method)*, 4 pp

Specifies three methods, A, B, and C, for latex, urethane foam, and open–cell PVC foam.

ISO 2440–83 *Polymeric Materials, Cellular Flexible — Accelerated Ageing Test*, 2 pp

Covers simulation of naturally occurring reactions such as oxidation or hydrolysis by humidity for open cellular latex and polyurethane foams.

ISO 2578–74 *Plastics — Determination of Time-Temperature Limits After Exposure to Prolonged Action of Heat*, 5 pp

Covers all plastic forms, not only cellular.

ISO 2581–75 *Rigid Cellular Plastics — Determination of Apparent Thermal Conductivity by Means of a Heat-Flow Meter*, 7 pp

ISO 2796–86 *Cellular Plastics, Rigid Test For Dimensional Stability*, 3 pp

Tests are carried out at specified conditions of temperature and humidity.

ISO 2896–87, *Cellular Plastics, Rigid — Determination of Water Absorption*, 7 pp

Measures the change in the buoyant force resulting from immersion of a specimen under a 50–mm head of water for four days.

ISO 3385–89 *Flexible Cellular Polymeric Materials — Deterioration of Fatigue by Constant–Load Pounding*, Third Edition, 7 pp

ISO 3386 *Polymeric Materials, Cellular Flexible — Determination of Stress/Strain Characteristics in Compression*

Part 1:1986. *Low–density materials*, 2 pp

Covers materials up to 250 kg/m^3 (15.6 lb/ft^3)

Part 2:1984. *High–density materials*, 2 pp

Covers materials over 250 kg/m^3 (15.6 lb/ft^3)

ISO 3582–78 *Cellular Plastic and Cellular Rubber Materials — Laboratory Assessment of Horizontal Burning Characteristics of Small Specimens Subjected to a Small Flame*, 7 pp

ISO 4589 *Plastics — Determination of Flammability by Oxygen Index*, First Edition, 17 pp

ISO 4590–81 *Cellular Plastics — Determination of Volume Percentage of Open and Closed Cells of Rigid Materials*, First Edition, 14 pp

This technique first measures the geometrical volume, and then the air–impenetrable volume of test specimens. The method provides for correcting the apparent open–cell volume by taking into account the surface cells opened by cutting during specimen preparation. Two alternative methods and corresponding apparatus are specified for the measurement of the impenetrable volume. The results obtained are to be used for comparison purposes only.

ISO 4638–84 *Polymeric Materials, Cellular Flexible — Determination of Air Flow Permeability*, 7 pp

ISO 4651-88 *Cellular Rubbers and Plastics — Determination of Dynamic Cushioning Performance*, 2nd Edition, 13 pp

This standard specifies a procedure for determining the dynamic cushioning performance of cellular rubber materials and rigid and flexible plastics by measuring the peak deceleration of a mass when it is dropped on a test piece. The test is intended primarily for quality assurance. However, since this type of test is also used to obtain design data notes are given in Annex A to assist in this objective. The method is applicable solely to packaging materials.

ISO 4897-85 *Cellular Plastics – Determination of the Coefficient of Linear Thermal Expansion of Rigid Materials at Sub-Ambient Temperatures*, 9 pp

Covers two methods, A and B.

ISO 4898-84 *Cellular Plastics – Specification for Rigid Cellular Materials Used in the Thermal Insulation of Buildings*, 6 pp

Covers polystyrene and polyurethanes made with isocyanates.

Addendum 1:   Phenol-formaldehyde cellular plastics (RC/PF) (1988), 1 p

Addendum 2:   Labelling and marking of products (1988), 3 pp

ISO 5999-82 *Polymeric Materials, Cellular Flexible — Polyurethane Foam for Load-Bearing Applications Excluding Carpet Underlay — Specification*, 11 pp

This specification primarily relates to the quality of polyurethane foam used for comfort-cushioning purposes. The foam is classified according to performance during a fatigue test, indentation hardness being used as a secondary means of grading the material.

ISO 6453-85 *Polymeric Materials, Cellular Flexible —Polyvinylchloride Foam Sheeting — Specification*, 9 pp

Covers two types, Type 1 – Open Cell and Type 2 – Closed Cell. The foam is graded according to hardness index, measured in accordance with ISO 2434.

ISO 6915–84 *Polymeric Materials, Cellular Flexible — Polyurethane Foam for Laminate Use — Specification*, 6 pp

Covers foams up to and including 20 mm thick intended for combination with suitable substrates such as non–woven, woven, or knitted fabrics to form a flexible laminate. There are three types.

ISO 7214–85 *Cellular Plastics — Polyethylene — Methods of Test*, 10 pp

ISO 7231–84 *Polymeric Materials, Cellular Flexible — Methods of Assessment of Air Flow Value at Constant Pressure–Drop*, 5 pp

ISO 7616–86 *Cellular Plastics, Rigid — Determination of Compressive Creep Under Specified Load and Temperature Conditions*, 4 pp

This is a revision of ISO/TR 2799.

ISO 7850–86 *Cellular Plastics, Rigid — Determination of Compressive Creep*, 3 pp

This standard specifies a method for the determination of compressive creep under various conditions of stress, temperature and relative humidity.

ISO 8067–89 *Flexible Cellular Polymeric Materials — Determination of Tear Strength*, 6 pp

ISO 8873–87 *Cellular Plastics, Rigid – Spray Applied Polyurathane Foam for Thermal Insulation of Buildings – Specification*, 8 pp

Lists requirements and test methods.

ISO 9054–90 *Cellular Plastics, Rigid — Test Methods for Self–Skinned High Density Materials*, 8 pp

This standard specifies the basic test procedures for the determi-nation of the physical properties of self–skinned (integral–skin), high–density (typically in excess of 100 kg/m^3 or 6.25 lb/ft^3) rigid cellular plastic materials. Test methods covered are:

- 9 Apparent Density – ISO 845
- Bending Strength – ISO 178
- Impact Strength – ISO 6603–1 and ISO 6603–2
- Tensile Test – ISO 1926
- Shear – ISO 1922 modified
- Screw Retention – described in Annex A
- Surface Hardness – ISO 868
- Thermal Conductance – ISO 8301 or ISO 8302
- Burning Behavior – as required by national practice
- Dielectric Strength – IEC 243
- Linear Thermal Expansion – ISO 4897
- Deflection Under Constant Load and Increasing Temperature – ISO 75

ISO/DIS 9772.2 *Cellular Plastics —Determination of Horizontal Burning Characteristics of Small Specimens Subjected to a Small Flame*, 15 pp

This Draft International Standard, circulated in April 1990, specifies a small–scale laboratory screening procedure for comparing the relative burning characteristics of horizontally oriented, small cellular plastic specimens having a density not less than 250 kg/m^3, determined in accordance with ISO 845, when exposed to a low–energy source of ignition.

ISO TR 9774–90 *Thermal Insulation Materials —Application Categories and Basic Requirements —Guidelines for the Harmonization of International Standards and Other Specifications*, 12 pp

This Technical Report applies only to prefabricated thermal insulation products. It consists largely of tables. Foams are not mentioned directly.

ISO/DIS 10351 *Plastics — Method of Test for the Determination of Combustibility of Specimens Using a 125 mm Flame Source*

This Draft International Standard, circulated in March 1990, specifies a small–scale laboratory screening procedure for comparing the relative burning characteristics and resistance to penetration by the ignition source of small plastic specimens exposed to a medium energy level (600 W) 125-mm lamp source of ignition.   This method is applicable to both solid materials and cellular plastic materials of density 250 kg/m^3 (3) or greater, determined in accordance with ISO 845. This method subjects small plastic specimens to a flame source approximately 10 times more severe than the flame source used in ISO 1210-82.

REFERENCES:

1.   Landrock, Arthur H. and Norman E. Beach, Chapter 11, "Commercial and Government Specifications and Standards", *Hand-Book of Plastics; and Elastomers*, C.A. Harper, ed., McGraw-Hill, New York, New York (1975).

2.   Landrock, A.H., *Standards and Specs: Are They Really Indigestible?", Plastics Design Forum*, 2 (6): 81-88 (November/December 1977).

3.   ASTM standards are available from the American Society for Testing and Materials, 1916 Race Street, Philadelphia, PA 19103. They may be purchased as "separates", or as published, in book form, under specialized headings.

4.   SAE-AMS specifications are available from the Society of Automotive Engineers, 400 Commonwealth Drive, Warrendale, PA 15096.

5.   Underwriters Laboratories standards are available from Underwriters Laboratories, Inc., 333 Pfingsten Road, Northbrook, Illinois 60062.

6.   Military and Federal Specifications, Standards and Handbooks and DOD Adopted industry standards may be obtained at no cost from the Standardization Documents Order Desk, Building 4D, 700 Robbins Avenue, Philadelphia, Pennsylvania 19111-5094.

7.   British Standards and ISO Standards in the United States are available from the American National Standards Institute (ANSI), 1430 Broadway, New York, New York 10018.

8.   An excellent reference, recently published, is the following hard–cover book. The author is currently with the U.S. Department of Defense and has worked for many years with military and industry standards in plastics and composites.

9.   Traceski, F.T., *Specifications & Standards for Plastics & Composites*, 1990, 224 pp. ISBN 087170–395–5. Published by ASM International, Materials Park, Ohio 44073, 800–368–9800.

GLOSSARY

Arthur H. Landrock

The definitions given in this glossary were taken from a number of sources, most of which are cited. The author's intention was to cover many of the terms used in the text which might not be familiar to some readers. The terms include definitions adopted by ASTM and ISO, along with definitions found in literature sources. Most of the ASTM definitions were taken from the 6th Edition of the *Compilation of ASTM Standard Definitions* published by the ASTM Standing Committee on Terminology (COT) (1). The author was a member of this committee for a number of years and was a member of the Working Group responsible for the Compilation. He has also been an active voting member of ASTM D-20's Subcommittee D 20.92 on Terminology and ASTM E 05 Subcommittee E 05.31 on Editorial and Terminology. Committee D 20 is the committee on Plastics and E 05 is the committee on Fire Standards.

The author is also a member of the USA Technical Advisory Group (TAG) for ISO Technical Committee (TC) 61 on Plastics, which group is also a subcommittee (D 20.61) of ASTM D 20. In TC 61 his primary responsibility has been the subcommittee on Terminology (SC 1).

It should be noted that the definitions listed in the ASTM Compilation (1) are those taken from ASTM Standards. Many of these definitions were originated in or adapted from ISO definitions. Most, however, were obtained from ASTM standards and, in such cases, the standard, and the ASTM committee responsible for the standard, are given in the Compilation. As in the Compilation many of the definitions provided in this glossary of 221 terms are not confined to a "bare–bones" statement identifying the concept, but have almost encyclopedic discussions appended to them. In a number of cases definitions or comments from more than one source are given for a single term.

In general, the ASTM definitions cited in this glossary were taken from terminology developed by the following ASTM committees: D–20 on Plastics, D–11 on Rubber and E–05 on Fire Standards. As noted above, the reader may observe a similarity between ASTM and ISO standard definitions since each of these organizations attempts to provide the best definitions available and readily adapt or use each other's definitions when desirable.

Literature sources have been used to provide definitions where no standard definitions could be found. In a very few cases the author has suggested his own definition where he felt qualified to do so.

abrasion – The surface loss of a material due to frictional forces; the wearing away of any part of a material by rubbing against another surface (1).

accelerator (promoter) – A substance used in small proportion to increase the reaction rate of a chemical system (reactants plus other additives) (2).

activator – A substance used in small proportion to increase the effectiveness of an accelerator (2). A material that speeds up a reaction in unison with a catalyst; an activator often starts the action of a blowing agent; used almost synonymously with *accelerator* or *initiator* (3).

additive – A substance used in minor amounts in the foam mixture, not required to produce foam but to modify properties (2–modified).

aging, air–oven – The process of exposing materials to the action of air at an elevated temperature at atmospheric pressure (1).

aging test, accelerated – An aging test in which the degradation of materials is intentionally accelerated to produce degradation more rapidly than expected in service (1–modified).

anisotropic – A term applied to foam having different properties in different directions, e.g., parallel to foam rise, as opposed to perpendicular to foam rise (3–modified).

antifoaming agent – An additive which reduces the surface tension of a solution or emulsion, thus inhibiting or modifying the formation of a foam (3).

antioxidant – A substance used to retard deterioration caused by oxidation (2).

antistatic agent – A chemical which imparts a slight to moderate degree of electrical conductivity to plastic compounds, thus preventing the accumulation of electrostatic charges in finished articles (3).

autoignition (self–ignition) – The ignition of a material caused by the application of pressure, heat, or radiation, rather than by an external source, such as a spark, flame or incandescent surface (1).

autoignition temperature (self–ignition temperature) – The minimum temperature at which autoignition occurs under specified conditions (1).

block (bun) – A cut–off segment of the continuously produced loaf of flexible or rigid foam made by the slab technique (3).

blocking – Relatively stiff materials used in packaging to immobilize materials or facilitate handling (1) (4).

blowing Agent (foaming agent) – A compounding ingredient used to produce gas by chemical or thermal action, or both, in the manufacture of hollow or cellular articles (1); blowing agents may be compressed gases, volatile liquids or chemicals that decompose or react to form a gas (2).

board stock – The flat sheets of rigid or flexible foam cut from blocks of foam; the term has recently been specifically applied to the product of a continuous lamination–foaming line on which the foam is sandwiched between two skins (3).

bonded foam – A product produced by the adhesion of small pieces of urethane foam with a suitable bonding agent (5).

bottoming point – That point in the stress–strain curve of a cushioning material where further increase in stress produces no increase in strain. For practical purposes it is often approximated.

bun – See *block*.

buoyancy factor – A number obtained by subtracting the density (in lb/ft^3) of an object floating in water from the density of water (64.3 lb/ft 3 for sea water and 62.4 lb/ft^3 for fresh water). An extra allowance is usually made for safety. The buoyancy factor number is the weight in pounds required to submerge a cubic foot of the foam.

catalyst – A substance that causes or accelerates chemical reaction. When added to the reactants in a minor amount the catalyst itself is not permanently affected by the reaction (3).

cell (bubble) (pore) – A single small cavity formed by gaseous displacement in a plastic or elastomer and surrounded completely by its walls (2) (3).

cell, closed – A cell totally enclosed by its walls and hence not interconnecting with other cells (1). (See *cell* and *cell, open*).

cell count – The number of cells or bubbles per linear inch or centimeter (3).

cell membrane – A thin intact film that forms the bubble walls in the final foam product (3).

cell, open – A cell not totally enclosed by its walls and hence interconnecting with other cells (1). (See *Cell* and *cell, closed*).

cell size – The average diameter of the pores (bubbles) in the final foam product (3).

cellular elastomer – A cured elastomeric material containing cells or small voids (1).

cellular material – A generic term for materials containing many cells (either open, closed, or both) dispersed through the mass (1).

cellular material, cored – A cellular material containing a multiplicity of holes (usually, but not necessarily, cylindrical in shape) molded or cut into the material in some pattern normally perpendicular to the largest surface and extending past or all the way through the piece (1).

cellular material, flexible – A cellular organic polymeric material which will not rupture when a specimen 200 by 25 by 25 mm (8 by 1 by 1 in.) is bent around a 25–mm (1–in.) diameter mandrel at a uniform rate of one lap in 5 seconds at a temperature between 18 and 29°C (1).

cellular plastics – Plastics containing numerous small cavities (cells), interconnecting or not, distributed throughout the mass (1).

cellular polymers – Two phase gas–solid systems in which the solid is a synthetic plastic or rubber and the solid phase is continuous (6).

cellular rubber – A general term covering all cellular materials that have an elastomer in the polymer phase (6).

cellular striation – A layer within cellular plastics that differs from the characteristic cell structure (2).

CFCs (chlorofluorocarbons) – As applied to polyurethane foams, blowing agents having chlorine and fluorine in their chemical structure. CFCs are gradually being replaced by other blowing agents because of concern about their adverse effect on the atmospheric ozone layer.

chemically foamed polymeric material – A cellular material in which cells are formed by gases generated from thermal decomposition or other chemical reaction (1).

clicking – The process of stamping out irregular–shaped articles from thin sheets of foam by means of a hammering medium and a steel–rule die; also called *die cutting* (3).

closed–cell cellular plastic – A cellular plastic in which almost all the cells are non–interconnecting (2).

CO_2*–blown foam* – A polyurethane foam in which all of the gas from expanding or blowing is carbon dioxide (CO_2) generated by the chemical reaction between water and isocyanates; also called *water–blown* foam (3). *cold–cure foam* – See *Cold molding.*

Cold molding – A special process of compression molding in which the molding is formed at room temperature and subsequently baked at elevated temperatures (1). Cold–molded (cold–cure) flexible urethane foam is now called *high–resilient foam* (HR), which see (1).

collapse (of cellular plastics) – Inadvertent densification of cellular plastics during manufacture resulting from breakdown of cell structure (1). Cell collapse is characterized by slumped and cratered surfaces, together with collapse of internal cells. It is caused by excessively rapid permeation of the blowing agent (gas) through the cell walls, or by weakening of the cell walls by plasticization (3).

combustible – Capable of undergoing combustion; the term is often delimited to specific fire–exposure conditions (7).

combustion – A chemical process of oxidation that occurs at a rate fast enough to produce heat and usually light, either as a glow or flame (7).

comfort cushioning – Cushioning used to provide more comfortable seating to human beings, as in upholstery; flexible foams with varying degrees of resilience are used.

compression–force deflection – The force necessary to produce a 50% compression over the entire top area of a foam specimen (3); also called *compression–load deflection (CLD)* (3).

compression set – A measure of the permanent deformation resulting from a fixed compression for a long period of time; the loss of thickness of a cushioning material after a specified time interval following removal of a compression load (4).

compression strength – The maximum compressive stress (nominal) carried by a test specimen during a compression test. It may or may not be the compressive stress (nominal) carried by the specimen at the moment of rupture (1). The maximum load that a certain cross–sectional area of a foam can sustain before losing a percentage of its thickness (generally 10%).

compressive strength at failure (nominal) – The compressive strength (nominal) sustained at the moment of failure of the test specimen if shattering occurs (1).

compressive stress – The compressive load per unit area of minimum original cross section within the gage boundaries, carried by the test specimen at any given moment. It is expressed in force per unit area (1).

core – The internal portion of a molded part, free of skin (5).

coupling agent – A substance that promotes or establishes a stronger bond at the interface of the resin matrix and the reinforcement (2); silanes and titanates are typical coupling agents (3).

cream time (initiation time) – The time between the start of the final mixing of foam ingredients and the point at which the clear mixture turns creamy or cloudy and starts to expand (3).

creep – The dimensional change with time of a material under load, following the initial instantaneous elastic deformation (3). The deformation of a material occurring with time and due to an externally applied constant stress. For cushioning materials specifically, it may be defined as the change in thickness of a cushion under static compressive load over a period of time (i.e., compressive creep) (1). It is usually measured as % deflection vs time, in hours (8).

crosslinking – The formation of a three–dimensional polymer by means of interchain reactions resulting in changes in physical properties (9). Crosslinking may be brought about by substances (crosslinking agents) or by radiation (2).

crumb – Finely divided pieces of flexible urethane foam that have been shredded, ground, milled or torn from foam trims or scraps (3).

cure, v. – To change the properties of a polymeric system into a more stable, usable condition by the use of heat, radiation, or reaction with chemical additives; curing can be accomplished, for example, by removal of solvent or by crosslinking (1).

curing agent – A substance that promotes or regulates a curing reaction (2)

cushion factor– The ratio of the maximum stress on a cushion to the energy absorbed by the cushion at the applied load. This concept is used to calculate cushion thickness required for a particular package.

cushioning, comfort – See *comfort cushioning.*

cushioning material – A material used to isolate or reduce the effect of externally applied shock or vibration forces, or both (1); resilient material used within a shipping container to reduce the shock transmitted to the product within the container (for example, foam plastic, corrugated fiberboard) (1).

cut–growth resistance – In rubber vulcanizates, the resistance to crack growth after repeated bend flexing under standardized conditions (ASTM D 1052). The initial cut is initiated by a special piercing tool.

damping – A general term applied to any mechanism that results in dissipation of energy from vibrating systems

decomposition temperature – The temperature range associated with decomposition of the polymer in the presence of oxygen (10).

density, apparent – The weight in air of a unit volume of material. This term is sometimes used synonymously with *bulk density* (1). The mass divided by the volume of a sample material, including both permeable and impermeable voids normally present in the material (1).

density, bulk – The weight per unit volume of a material, including voids inherent in the material as tested. See *density, apparent* (1).

density, core – The relatively uniform density of the interior portion of a foamed section or item, excluding any collapsed foam skins. This term is synonymous with *apparent density* (1).

density, overall – The average density of the entire foam item, including any molded skins and variations within the item.

elastic modulus – The ratio of stress to strain below the proportional limit (which see).

elastomer – A macromolecular material (i.e. polymer) that returns rapidly to approximately its initial dimensions and shape after substantial deformation by a weak stress and subsequent release of the stress (2–modified).

emulsifier (emulsifying agent) – A surface–active substance that promotes and maintains the dispersion of two incompletely miscible liquids, or of a solid and a liquid, by reducing the interfacial tension between the two phases (2).

encapsulation, complete – A cushioning method in which a packaged item to be protected is completely surrounded by a homogeneous mass of foam. The item is usually protected by a cocoon of plastic film before the foam is applied (2).

energy absorption – The ability of cushioning materials to absorb energy under static conditions, calculated as the area beneath the stress–strain curve for the loading and unloading condition.

exotherm – In foam formation, the heat liberated by some of the chemical reactions occurring in the foaming mass; this heat accelerates the foaming processes (3).

expandable plastic – A plastic in a form that can be made cellular by thermal, chemical, or mechanical means (2) (11).

expandable polystyrene – Free–flowing round beads or strands of polystyrene (PS) containing an expanding (blowing) agent, such as pentane, which is converted to closed–cell polystyrene foam by preexpansion and molding (3–modified).

expanded plastics – See *cellular plastics.*

expanded rubber – Cellular rubber having closed cells made from a solid rubber compound (13).

fascia – Elastomeric coverings for the energy–absorbing system of an automobile, such as the bumper, or even the entire front and rear ends. Microcellular urethane foams are frequently used for fascia (3).

filler – A relatively inert material added to a foam formulation to modify its strength, permanence, working properties, or other qualities, or to lower costs. If it enhances mechanical properties it is called a *reinforcing*

filler or reinforcement (1–modified).

filler, reinforcing – A fibrous additive that increases stiffness, strength, impact resistance, and hardness (3–modified).

fire gases – The airborne products emitted by a material undergoing pyrolysis or combustion, which, at the relevant temperature, exist in the gas phase (7).

fire–retardant adj. – The quality of a substance of suppressing, reducing, or delaying markedly the combustion of certain materials (14). Fire retardants are usually incorporated as additives during compounding, but are sometimes applied to surfaces of finished articles (15). *Fire retardants* cause a material to resist burning when exposed to high–energy sources, such as a burning sofa (16).

flame–retardant adj. – The quality of a substance of resisting burning when exposed to a relatively low–energy source, such as a lighted cigarette, match, candle, cigarette lighter, or stove burner (16).

flame–spread index – A number or classification indicating a comparative measure derived from observations made during the progress of the boundary of a zone of flame under defined test conditions (7). See ASTM E 84 and E 162 in Chapter 11).

flash fire – A fire (flame front) that spreads with extreme rapidity, such as one that races over flammable liquids and through gases (17). Extreme rapidity of flame propagation is generally considered as characteristic of a flash fire. A flash fire is not the same as a *flashover* (which see).

flashover – The point in a fire at which sufficient flammable gas concentration has been reached to cause all combustibles in an enclosure to burst into flaming (18).

flash point – The lowest temperature, corrected to 101.3 kPa (1.0 atmosphere) of pressure, of a sample at which application of an ignition source causes the vapor of the sample to ignite momentarily under specified conditions of test (7). (See *flash ignition temperature*).

flash–ignition temperature (flash temperature) – The lowest temperature of air passing around a specimen at which a sufficient amount of combustible gas is developed to be ignited by an external heat source, such as a flame (19). The temperature at which decomposition gases can be ignited by a spark or flame (20). (see *flash point*).

flexible – A descriptive term applied to a cellular organic polymeric material that will not rupture when a specimen 200 by 25 by 25 mm is bent around a 25–mm diameter mandrel at a uniform rate of one lap in 5 seconds in the form of a helix at a temperature between 18 and 29°C (1).

flexible polyurethane foams – Blown polymers made by reacting active hydrogen–ended, usually hydroxyl group–ended, polymers of molecular weights from about 1000 to about 10,000, with diisocyanates or polyisocyanates (21). Open–cell structures through which liquids or gases can freely pass; characterized by relatively low compression strength and ability to recover after loading; used for comfort cushioning, filters, etc.

flexural modulus – A measure of the ability of a foam to bend easily around objects, or to conform to surfaces. Flexural modulus is dependent on the tensile modulus of the polymer phase in plastic foams. It is also dependent on density and skins on the foam surface.

fluorocarbon–blown foam – Foam produced entirely by the gas generated from the boiling of a fluorocarbon (3).

foam – (A deprecated term applying to cellular polymeric materials, plastics or elastomers). A product, either flexible or rigid, that has been produced by the internal generation or liberation of a gas in a fluid medium that is polymerizing while expanding in volume. The final result is either an open– or closed–cell product (3). NOTE – This definition does not cover *syntactic foams* (author).

foamed plastics – See *cellular plastics*.

foaming agent –   See *blowing agent*.

foam–in–bag packaging – A special form of foam–in–place packaging (which see) in which the liquid polyurethane components are poured into plastic bags before rigidifying and taking the shape of the items to be protected. The function of the foam is that of blocking and bracing, since the foam is rigid.

foam–in–place packaging – A system of cushioning or blocking and bracing of packaged items in which the liquid urethane components are poured into the container around the item to be protected and then rigidified to more or less take the shape of the item. There are a number of techniques, including "foam–in–bag" and "encapsulation".

fragility rating (G–Factor G–value) – The ratio of the maximum acceleration that an object can safely withstand to the acceleration of gravity (4).

free rise – The unhampered expansion of a foam sample or product in a container with no top and a height of sidewall not greater than twice the diameter (3).

friability – The degree of dusting or surface susceptibility to particle loss of certain relatively brittle foams.

frothing – A technique of applying foam by introducing blowing agents as small air bubbles, under pressure, into the liquid mixture of foam ingredients (3). The froth stream is quite fluid, resembling shaving cream in appearance, and flows readily (12).

fuel value – See *heat of combustion.*

glass transition – The reversible change in an amorphous polymer, or in amorphous regions of a partially crystalline polymer from (or to) a viscous or rubbery condition to (or from) a hard and relatively brittle one (1).

glass–transition temperature (T_g) – The approximate midpoint of the temperature range over which the glass transition takes place (1).

hardening agent (hardener) – A curing agent that promotes or regulates the curing reaction of resins that yield rigid (hard) products (2).

hardness – The resistance to indentation, as measured under specific conditions (1).

hardness index – The term used in some specifications for the 50% indentation force deflection (IFD) value (which see).

heat of combustion (fuel value) (heat value) – The amount of heat released by the oxidation of 1 mole of a substance at constant pressure or constant volume (22). The heat of combustion must be in excess of the ignition point, or the flame will extinguish itself (23). In the literature the values are frequently reported in Kcal/g mol and BTU/ft^3, and BTU/board ft.

heat value – See *Heat of combustion.*

high–resilient (HR) foam – In trade practice, a foam that has a sag factor (which see) of 2.7 and above. Such foams are made by the *cold–curing process*. With high sag a better cushion is obtained for seating applications. See also *cold molding.*

high–pressure impingement mixing (HPIM) – A form of molding in which two or more pressurized reactive streams are impinged in the mixing chamber prior to injection in the mold. The reactants are delivered through nozzles to the chamber at 2000 to 3000 psi (13.6 – 20.7 MPa), although pressures in the chamber and mold remain quite low, around 15 to 75 psi (0.10 – 0.51 MPa). HPIM is usually classified as a form of reaction injection molding (RIM) (8).

hydroxyl number – A measure of the length of polymer chains in polyether polyols. The lower the hydroxyl number, the longer is the chain. Hydroxyl numbers between 40–75 give *flexible* foams, and between 350–600 give *rigid* foams, with *semi–rigid* foams in between.

hygrothermal stability test – A test to determine the resistance of a material to a combination of elevated temperature and humidity (4).

hysteresis – A measure of the ability of a foam to dampen vibrations. The difference in area under a load indentation (stress–strain) curve as the load is applied and then released.

ignition – Initiation of combustion (which see). The combustion may be evidenced by glow, flame, detonation, or explosion (7).

ignition point (ignition temperature) – The lowest temperature at which a solid can ignite, either directly, or at the evolution of combustible gas which will ignite and sustain its combustion process (14). There are two categories, *flash–ignition temperature* and *self–ignition temperature*, which see.

indentation–force deflection (IFD) (indentation–load deflection (ILD)) – A measure of the force necessary to produce 25% and 65%, or other designated indentation, in a foam product (5).

initiator – A substance used in small proportion that starts a chemical reaction, for example, by providing free radicals (2).

integral–skin foam – A molded flexible urethane foam product having a dense, tough, outer foam skin and a lower–density core (3). Also called "self–skinning foam".

intumescence – The expansion of a fire–retardant coating into a foamlike carbonaceous char on heating; continued heating of the coating thereafter converts it by pyrolysis into a heat–resistant carbonaceous foam. The insulating qualities of such a foam then protect the substrate from the effects of a flame (19).

isocyanate – A chemical compound having one or more reactive NCO radicals or groups attached to the main chain (3). A starting material for polyurethane foam.

isocyanate plastic – A plastic based on polymers made by the reaction of polyfunctional isocyanates with other compounds (2).

isocyanurate plastic – A plastic based on polymers in which trimerization of isocyanates incorporates six–membered isocyanurate ring groups in a chain (2). Such groups impart a higher degree of thermal stability and lower combustibility characteristics than is obtained with urethane plastics.

isotropic – Having the same properties in all directions.

"K"–factor – A measure of the insulating ability or thermal conductivity of a foam or other material. It is usually expressed as $BTU/hr/ft^2/°F/in$ $(cal/sec/cm^2/°C/cm)$ (3).

lamination – The process of adhering thin sheets of flexible foam and fabric together. Two common processes are *flame lamination* and *adhesive lamination* (3).

latex – A stable aqueous dispersion of a polymeric material having particle diameters in the range of about 500 to 50,000 (0.05μ to 5μ)(24).

latex foam – Cellular rubber made from rubber latex (1).

latex foam rubber – A cellular rubber produced by frothing a rubber latex or liquid rubber, gelling the frothed latex and then vulcanizing it in the expanded state (6).

limiting oxygen index (LOI) – A measure of the minimum concentration of oxygen in an oxygen–nitrogen atmosphere that is necessary to support a flame for at least 3 minutes under specified test conditions (see ASTM D 2863). The LOI is used to illustrate the relative flammability of polymeric materials (25).

liquid injection molding (LIM) – A variation of the RIM process of molding in which the reactive components are mixed in a large chamber by mechanical stirring before injection of the homogeneous reaction mixture into the mold; also called *liquid reaction molding (LRM)* (26).

liquid reaction molding (LRM) – See Liquid injection molding (21).
lost–foam casting – A sand–casting method using loose, unbonded sand compacted through vibration around an expanded polystyrene (EPS) foam pattern. As the metal is poured into the foam, the EPS pattern evaporates (i.e. lost foam), leaving a casting that exactly duplicates the pattern.

mechanically foamed plastic – A cellular plastic in which the cells are formed by the physical incorporation of gases (1).

microcellular foams – Flexible polyurethane foams varying in density from about 30–60 lb/ft^3 (481–961 kg/m^3). They are similar in properties to solid urethanes, but are lighter in weight. Applications are for shoe soles, automotive bumpers, seals, gaskets, and vibration pads (8).
mixing ratio – The ratio required of the individual foam components to achieve maximum mechanical properties. The ratio may be expressed as parts by weight or by volume.

modulus of elasticity (E) (elastic modulus) (Young's modulus) – The ratio of stress (nominal) to corresponding strain below the proportional limit of a material (which see). It is expressed in force per unit area, usually in kg force/cm^2 (psi), based on the average initial cross–sectional area (1).

molded foam – A cellular product having the shape of the enclosed chamber in which it is produced by foaming (5).

mold–release agent (lubricant) – A material applied to the surface of a mold to facilitate removal of a molded item from the mold.

nonrigid plastic – A plastic that has a modulus of elasticity, E, either in flexure or in tension, of not over 70 MPa (700 kgf/cm^2) (10,000 psi) at 23°C and 50% RH (1).

nucleating agent – A chemical substance which, when incorporated in polymers, forms nuclei for the growth of crystals in the reaction; in foams, nucleating agents act by forming many small bubbles, rather than fewer large bubbles (3). *Colloidal silicas* and *micro–expanded silicas* are commonly used nucleating agents, as are *dry air* and *nitrogen* (21).

open–cell cellular plastics – Cellular plastics in which almost all the cells are interconnecting throughout (2). Flexible foams should have a high predominance of open cells for best quality.

optical density of smoke (D) – A measure of the attenuation of a light beam passing through smoke (7).

perm – A unit of measurement of water vapor permeance; a metric perm, or 1 g/24 h·m^2·mm Hg (1 grain/h·ft2·in Hg) (1).

permanent set –The permanent change in thickness of an unloaded cushion as a result of an applied compressive load for any given time interval and any given unloaded recovery time period (1).

permeability – The property of a material that permits a fluid or gas to pass through it; in construction, the term commonly refers to water vapor permeability of a sheet material or assembly and is defined as water vapor permeance per unit thickness (1–modified).

permeance (water vapor) – The ratio of the rate of water vapor transmission (WVT) through a material or assembly between its two parallel surfaces to the vapor pressure differential between the surfaces. The metric unit of measure is the metric perm, 1 g/24 h·m^2·mm Hg (1).

plastic (n) – A material that contains as an essential ingredient one or more organic polymeric substances of large molecular weight, is solid in its finished state, and, at some stage in its manufacture or processing into finished articles, can be shaped by flow (excludes rubber, textiles, adhesives and paints (1).

plastic foam – See Cellular plastics.

plasticizer – A substance incorporated in a material to increase its workability, flexibility, or distensibility; a compounding material used to enhance the deformability of a polymeric compound (1).

polyester foams – Urethane foams made by reacting the isocyanate with a polyester (3).

polyether foams – Urethane foams made by reacting the isocyanate with a polyether (3).

polyisocyanurate – See Isocyanurate plastic.

polymer – A high–molecular–weight compound, usually organic, but also including silicone compounds, natural or synthetic, whose structure can be represented by a repeated small unit, the *mer*; if two or more monomers are involved a *copolymer* is obtained (3–modified).

polyol – A organic compound with more than one hydroxyl group attached to the molecule (3). A starting material for polyurethane foam.

polyurethane foam (flexible) – See Flexible polyurethane foams.

polyurethanes – A large family of polymers based on the reaction products of an organic isocyanate with compounds containing a hydroxyl group (3). Polymers in which the repeated structural unit is of the urethane type (2).

premix (masterbatch) (polyol mix) – In urethane foams, the mixture resulting from blending many of the minor ingredients in with the polyol in an effort to reduce the final number of components, or to allow more time for mixing or blending those chemicals that may or may not be readily miscible in the short period of exposure to the final mixing (3–modified).

prepolymer – A chemical intermediate used at times in the production of urethane foams. It is normally manufactured by prereacting all of the isocyanate with part or all of the polyol (3). A polymer of degree of polymerization (DP) between that of the monomer or monomers and that of the final polymer (1).

processing aids – Additives such as viscosity depressants, mold–release agents, emulsifiers, lubricants, and antiblocking agents (3).

promoter – See *accelerator.*

proportional limit – The greatest stress that a material is capable of sustaining without any deviation from proportionality of stress to strain (Hooke's Law). It is expressed in force per unit area (1).

reaction injection molding (RIM) – A process involving the fast metering and mixing of two or more reactive chemicals and the injection of the resulting reaction mixture into a closed mold (21). RIM most frequently involves polyurethane.

release agent – See *mold release agent.*

resilience – A material characteristic indicating an ability to withstand temporary deformation without permanent deformation or rupture (4). The ratio of energy output to energy input in a rapid (or instantaneous) full recovery of a deformed specimen (1).

resin – A solid or pseudosolid organic material, often of high molecular weight, which exhibits a tendency to flow when subjected to stress, usually has a softening or melting point, and usually fractures conchoidally. The term is ordinarily applied to designate any polymer that is a basic material for plastics, including non–organic silicones (1–modified).

reticulated urethane foam – Very low–density urethane foams character-ized by a three–dimensional skeletal structure of ribs or struts (which see) with few or no membranes between the strands. Such foams contain approximately 97% void space (3).

R–factor – The reciprocal of "K"–factor. This concept is currently in wide use in selling thermal insulation because the better insulations have higher values and the various elements comprising an insulation system can be added arithmetically to provide an overall R–factor.

ribs (struts) – Thread–like structures formed at the joint between adjacent bubbles in a foam which becomes open–celled. The ribs are usually reinforced by the remains of the cell membranes in good–quality foam (3). *rigid plastic* – A plastic that has a modulus of elasticity, E, either in flexure or in tension, greater than 700 MPa (7000 kgf/cm^2) (100,000 psi) at 23°C and 50% RH (1).

rigid polyurethane foams – Closed–cell polyurethane foams having high compressive strength–to–weight ratios and outstanding thermal–insulation properties.

rigid, self–skinning foam – A rigid cellular foam having one or more surface zones with an apparent density substantially higher than that of the core material. This surface zone density approaches that of the unblown polymeric material. Also called "integral–skin foam".

RIM – See *reaction injection molding.*

rise time (pulse rise time) – In cushioning, the interval between a point near the start of the pulse equal to 10% of the peak acceleration and a point just before peak which is 90% of peak (1).

RRIM – Reinforced–high–modulus RIM (which see) (21).

rubber – A material that is capable of recovering from large deformations quickly and forcibly, and can be, or already is, modified to a state in which it is essentially insoluble (but can swell) in boiling solvent, such as benzene, methyl ethyl ketone, and ethanol–toluene azeotrope. A rubber in its modified state, free of diluents, retracts within 1 minute to less than 1.5 times its original length after being stretched at room temperature (18 to 29°C) to twice its length and held for 1 minute before release (1).

rubber, expanded – Cellular rubber having closed cells made from a solid rubber compound.

rubber latex – See *latex*.

sag factor – The ratio of the load needed to compress a foam by 65% to the load needed to compress the foam by 25%. Sag factor is a measure of desirability of comfort cushioning (8–modified).

self–extinguishing – A term no longer in use.

self–ignition – See *autoignition*.

semi–rigid plastic – A plastic that has a modulus of elasticity, E, either in flexure or in tension, of between 70 and 700 MPa (700 and 7000 kgf/cm^2) (10,000 and 100,000 psi) at 23°C and 50% RH (1). Practically all "semi–rigid" foams are now considered flexible foams.

shrinkage (of cellular plastics) – An inadvertent dimensional decrease of cellular plastics without breakdown of cell structure (2).

skin – A relatively dense layer at the surface of a cellular polymeric material (1). See also *integral skin foam*.

slab foam – Foam made by the continuous pouring of mixed liquids onto a conveyor, and generating a continuous loaf of foam for as long as the machine is operating. It is generally classed as free–rise or unconfined, although fixed side guides give the loaf a generally rectangular cross section with a slightly rounded top (3).

smoke – The airborne solid and liquid particulates and gases evolved when a material undergoes pyrolysis or combustion. So–called chemical smokes are excluded from this definition (1).

smoldering – Combustion of a solid without flame, often evidenced by visible smoke (1). A very small air supply is required for smoldering to take place, and the gaseous products it produces are frequently toxic (26).

softness – A subjective characterization usually determined by compressing a small sample of foam with the fingers or hands. The physical property most important in determining whether a foam is soft to the "feel" is the initial compression modulus. Soft foams generally have open–cell structures and small cell size within cell walls.

solvent cement – An adhesive made by dissolving a thermoplastic resin or compound in a suitable solvent or mixture of solvents. When the solvent evaporates the drying resin acts as an adhesive (1–modified).

sponge rubber – Cellular rubber consisting predominantly of open cells and made from a dry rubber compound (1).

spontaneous ignition – Initiation of combustion caused by an internal chemical exothermic reaction without externally applied heat (1–modified).

spray–up – In processing cellular plastics such as epoxy and polyurethane types, the spraying of fast–reacting resin catalyst systems onto a surface where they react to foam and cure. The resin and catalyst are usually sprayed through separate nozzles so that they are mixed during the sprayup operation (2).

stabilizer – An ingredient added to a plastic to retard degradation due to the effects of heat, light or oxidation (1). An agent used in compounding to assist in maintaining the physical and chemical properties of a material throughout its processing and service life. The term covers emulsion stabilizers, viscosity stabilizers, antioxidants, and antiozonants (3).

steam molding – A process used to mold parts from preexpanded beads of polystyrene, using steam as a source of heat to expand the blowing agent in the material (3).

structural foam – A term originally used for cellular thermoplastic foams with integral solid skins and with high strength–to–weight ratios, but more recently sometimes used for high–density cellular plastics strong enough for structural applications (15). The latter include thermosetting foams such as polyurethane and polyisocyanurate.

surface flame spread – The propagation of a flame away from the source of ignition across the surface of a liquid or solid (1). See ASTM E 84 for a test method on flame–spread ratings.

surfactant – A contraction of the term "surface–active agent"; a material that improves the emulsifying, dispersing, spreading, wetting, or other surface–modifying properties of liquids (1). In foams, surfactants are widely used to control the size and shape of the bubbles or cells in the final product (3).

syntactic cellular plastic (syntactic foam) – A cellular plastic in which hollow microsphere fillers are used as the low–density elements (2). A material consisting of hollow–sphere fillers in a resin matrix (1). The most common matrix resins are epoxies and polyesters (12).

tack–free time – The time elapsed between the initial mixing of foam in–gredients and the point at which the foam crown may be touched lightly with no perceptible adherence of the foam to the fingers. When the foam is tack–free complete gelation or hardening has been achieved.

Talalay foam process – A method of latex foam production producing relatively large spherical, fairly uniform–sized thick–wall interconnecting cells and almost skinless surface. The result is a highly efficient load–supporting structure whose resilience is improved by the rapid expelling and taking in of air during compression and release (24).

TDI – A common symbol for *toluene diisocyanate*, particularly the 80–20 isomer blend (3).

thermal conductivity – For a homogeneous material not affected by thickness, the rate of heat flow under steady conditions through unit area, per unit temperature gradient in the direction perpendicular to the area (1) (2). See *K–factor*.

thermally foamed plastic – A cellular plastic produced by applying heat to effect gaseous decomposition or volatilization of a constituent

thermoplastic adj. – Capable of being repeatedly softened by heating and hardened by cooling through a temperature range characteristic of the plastic, and, in the softened state, capable of being shaped by flow into

articles by molding or extrusion. The change upon heating of thermoplastic materials is substantially physical (1–modified).

thermosetting adj. – Capable of being changed into a substantially infu-
sible or insoluble product when cured by heat or other means (1).

toxicity – The amount of a substance which produces detrimental effects
in an animal or part of an animal. It is expressed as a dosage unit over
a dimensional feature of the biologic system, such as mg of the material
divided by the body weight, as mg/kg (27–modified).

toxicology – The study of poisons (27).

ultraviolet absorber (stabilizer) (inhibitor) – An additive used to minimize or prevent the degradation of polymers by ultraviolet light. These
materials screen out radiation beyond the violet end of the spectrum (3)

unicellular plastic – Synonymous with *closed–cell cellular plastic*, which
see.

urethane foam – A cellular product produced by the interaction of active
hydrogen compounds, water, and isocyanates; may be rigid, flexible, or
semi–rigid (5–modified).

urethane foam, flexible – Cellular elastomeric material made by the condensation of a polyol with an organic isocyanate and containing a
preponderence of open cells (1–modified).

urethane linkage – A chemical group formed by the reaction of an
isocyanate with a hydroxyl group (8).

void – A cavity unintentionally produced in a cellular material and substantially larger than the characteristic individual cells of the material
(1–modified).

water–blown foam – A polyurethane foam in which the gas for expansion was generated by the reaction between water and an isocyanate-
bearing material (3).

water–vapor transmission rate (WVTR) – The rate of water–vapor flow,
under steady specified conditions, through a unit area of a material,

between its two parallel surfaces and normal to the surfaces. The metric unit of measure is 1 g/24 hr· m² (1).

REFERENCES

1. American Society for Testing and Materials (ASTM), Compilation of ASTM Standard Definitions, sponsored by the ASTM Committee on Terminology, 6th Edition, (1986), 907 pp.

2. International Organization for Standardization (ISO), International Standard 472–(1979), Plastics – Vocabulary, 1st edition, 93 pp. Addendum 1 (1982); Addendum 2 (1983); Addendum 3 (1984).

3. Foams – desk–top data bank[R], edition 2, M.J. Howard, ed., published by the International Plastic Selector, Inc., San Diego, CA, a subsidiary of Cordura Publications, (1980), 734 pp.

4. U.S. Department of Defense, Military Standardization Handbook MIL–HDBK–304B, Packaging Cushioning Design, (31 October 1978), 518 pp.

5. American Society for Testing and Materials, ASTM D 3574–86, Standard Methods for Testing Flexible Cellular Materials – Slab, Bonded and Molded Urethane Foams, (1986) pp. 22

6. Skochdopole, R.E., article on "Cellular Materials", pp 80–130, in Encyclopedia of Polymer Science and Technology, Vol. 3, Wiley–Interscience, New York, (1965).

7. American Society for Testing and Materials, ASTM E 176–86, Standard Terminology Relating to Fire Standards, (1986) 5 pp.

8. Chapter 20 – "Cellular Plastics", in Plastics Engineering Handbook of the Society of the Plastics Industry, Inc., 4th edition, J. Frados, ed., Van Nostrand Reinhold, New York, (1976).

9. American Society for Testing and Materials, ASTM D 883–86b, Standard Definitions of Terms Relating to Plastics, (1986) 16 pp.

10.   Mark, H.F., Atlas, S.M., Shalaby, S.W. and Pearce, E.M., Chapter 1, "Combustion of Polymers and its Retardation", in *Flame-Retardant Polymeric Materials* (M. Lewin, S.M. Atlas, and E.M. Pearce, eds.), Plenum Press, New York (1975). Abridged version in *Polymer News* 2 (5/6):3–12 (1975).

11.   International Organization for Standardization (ISO), International Standard ISO 472/Addendum 1, Plastics – Vocabulary, (1982), 7 pp.

12.   U.S. Department of Defense, Plastics Technical Evaluation Center, U.S. Army ARDEC, Dover, N.J., PLASTEC Report R52 – *Handbook of Plastic Foams*, by Arthur H. Landrock, (February 1985), 86 pp. Available from National Technical Information Service (NTIS) as AD A156 758.

13.   American Society for Testing and Materials, ASTM D 1056–85, Standard Specification for *Flexible Cellular Materials – Sponge or Expanded Rubber*, (1985) 13 pp.

14.   International Organization for Standardization (ISO), International Standard ISO 3261–1975, *Fire Tests – Vocabulary*, (1975) 11 pp.

15.   Whittington, L.R., *Whittington's Dictionary of Plastics*, 2nd Edition, Technomic Publishing Co., Lancaster, PA (1978).

16.   Sanders, M.J., "Flame Retardants – Special Report", *Chemical and Engineering News*, 56 (17): 22–28 (April 24, 1978).

17.   Kunshinoff, B.W., Fristrom, R.M., Ordway, G.L. and Tave, R.L., *Fire Sciences Dictionary*, John Wiley and Sons, New York (1977).

18.   *Precautions for the Proper Usage of Polyurethane, Polyisocyan- urates and Related Materials*, 2nd Edition, prepared by the Chemical Division of the Upjohn Company, Technomic Publishing Co., Lancaster, Pennsylvania (1980).

19.   Hendersinn, R., "Fire Retardancy (Survey)", in: *Encyclopedia of Polymer Science and Technology*, (N.B. Bikales, ed), Supplement, Vol. 2, pp 270–339, Wiley–Interscience, New York (1977).

20.   Pearce, E.M., Shalaby, S.W. and Baker, P.M., Chapter 6, "Retardation of Combustion of Polyamides", in *Flame Retardant Polymeric Materials*, (M. Lewin, S.M. Atlas and E.M. Pearce, eds), Plenum Press, New York (1974).

21.   Woods, G., *Flexible Polyurethane Foams – Chemistry and Technology*, Applied Science Publishers, London and New York, (1982).

22.   *Dictionary of Scientific and Technical Terms*, (D.N. Lapedes, editor in chief) McGraw–Hill, New York (1974).

23.   Einhorn, I.N., "Fire Retardance of Polymeric Materials", in *Reviews in Polymer Technology*, (I. Skeist, ed), Vol. 1, pp 113,184, Marcel Dekker, New York (1977).

24.   Rogers, T.H. and Hecker, K.C., Chapter 10 – "Latex and Foam Rubber", in *Rubber Technology*, 2nd Edition, (M. Morton, ed) Van Nostrand Reinhold, New York (1973).

25.   Landrock, A.H., *Handbook of Plastics Flammability and Combustion Technology*, Noyes Publications, Park Ridge, New Jersey, (1983).

26.   Ohlemiller, T.J., Rogers, F.F., and Davidson, B., "A survey of Several Factors Influencing Smoldering Combustion in Flexible and Rigid Polymer Foams", American Institute of Chemical Engineers, 85th National Meeting and Chemical Plant Equipment Exposition, (4–8 June 1978), Philadelphia, PA.

27.   Doull, J., Klaasen, G.D., and Amdur, M.O., editors, *Casarett and Doull's Toxicology: The Basic Science of Poisons*, Macmillan, N.Y., (1980).

INDEX